Topoisomerases as Targets for Novel Drug Discovery

Topoisomerases as Targets for Novel Drug Discovery

Guest Editor

Andrej Perdih

MDPI

Basel • Beijing • Wuhan • Barcelona • Belgrade • Novi Sad • Cluj • Manchester

Guest Editor
Andrej Perdih
Theory Department
National Institute of
Chemistry
Ljubljana
Slovenia

Editorial Office
MDPI AG
Grosspeteranlage 5
4052 Basel, Switzerland

This is a reprint of the Special Issue, published open access by the journal *Pharmaceuticals* (ISSN 1424-8247), freely accessible at: https://www.mdpi.com/journal/pharmaceuticals/special_issues/drug_topoisomerases.

For citation purposes, cite each article independently as indicated on the article page online and as indicated below:

Lastname, A.A.; Lastname, B.B. Article Title. *Journal Name* **Year**, *Volume Number*, Page Range.

ISBN 978-3-7258-6081-4 (Hbk)
ISBN 978-3-7258-6082-1 (PDF)
https://doi.org/10.3390/books978-3-7258-6082-1

Contents

About the Editor . . . . **vii**

Andrej Perdih
Topoisomerases as Targets for Novel Drug Discovery
Reprinted from: *Pharmaceuticals* **2025**, *18*, 1693, https://doi.org/10.3390/ph18111693 . . . . **1**

Prasanna Anjaneyulu Yakkala, Naveen Reddy Penumallu, Syed Shafi and Ahmed Kamal
Prospects of Topoisomerase Inhibitors as Promising Anti-Cancer Agents
Reprinted from: *Pharmaceuticals* **2023**, *16*, 1456, https://doi.org/10.3390/ph16101456 . . . . **4**

Victor M. Matias-Barrios and Xuesen Dong
The Implication of Topoisomerase II Inhibitors in Synthetic Lethality for Cancer Therapy
Reprinted from: *Pharmaceuticals* **2023**, *16*, 94, https://doi.org/10.3390/ph16010094 . . . . **28**

Scott Grossman, Colin W. G. Fishwick and Martin J. McPhillie
Developments in Non-Intercalating Bacterial Topoisomerase Inhibitors: Allosteric and ATPase Inhibitors of DNA Gyrase and Topoisomerase IV
Reprinted from: *Pharmaceuticals* **2023**, *16*, 261, https://doi.org/10.3390/ph16020261 . . . . **48**

Xiaoqin Yang, Jiamei Chen, Yitao Wang, Yihan Wu and Jinming Zhang
Managing Irinotecan-Induced Diarrhea: A Comprehensive Review of Therapeutic Interventions in Cancer Treatment
Reprinted from: *Pharmaceuticals* **2025**, *18*, 359, https://doi.org/10.3390/ph18030359 . . . . **65**

Jessica Ceramella, Domenico Iacopetta, Anna Caruso, Annaluisa Mariconda, Anthi Petrou, Athina Geronikaki, et al.
5,8-Dimethyl-9*H*-carbazole Derivatives Blocking hTopo I Activity and Actin Dynamics
Reprinted from: *Pharmaceuticals* **2023**, *16*, 353, https://doi.org/10.3390/ph16030353 . . . . **97**

Ilija N. Cvijetić, Barbara Herlah, Aleksandar Marinković, Andrej Perdih and Snežana K. Bjelogrlić
Phenotypic Discovery of Thiocarbohydrazone with Anticancer Properties and Catalytic Inhibition of Human DNA Topoisomerase IIα
Reprinted from: *Pharmaceuticals* **2023**, *16*, 341, https://doi.org/10.3390/ph16030341 . . . . **116**

Mohamed Badr, Elshaymaa I. Elmongy, Doaa Elkhateeb, Yasmine S. Moemen, Ashraf Khalil, Hadeer Ali, et al.
In Silico and In Vitro Investigation of Cytotoxicity and Apoptosis of Acridine/Sulfonamide Hybrids Targeting Topoisomerases I and II
Reprinted from: *Pharmaceuticals* **2024**, *17*, 1487, https://doi.org/10.3390/ph17111487 . . . . **135**

Syed Nasir Abbas Bukhari, Mohamed Abdelwahab Abdelgawad, Naveed Ahmed, Muhammad Wahab Amjad, Muhammad Ajaz Hussain, Mervat A. Elsherif, et al.
Synthesis, Characterization, and Biological Evaluation of Meldrum's Acid Derivatives: Dual Activity and Molecular Docking Study
Reprinted from: *Pharmaceuticals* **2023**, *16*, 281, https://doi.org/10.3390/ph16020281 . . . . **148**

Josephine Geertsen Keller, Kamilla Vandsø Petersen, Karol Mizielinski, Celine Thiesen, Lotte Bjergbæk, Rosa M. Reguera, et al.
Gel-Free Tools for Quick and Simple Screening of Anti-Topoisomerase 1 Compounds
Reprinted from: *Pharmaceuticals* **2023**, *16*, 657, https://doi.org/10.3390/ph16050657 . . . . **161**

Enas S. Gad, Ahmed M. Ashour, Amany M. Gad, Ali Khames, Shaimaa G. Ibrahim, Mohamed H. A. Gadelmawla and Mona Mansour
Hepatoprotection by Methylene Blue Against Doxorubicin Toxicity Through Coordinated Modulation of Oxidative Stress, ER Stress, and Apoptotic Pathways
Reprinted from: *Pharmaceuticals* **2025**, *18*, 1625, https://doi.org/10.3390/ph18111625 . . . . **175**

About the Editor

Andrej Perdih

Andrej Perdih is a Research Associate Professor in the Laboratory for Computational Biochemistry and Drug Design at the National Institute of Chemistry in Ljubljana, Slovenia. He completed his PhD studies in Medicinal Chemistry at the University of Ljubljana. Following that, he undertook two postdoctoral stays at Freie Universität Berlin, Germany, and at the University of Michigan, USA. His main research interests are computational biochemistry and computational molecular design. In particular, he focuses on the study of type IIA topoisomerases and the design of catalytic inhibitors as novel anticancer agents.

pharmaceuticals

MDPI

Editorial

Topoisomerases as Targets for Novel Drug Discovery

Andrej Perdih [1,2]

[1] National Institute of Chemistry, Hajdrihova 19, SI 1000 Ljubljana, Slovenia; andrej.perdih@ki.si; Tel.: +386-1-4760-376

[2] Faculty of Pharmacy, University of Ljubljana, Aškerčeva 7, SI 1000 Ljubljana, Slovenia

DNA topoisomerases (topo) are essential enzymes that maintain the integrity of the genome by regulating the topological state of DNA during replication, transcription, recombination and repair [1]. These enzymes, which are found in bacteria, archaea, eukaryotes and even some viruses, are categorized into type I and type II topoisomerases, which are distinguished by their mechanisms of transient single- or double-strand DNA cleavage and strand passage that allows DNA topology to be altered [2]. Due to their unique ability to manipulate DNA topology, topoisomerases are often referred to as the "Magicians of the DNA world", as "in their presence, DNA strands or double helices can pass through each other as if all physical boundaries had disappeared" [2].

The first member of this protein family, the protein ω, now known as *Escherichia coli* topo I, was discovered by James Wang and published in 1971 [3]. This enzyme became the founding member of the type I topoisomerase family and marked the beginning of a rapidly growing field of research. Since then, numerous members have been identified and characterized, leading to vibrant research activity over the past five decades [4].

The ability to manipulate DNA topology has made topoisomerases key targets for antibiotic and cancer therapy, and many drugs acting on them have a long history of success in clinical practice. In the field of infectious diseases, nalidixic acid, introduced in the 1960s, was the starting point for the development of fluoroquinolones, such as ciprofloxacin and levofloxacin, and newer agents such as moxifloxacin, which target bacterial DNA gyrase and topoisomerase IV. In the field of chemotherapeutics, etoposide and teniposide, semi-synthetic derivatives of the natural product podophyllotoxin, were introduced in the 1980s. They inhibit human topoisomerase II by stabilizing the transient covalent complex between topo II and DNA. Anthracyclines, including doxorubicin, daunorubicin and epirubicin, are another established group with a similar mode of action on topo II. Camptothecin, a natural product of *Camptotheca acuminata* discovered in the 1960s, led to the development of the clinically approved topo I inhibitors topotecan and irinotecan in the 1990s, which are used to treat various types of cancer [5].

Despite decades of clinical use, these topoisomerase-targeting agents have several limitations, including antibiotic resistance, off-target toxicity, induction of secondary malignancies, cardiotoxicity and limited tumor selectivity [5,6]. This Special Issue presents a collection of review and research articles that highlight recent progress in overcoming these challenges through the development of new agents targeting these established targets as well as novel insights into the management of chemotherapy-induced toxicities.

A large portion of the review articles highlight the strategies currently being developed for novel topoisomerase inhibitors in oncology and infectious diseases. One of these provides a comprehensive perspective on topoisomerase inhibitors in cancer, covering both topo I and topo II agents, and highlights novel chemical scaffolds, hybrid molecules and clinical candidates that could overcome resistance and toxicity, supported by structural

 https://doi.org/10.3390/ph18111693

insights for rational drug design [7]. Complementing this, another review looks at the development of bacterial gyrase inhibitors that circumvent fluoroquinolone resistance by targeting either allosteric or ATP-binding sites [8]. The next review focuses on synthetic lethality strategies that combine topoisomerase II inhibition with disruption of DNA repair pathways or oncogenic drivers, offering improved therapeutic selectivity and efficacy [9].

The next section of the Special Issue presents original research contributions. These include 5,8-dimethyl-9*H*-carbazole derivatives with dual topo I/II inhibition, which illustrates the potential of multitarget strategies [10]. The phenotypically discovered thiocarbohydrazone compound, which acts as a catalytic topo II inhibitor, shows cytotoxic activity against cancer cells [11]. Complementing these reports is a newly developed rolling circle amplification assay for rapid, gel-free screening of topo I inhibitors, which enables efficient identification of active compounds [12]. These and other studies in this Special Issue demonstrate how novel chemical scaffolds and innovative screening platforms are helping to advance the next generation of topoisomerase therapeutics.

Overall, the Special Issue *Topoisomerases as Targets for Novel Drug Discovery* demonstrates that successfully tackling these established targets requires translational strategies that combine structural biology [13], medicinal chemistry, computational science [14] and other interdisciplinary approaches to develop next-generation topoisomerase inhibitors that can overcome long-standing clinical challenges.

Acknowledgments: As Guest Editor of this Special Issue, I would like to thank all the authors for their valuable contributions. I would also like to thank the reviewers for their constructive feedback and the dedicated staff of the *Pharmaceuticals* Editorial Office, for their exceptional cooperation and support throughout the publication process.

Conflicts of Interest: The author declares no conflict of interest.

References

1. Pommier, Y.; Nussenzweig, A.; Takeda, S.; Austin, C. Human Topoisomerases and Their Roles in Genome Stability and Organization. *Nat. Rev. Mol. Cell Biol.* **2022**, *6*, 407–427. [CrossRef] [PubMed]
2. Wang, J.C. Cellular Roles of DNA Topoisomerases: A Molecular Perspective. *Nat. Rev. Mol. Cell Biol.* **2002**, *6*, 430–440. [CrossRef] [PubMed]
3. Wang, J.C. Interaction Between DNA and an *Escherichia coli* Protein Omega. *J. Mol. Biol.* **1971**, *55*, 523–533. [CrossRef] [PubMed]
4. McKie, S.J.; Neuman, K.C.; Maxwell, A. DNA Topoisomerases: Advances in Understanding of Cellular Roles and Multi-Protein Complexes via Structure-Function Analysis. *Bioessays* **2021**, *43*, e2000286. [CrossRef] [PubMed]
5. Bondarev, A.D.; Jonsson, J.; Chubarev, V.N.; Tarasov, V.V.; Lagunas-Rangel, F.A.; Schiöth, H.B. Recent Developments of Topoisomerase Inhibitors: Clinical Trials, Emerging Indications, Novel Molecules, and Global Sales. *Pharmacol. Res.* **2024**, *209*, 107431. [CrossRef] [PubMed]
6. Nitiss, K.C.; Nitiss, J.L.; Hanakahi, L.A. DNA Damage by an Essential Enzyme: A Delicate Balancing Act on the Tightrope. *DNA Repair* **2019**, *82*, 102639. [CrossRef] [PubMed]
7. Yakkala, P.A.; Penumallu, N.R.; Shafi, S.; Kamal, A. Prospects of Topoisomerase Inhibitors as Promising Anti-Cancer Agents. *Pharmaceuticals* **2023**, *16*, 1456. [CrossRef] [PubMed]
8. Grossman, S.; Fishwick, C.W.G.; McPhillie, M.J. Developments in Non-Intercalating Bacterial Topoisomerase Inhibitors: Allosteric and ATPase Inhibitors of DNA Gyrase and Topoisomerase IV. *Pharmaceuticals* **2023**, *16*, 261. [CrossRef] [PubMed]
9. Matias-Barrios, V.M.; Dong, X. The Implication of Topoisomerase II Inhibitors in Synthetic Lethality for Cancer Therapy. *Pharmaceuticals* **2023**, *16*, 94. [CrossRef] [PubMed]
10. Ceramella, J.; Iacopetta, D.; Caruso, A.; Mariconda, A.; Petrou, A.; Geronikaki, A.; Rosano, C.; Saturnino, C.; Catalano, A.; Longo, P.; et al. 5,8-Dimethyl-9H-Carbazole Derivatives Blocking hTopo I Activity and Actin Dynamics. *Pharmaceuticals* **2023**, *16*, 353. [CrossRef] [PubMed]
11. Cvijetić, I.N.; Herlah, B.; Marinković, A.; Perdih, A.; Bjelogrlić, S.K. Phenotypic Discovery of Thiocarbohydrazone with Anticancer Properties and Catalytic Inhibition of Human DNA Topoisomerase IIα. *Pharmaceuticals* **2023**, *16*, 341. [CrossRef] [PubMed]

12. Keller, J.G.; Petersen, K.V.; Mizielinski, K.; Thiesen, C.; Bjergbæk, L.; Reguera, R.M.; Pérez-Pertejo, Y.; Balaña-Fouce, R.; Trejo, A.; Masdeu, C.; et al. Gel-Free Tools for Quick and Simple Screening of Anti-Topoisomerase 1 Compounds. *Pharmaceuticals* **2023**, *16*, 657. [CrossRef] [PubMed]
13. O'Donnell, A.C.; Berger, J.M. Structural Mechanisms of Topoisomerase-Targeting Drugs. *Annu. Rev. Biochem.* **2025**, *94*, 223–251. [CrossRef] [PubMed]
14. Radaeva, M.; Dong, X.; Cherkasov, A. The Use of Methods of Computer-Aided Drug Discovery in the Development of Topoisomerase II Inhibitors: Applications and Future Directions. *J. Chem. Inf. Model.* **2020**, *60*, 3703–3721. [CrossRef] [PubMed]

pharmaceuticals

Review

Prospects of Topoisomerase Inhibitors as Promising Anti-Cancer Agents

Prasanna Anjaneyulu Yakkala [1], Naveen Reddy Penumallu [2], Syed Shafi [3] and Ahmed Kamal [1,4,5,*]

[1] Department of Pharmaceutical Chemistry, School of Pharmaceutical Education and Research, Jamia Hamdard, New Delhi 110062, India; prasannayakkala@jamiahamdard.ac.in
[2] Department of Pharmacology, School of Pharmaceutical Education and Research, Jamia Hamdard, New Delhi 110062, India; naveenjnv97@gmail.com
[3] Department of Chemistry, School of Chemical and Life Sciences, Jamia Hamdard, Hamdard Nagar, New Delhi 110062, India; syedshafi@jamiahamdard.ac.in
[4] Department of Pharmacy, Birla Institute of Technology and Science (BITS) Pilani, Hyderabad Campus, Dist. Medchal, Hyderabad 500078, India
[5] Telangana State Council of Science & Technology, Environment, Forests, Science & Technology Department, Hyderabad 500004, India
* Correspondence: ahmedkamal@iict.res.in or ahmedkamal@hyderabad.bits-pilani.ac.in

Abstract: Topoisomerases are very important enzymes that regulate DNA topology and are vital for biological actions like DNA replication, transcription, and repair. The emergence and spread of cancer has been intimately associated with topoisomerase dysregulation. Topoisomerase inhibitors have consequently become potential anti-cancer medications because of their ability to obstruct the normal function of these enzymes, which leads to DNA damage and subsequently causes cell death. This review emphasizes the importance of topoisomerase inhibitors as marketed, clinical and preclinical anti-cancer medications. In the present review, various types of topoisomerase inhibitors and their mechanisms of action have been discussed. Topoisomerase I inhibitors, which include irinotecan and topotecan, are agents that interact with the DNA-topoisomerase I complex and avert resealing of the DNA. The accretion of DNA breaks leads to the inhibition of DNA replication and cell death. On the other hand, topoisomerase II inhibitors like etoposide and teniposide, function by cleaving the DNA-topoisomerase II complex thereby effectively impeding the release of double-strand DNA breaks. Moreover, the recent advances in exploring the therapeutic efficacy, toxicity, and MDR (multidrug resistance) issues of new topoisomerase inhibitors have been reviewed in the present review.

Keywords: cancer; clinical; marketed drugs; pre-clinical; topoisomerase

1. Introduction

Topoisomerases are a crucial class of enzymes that play a pivotal role in various DNA-related processes, including DNA replication and transcription. They intricately control DNA topology, enabling the strands to unwind, separate, and rejoin seamlessly during essential cellular processes [1]. There are mainly two categories of topoisomerases (TOPO): topoisomerase I (TOPO I) and topoisomerase II (TOPO II). Further, there are three subcategories of type I topoisomerases: type IA, IB, and IC [2]. Type IA topoisomerases, also known as bacterial topoisomerase I, require a single-stranded region in the DNA to bind [3]. They cleave one of the strands in double-stranded DNA and process a transient covalent bond (*via* tyrosine residue) with the 5′-phosphoryl group of the DNA [4,5]. This covalent attachment allows the enzyme to pass the unbroken DNA strand through the nick, altering the DNA topology. The process by which the enzyme allows the DNA to pass through is referred to as the "strand passage" mechanism. Type IA topoisomerases relax negative supercoils, resolve DNA knots, and facilitate DNA replication and transcription [6].

Whereas, type IB topoisomerases are discovered in both prokaryotes and eukaryotes, including humans. This subtype includes human topoisomerase I. Type IB topoisomerases break one strand of the DNA, similar to type IA enzymes [1]. However, they differ in the attachment of the catalytic site of tyrosine residue [7]. Type IB topoisomerases covalently link the tyrosine to the 3′-phosphoryl group of the DNA [8]. They use a mechanism called the "swivel" mechanism to relax the DNA supercoils. In this mechanism, the enzyme interacts with the DNA, breaks one strand, allows the DNA to rotate around the intact strand, and finally reveals the nick [9]. Type IB topoisomerases are involved in DNA replication, transcription, repair, and chromatin remodeling. Type IC topoisomerases are structurally similar to type IB enzymes and share some functional characteristics. They break a single strand in the DNA and covalently link with the catalytic-pocket tyrosine to the 3′-phosphoryl end of the DNA [10]. Type IC topoisomerases follow the swivel mechanism to relax the DNA supercoils, similar to type IB topoisomerases. These enzymes were primarily discovered in archaea, but they are also present in some bacteria and viruses. Their specific functions and role in DNA metabolism are still being studied [11].

Similarly, TOPO II also consists of two subtypes: type IIA and IIB. It is probable that TOPO IIA plays a key function in the cell division process. TOPO IIB is expressed in post-mitotic cells and may be important in regulating the expression of long genes, even at this early stage [12].

1.1. Topoisomerase I Mechanism of Action

The general mechanism of action (MOA) of TOPO I involves the relaxation of DNA supercoiling and the prevention of DNA torsional stress at the outset; TOPO I recognizes and binds to a specific DNA sequence, usually regions with single-stranded DNA, through non-covalent interactions [13]. Then, TOPO I introduces a reversible cleave in one of the DNA strands [14]. This break occurs by forming a covalent linkage between the active site or catalytic site of tyrosine (TYR) residue and the 3′-phosphoryl group of the DNA [15]. Inhibitors of TOPO I interfere with the denaturation and annealing of the DNA backbone, which is important for DNA replication and transcription. By blocking the action of TOPO enzymes, it is possible to disrupt the ligation step in DNA during replication or transcription, leading to DNA cleavage [16]. These breakdowns can disrupt the genomic integrity of cells and activate mechanisms, such as apoptosis or necrosis (Figure 1) [17].

Mammalian cells typically possess seven different active sites in topoisomerase enzymes (Figure 1), with four encoding type I enzymes and three encoding type II enzymes. Type I topoisomerases are monomeric enzymes discovered in mammalian cells [18]. They bind to double-stranded DNA and relieve torsional stress during transcription by cleaving one strand of DNA. TOPO I drugs or inhibitors are primarily employed in colorectal and ovarian cancer treatments [19]. Examples of topoisomerase I inhibitors include camptothecin derivatives, particularly camptothecin **9**, topotecan **10**, irinotecan **11**, and belotecan **12**, which are accepted by the U.S. Food and Drug Administration (FDA) for colon, lung, and ovarian cancer treatments, however, these drugs have certain limitations [20,21]. They can undergo spontaneous inactivation in the blood, requiring longer infusion times due to rapid drug reversal, and some cancer cells with increased membrane transporters may develop resistance to these drugs [11,13].

1.2. Mechanism of Action of Topoisomerase II

TOPO II is an enzyme that recognizes specific DNA sequences, usually regions with double-stranded DNA, during its mechanism of action (MOA) [22]. It cleaves both strands of the double helix, creating double-strand breaks in a coordinated manner [23,24]. Unlike type I topoisomerases, type II topoisomerase enzymes exist as multimeric complexes with α and β domains. Inhibitors of TOPO II induce double-strand breaks, leading to the activation of apoptosis in cells [6,25]. The capability of TOPO II in the denaturation and annealing of DNA is central to its active site. All topoisomerases utilize catalytic site tyrosine amino acid residues to facilitate DNA distraction and ligation. Type II enzymes, which break both strands of the double helix, have one of these tyrosine residues in each protomer subunit [11,26]. TOPO II starts DNA breakdown through the nucleophilic attack of the catalytic position of the tyrosine amino acid residue on the 3′ end of the phosphate group in the nucleic acid backbone [24].

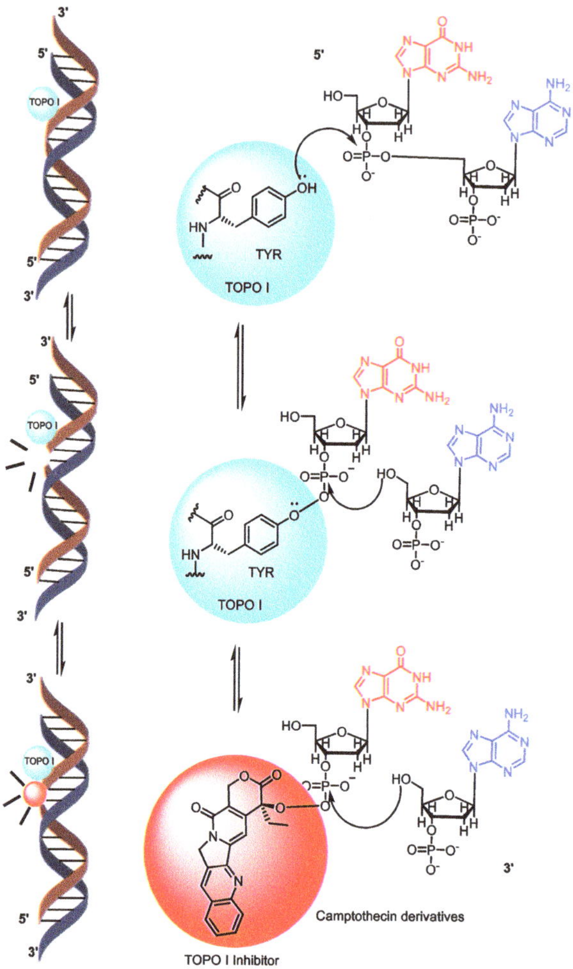

Figure 1. Mechanism of action of topoisomerase I.

As a result, a covalent phosphotyrosine bond is formed that binds the protein to the freshly generated 5′ end of the DNA chain [27,28]. Simultaneously, a 3′-hydroxyl group is generated on the opposite end of the cleaved strand. The cleavage occurs at staggered scissile bonds across the major groove in the double helix, generating four-base 5′ single-stranded cohesive ends that are covalently linked to separate protomer subunits of the enzyme. TOPO II inhibitors used in anti-cancer drugs exert their effects by interfering with the enzyme's mechanism of action. They induce double-strand breaks, leading to DNA damage and the activation of apoptosis in cancer cells [29]. Drugs, such as anthracycline, anthracenedione, acridine, and epipodophyllotoxin derivatives, are well-known inhibitors of topoisomerase II [10,30,31] as shown in Figure 2.

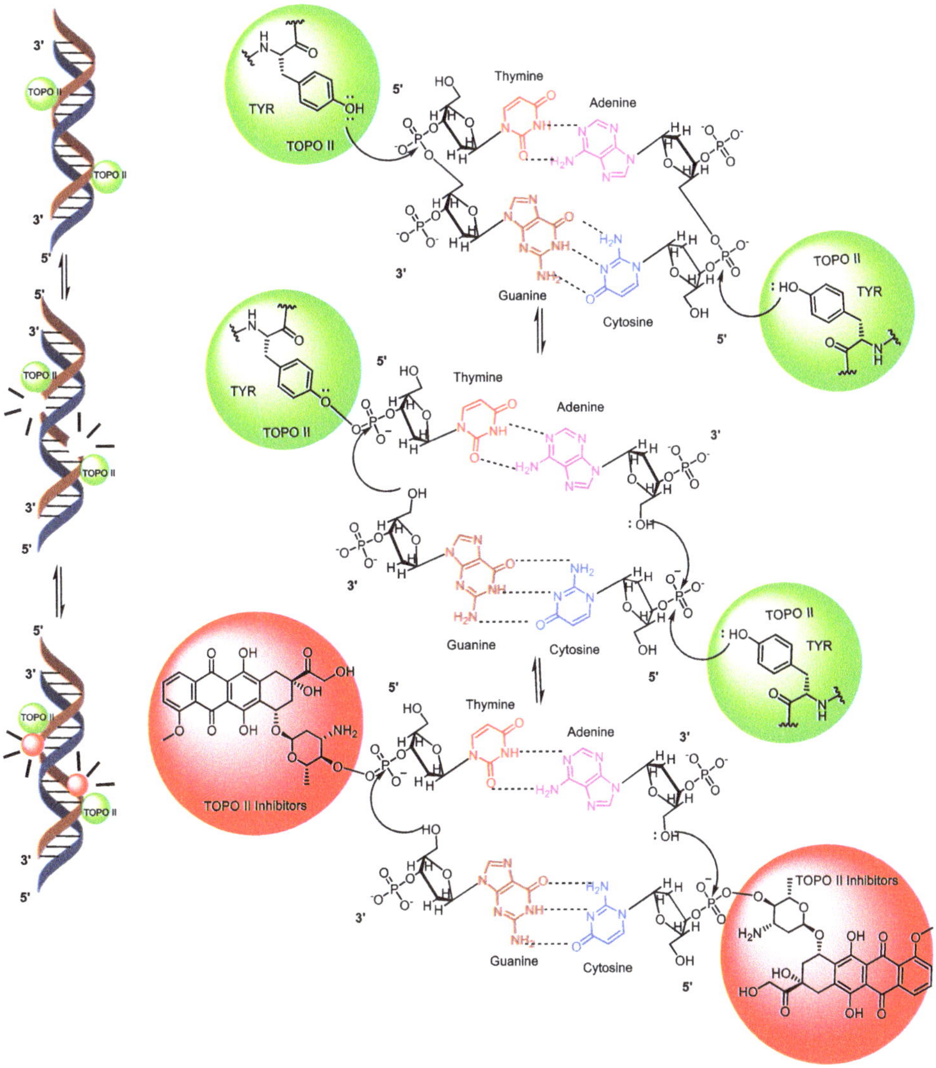

Figure 2. Topoisomerase II mechanism of action.

In recent literature, several noteworthy reviews have been published detailing the crystal structures of topoisomerase enzymes in a complex with their corresponding inhibitors. Specifically, these reviews have shed light on the structural characteristics of both topoisomerase I (TOPO I) and topoisomerase II (TOPO II) proteins. Here, we provide a comprehensive overview of the protein structures complex with TOPO, as revealed by these crystallographic studies [1,32,33] as depicted in Figure 3.

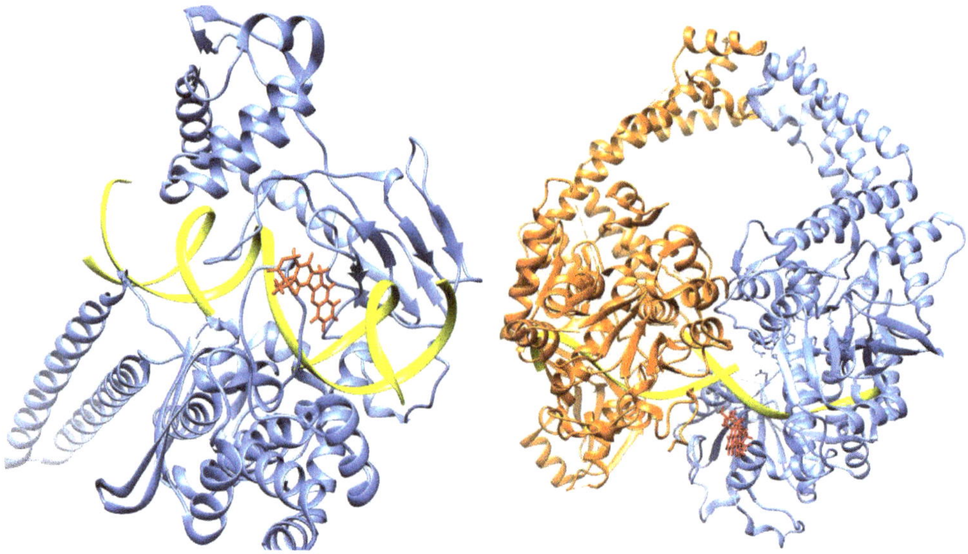

Figure 3. Structures of human topoisomerases. Shown are the structures of the full-length human topoisomerase I (left, PDB ID 1k4t) and topoisomerase IIβ (right, PDB ID 5gwi) enzymes with bound DNA, representative of the overall structure and domains of the two sub-family types.

2. Drugs in Clinical Usage

2.1. Anthracycline-Based Clinically Used DNA-Topoisomerase Inhibitors

The anthracycline agents were the primary class of TOPO inhibitors used in cancer treatment. They were initially discovered in bacterial Streptomyces species and found to have anti-proliferative and antibiotic properties [32]. Some clinically marketed anthracycline congeners are doxorubicin **1**, epirubicin **2**, valrubicin **3**, daunorubicin **4**, and idarubicin **5**. Doxorubicin, in particular, is used for various cancers, such as breast cancer, leukemia, lymphoma, sarcomas, carcinomas, and other tumors [34]. Daunorubicin and idarubicin are used for leukemia [35], while epirubicin is used after breast cancer surgery and valrubicin for urinary bladder carcinoma [36–38]. Anthracycline agents exert their cytotoxic effects by acting as "poisons" for topoisomerases, specifically type IIα topoisomerases (TOPO IIα and TOPO IIβ). They intercalate into DNA, bind to it, and are subsequently broken down by topoisomerases. This interaction stabilizes the DNA–TOPO complex, inhibiting DNA re-ligation [39,40].

This mechanism contributes to their ability to induce cell death. Additionally, anthracyclines generate free radicals in an iron-dependent manner, which further enhances their cytotoxic effects on cancer cells [41]. Recent studies advise that the inhibition of TOPO IIβ, which is excessively expressed in cardium, by anthracycline congeners may lead to cardiotoxicity through apoptosis and reactive oxygen species (ROS) production [42] (Figure 4, Table 1).

2.2. Anthracenedione and Acridine Derivatives

Anthracenedione agents, including mitoxantrone **6** and pixantrone **7**, are synthetic drugs that mimic the action of anthracycline drugs (Figure 5). Analogous to anthracyclines, anthracenedione agents act as topoisomerase poisons, primarily targeting TOPO II. Mitoxantrone intercalates into DNA bound by topoisomerase, avoiding DNA re-ligation and finally causing DNA breakage and interruption of the DNA repair processes. Pixantrone, an aza-anthracenedione, was accepted for the treatment of non-Hodgkin B-cell lymphoma. It depicts cytotoxic effects through DNA intercalation, similar to anthracyclines. Animal models have shown that treatment with doxorubicin leads to increased heart weight, whereas treatment with pixantrone does not have the same effect [43] (Table 1).

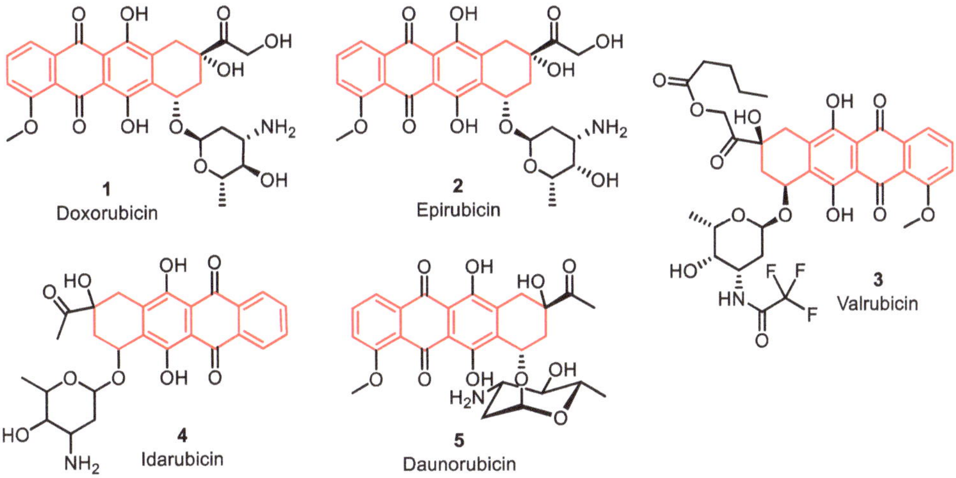

Figure 4. Anthracycline derivatives as DNA–topoisomerase inhibitors.

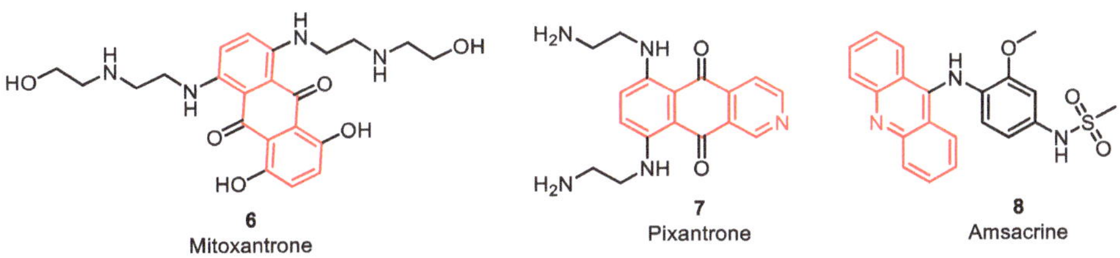

Figure 5. Anthracenedione and acridine analogues as DNA–TOPO inhibitors.

Amsacrine (m-AMSA **8**) is the only marketed agent in its chemical class. It is a synthetic drug that consists of an acridine ring (Figure 5). Similar to the previously discussed agents, amsacrine acts as a TOPO inhibitor, specifically aimed at type II TOPO. Notably, amsacrine was the first drug recognized to inhibit eukaryotic TOPO II. The acridine ring of the amsacrine is responsible for its intercalation into DNA and contributes to its activity. On the other hand, the non-interactive m-AMSA head group provides specificity for the DNA–TOPO cleavage complex. This combination of structural components enables amsacrine to exert its cytotoxic effects [44] (Table 1).

2.3. *Camptothecin Analogues*

Several clinically used camptothecin derivatives are available in the market, including camptothecin **9**, topotecan **10**, irinotecan **11**, and belotecan **12** (Figure 6). These derivatives stem from the camptothecin alkaloid, initially obtained from the Chinese tree *Camptotheca acuminata*. Camptothecin-derived congeners act initially as topoisomerase inhibitors affecting TOPO I. Many researchers have demonstrated that camptothecin inhibits TOPO I, leading to DNA strand breaks and inhibition of DNA replication. Irinotecan and topotecan are clinically accessible hydrophilic versions of camptothecin. They reversibly coordinate and form a ternary complex with TOPO I and DNA, as mentioned earlier [1,41]. The FDA approved it as a second-line treatment for lung cancer, with a cisplatin **27** combination for stage IV-B cervical carcinoma patients instead of surgery or radiation [42,43]. Irinotecan is used for metastatic colon or rectal carcinoma following treatment failure or progression with fluorouracil **26** [45–48]. It can also be given in combination with leucovorin **28** and 5-fluorouracil [49] (Figure 9). Belotecan, one of the comparatively newer camptothecin congeners, has been approved in South Korea for lung and ovarian cancer treatment since 2003. It shares the same MOA as other agents in its class. In comparison to previously known camptothecin-based agents, belotecan demonstrates similar efficacy with decreased toxicities [50] (Figure 6, Table 1).

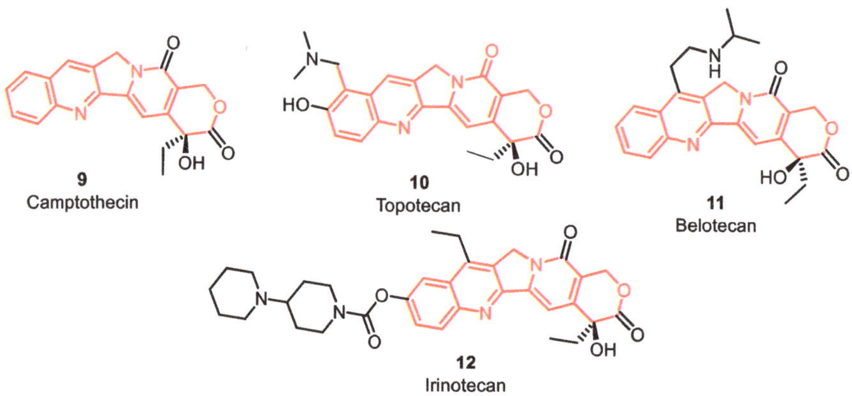

Figure 6. Camptothecin derivatives as DNA–topoisomerase inhibitors.

2.4. Epipodophyllotoxin Derivatives

Etoposide **13** and teniposide **14** (Figure 7) were obtained from epipodophyllotoxins, which are natural sources obtained from the mayapple plant (*Podophyllum peltatum*). These drugs act as topoisomerase poisons by binding to type II TOPO, similar to the derivatives discussed earlier. Etoposide has been used in combination with chemotherapy regimens for refractory testicular tumors and as part of the treatment for lung cancer in combination with cisplatin [51]. Teniposide is approved for use in combination with other chemotherapy medications to treat refractory pediatric acute lymphoblastic leukemia [52].

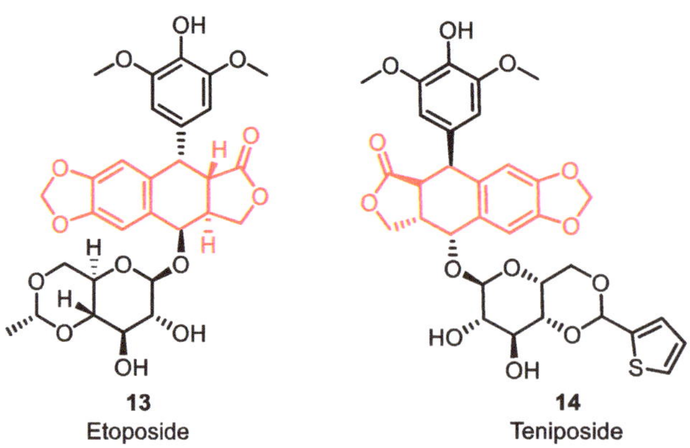

Figure 7. Epipodophyllotoxin derivatives as DNA–topoisomerase inhibitors.

Table 1. Marketed topoisomerase inhibitors.

Sl.	Drug	Class	Mechanism/Target	Indications	Adverse Drug Reactions	Ref.
1	Doxorubicin	Anthracycline	Topoisomerase IIα and IIβ poison, intercalation, free radicals	Various cancers	Cardiotoxicity, myelosuppression, nausea, potential for cumulative toxicity	[53,54]
2	Epirubicin	Anthracycline	Topoisomerase IIα and IIβ poison, intercalation, free radicals	Breast cancer	Cardiotoxicity, myelosuppression, nausea, potential for cumulative toxicity	[36]
3	Valrubicin	Anthracycline	Topoisomerase IIα and IIβ poison, intercalation, free radicals	Urinary bladder carcinoma	Local irritation, urinary symptoms, myelosuppression	[55]
4	Daunorubicin	Anthracycline	Topoisomerase IIα and IIβ poison, intercalation, free radicals	Leukemia	Cardiotoxicity, myelosuppression, nausea, potential for cumulative toxicity	[56]
5	Idarubicin	Anthracycline	Topoisomerase IIα and IIβ poison, intercalation, free radicals	Leukemia	Cardiotoxicity, myelosuppression, nausea, potential for cumulative toxicity	[57]
6	Mitoxantrone	Anthracenedione	Topoisomerase II poison, intercalation	Leukemia, prostate cancer, MS	Myelosuppression, potential for cumulative toxicity, potential for myelosuppression	[58]
7	Pixantrone	Anthracenedione	Topoisomerase II poison, intercalation	Non-Hodgkin B-cell lymphoma	Myelosuppression, potential for cumulative toxicity	[43]
8	Amsacrine	Acridine	Topoisomerase II poison, intercalation	Acute leukemia	Myelosuppression, potential for cumulative toxicity	
9	Camptothecin	Camptothecin	Topoisomerase I inhibitor, DNA strand breaks	Not specified	Gastrointestinal toxicity, myelosuppression, potential for cumulative toxicity	[59]
10	Topotecan	Camptothecin	Topoisomerase I inhibitor, DNA strand breaks	Small-cell lung cancer	Myelosuppression, gastrointestinal toxicity, potential for cumulative toxicity	[60,61]
11	Irinotecan	Camptothecin	Topoisomerase I inhibitor, DNA strand breaks	Colon and rectal carcinoma	Diarrhea, myelosuppression, potential for cumulative toxicity	[62]
12	Belotecan	Camptothecin	Topoisomerase I inhibitor, DNA strand breaks	Non-small-cell lung cancer, ovarian cancer	Myelosuppression, gastrointestinal toxicity, potential for cumulative toxicity	[50]
13	Etoposide	Epipodophyllotoxin	Topoisomerase II poison, intercalation	Testicular tumors, small-cell lung cancer	Myelosuppression, gastrointestinal toxicity, potential for cumulative toxicity	[63]
14	Teniposide	Epipodophyllotoxin	Topoisomerase II poison, intercalation	Childhood acute lymphoblastic leukemia	Myelosuppression, gastrointestinal toxicity, potential for cumulative toxicity	[64]

3. TOPO Inhibitors in Clinical Trials

This segment is focused on novel and new TOPO inhibitor scaffolds that were examined in human clinical trials, with a summary provided in Table 2. The clinical trials discussed in this section are categorized based on their phase (1, 2, or 3) and include relevant identifiers, such as National Clinical Trial (NCT) numbers or other regulatory agency identifiers. The information is sourced from the U.S. National Library of Medicine's clinical trials database and the WHO/ICMJE ISRCTN registry.

Table 2. Clinical trials on DNA–topoisomerase inhibitors (clinicaltrials.gov, accessed on 17 August 2023).

Sl.	Drug	Class	Mechanism/Target	Study Phase	NCT Number
1	Indenoisoquinolines 15, 16, 17	Non-camptothecin type I inhibitors	Stabilize DNA–topoisomerase cleavage complex, preferential DNA cleavage sites	Phase 1	NCT-01794104
2	Namitecan (ST1968)	Topoisomerase I inhibitor	Inhibits topoisomerase I, demonstrated anti-tumor activity	Phase 1	Not specified
3	Vosaroxin	Anti-cancer quinolone derivative (AQD)	Targets type II topoisomerases, induces DNA damage and apoptosis	Phase 2	NCT-02658487
4	Cytarabine	Nucleoside analogue	Incorporates into DNA, inhibits DNA synthesis	Phase 2	NCT-02658487
5	CRLX101	Camptothecin nanoparticle conjugate	Increases tumor cell exposure to camptothecin, tumor-specific targeting	Phase 2	NCT-01380769
6	Pixantrone	Anthracenedione	Induces DNA damage, anti-tumor activity	Phase 3	NCT-01321541
7	Aldoxorubicin	Pro-drug of doxorubicin	Delivers doxorubicin directly to tumor tissue	Phase 3	NCT-02049905
8	Silatecan	Silicon-containing camptothecin derivative	Inhibits topoisomerase I, being evaluated for gliosarcoma	Phase 2	NCT-01124539
9	Becatecarin	Rebeccamycin analogue	Dual topoisomerase I and II poison, clinical development ceased	Phase 2	NCT-00132600
10	Edotecarin	Rebeccamycin analogue	Dual topoisomerase I and II poison, clinical development ceased	Phase 2	NCT-02310763

3.1. Topoisomerase Inhibitors in Phase 1 Clinical Trials

The U.S. National Cancer Institute (NCI) conducted a phase 1 clinical trial (NCT-01794104, (Figure 8, Table 2) investigating a novel class of non-camptothecin type I topoisomerase poisons, called indenoisoquinolines, indotecan **15**, inimitecan **16**, and indenoisoquinoline NSC 314622 **17**, for neoplasm lymphoma [65]. Indenoisoquinolines create a stable DNA–TOPO cleavage complex like camptothecin analogues, but exhibit a preference for specific DNA cleavage sites. This preference enables them to effectively target camptothecin-resistant cell lines. These derivatives are chemically stable and target their action on cells that overexpress ATP-binding cassette transporters ABCG2 and P-glycoprotein (MDR1). By stabilizing the breakdown complex, they induce DNA destruction, representing their efficacy as potent anti-cancer treatments. Furthermore, indenoisoquinolines delay DNA repair, leading to apoptosis. After five days of administration, LMP400 has linear pharmacokinetics, at which point drug aggregation is seen. Weekly dosing is thought to raise the drug's peak levels, resulting in improvements in safety and efficacy [66].

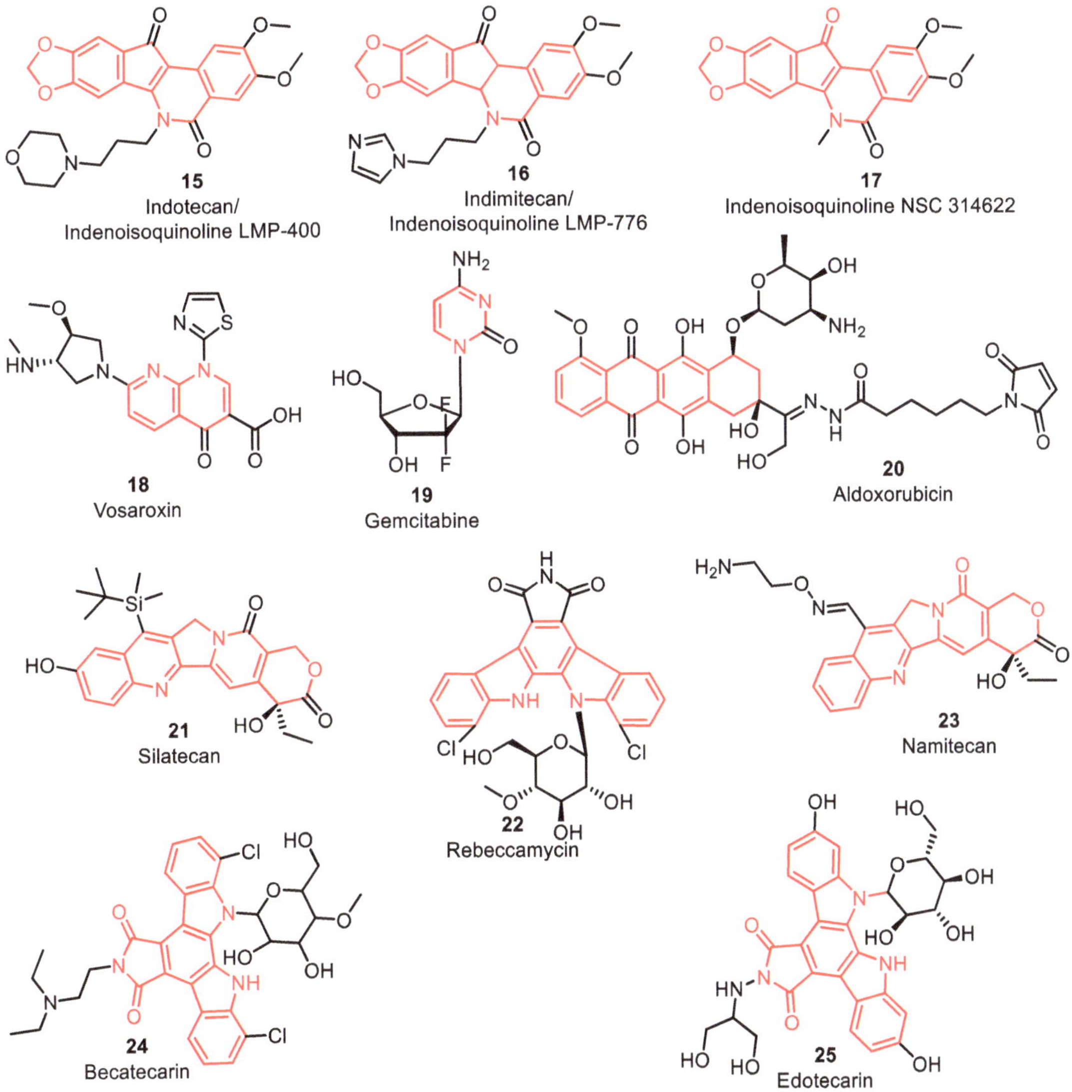

Figure 8. Representative DNA–TOPO inhibitors in clinical trials.

Namitecan (ST1968) **23** is a TOPO I inhibitor that has demonstrated good anti-proliferative activity and has a far better safety profile compared to topotecan **10** and irinotecan **12**. Pharmacokinetic studies using repeated dosing schedules have shown no production or accumulation of metabolites due to its short half-life. Current research has validated the safety and pharmacokinetic profile of namitecan, including manageable neutropenia, and has shown efficacious anti-proliferative activity, with positive responses observed in endometrium and bladder cancers [67].

3.2. Topoisomerase Inhibitors in Phase 2 Clinical Trials

The Vanderbilt-Ingram Cancer Center enrolled subjects for phase 2 clinical trials (NCT-02658487) to evaluate the effectiveness of vosaroxin **18** and cytarabine **29** (Figure 9) in treating patients with untreated acute myeloid leukemia. Vosaroxin is an anti-proliferative quinolone conjugate that targets type II TOPO (Figure 8). Quinolone conjugates act on the DNA–TOPO to breakdown complex and intercalate DNA at exact GC-rich sites to avoid DNA annealing by TOPO. It leads to specific site DNA destruction, prolongation of the S-phase, G2-phase cell cycle arrest, and finally leads to apoptosis [32]. Vosaroxin consists of a quinolone core, which makes it less active compared to the remaining classes of TOPO inhibitors. It produces fewer toxic metabolites, ROS and cardiotoxicity. Furthermore, vosaroxin can induce p53-independent apoptosis, making it effective against drug resistance mechanisms associated with p53 inactivation. Its stable quinolone structure is not extensively metabolized by enzymes or capable of inducing or inhibiting p450 activity, minimizing the potential for drug–drug interactions and even enhancing the activity of other anti-cancer drugs, such as cytarabine [68].

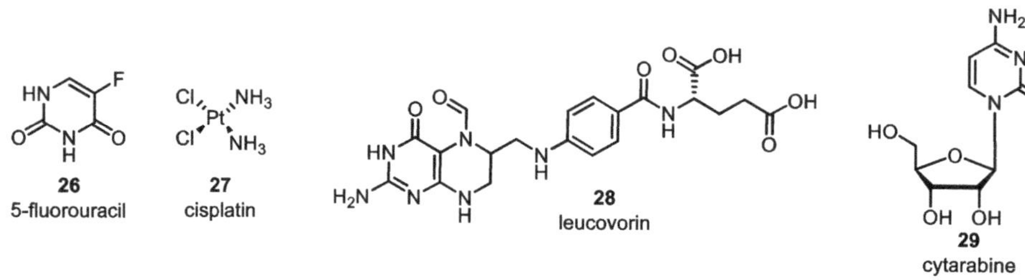

Figure 9. Miscellaneous DNA–topoisomerase inhibitors.

The primary goal of this research is to assess the proportion of patients achieving complete remission after undergoing initial treatment with a combination of vosaroxin and cytarabine. This study focuses on individuals who have recently been diagnosed with acute myelogenous leukemia and have not received any prior treatment. The aim is to evaluate the effectiveness of the vosaroxin and cytarabine combination during the induction therapy phase [69]. In another phase 2 clinical trial completed by NewLink Genetics Corporation (NCT-01380769), the impact of CRLX101 on the average survival of patients with advanced non-small-cell lung cancer (NSCLC) was studied. CRLX101 is a camptothecin nanoparticle conjugated to a cyclodextrin-based polymer, designed to increase tumor cell exposure to camptothecin, while minimizing side effects. The nanoparticle size facilitates tumor-specific targeting by extravasating from the leakier blood vessels found in tumors [70] (Figure 8).

3.3. *Topoisomerase Inhibitors in Phase 3 Clinical Trials*

CTI BioPharma performed a phase 3 study (NCT-01321541) comparing the ability of pixantrone **7** with rituximab to gemcitabine **19** with rituximab in 260 patients with relapsed or refractory diffuse large B-cell lymphoma or follicular grade 3 lymphoma. The major results measure in the study was progression-free survival (PFS), with secondary outcome measures including total survival, and complete and total response rate, as well as safety evaluations, such as adverse events and laboratory values falling outside predetermined ranges [71]. Aldoxorubicin **20**, a pro-drug of doxorubicin, is considered a promising option for the treatment of soft tissue sarcomas [72]. It contains a carboxylic hydrazine that binds to albumin in the blood and is then released in the acidic tumor environment, delivering doxorubicin directly to the tissue. A phase 3 study, sponsored by CytRx (NCT-02049905), investigated the administration of aldoxorubicin to patients with soft tissue sarcomas, with the active comparator being the investigator's choice among various treatment options. The study assessed the overall survival, safety parameters, and tumor response. Other notable topoisomerase inhibitors undergoing clinical trials include silatecan **21**, a silicon-containing camptothecin analogue [73], and rebeccamycin analogs obtained from the natural product rebeccamycin **22** [74]. Silatecan is being evaluated in a phase 2 study for gliosarcoma (NCT-01124539), while rebeccamycin analogs, such as becatecarin **24** and edotecarin **25**, have progressed to phase 2 trials. These compounds exhibit dual TOPO I and II inhibitors.

However, the clinical development of becatecarin and edotecarin has stopped, and it remains uncertain whether additional rebeccamycin compounds will be tested clinically (Figure 8) [75].

4. Topoisomerase Inhibitors in Preclinical Studies

4.1. Naphthalimide–Benzothiazole Derivatives

Rao et al. reported novel naphthalimide–benzothiazole derivatives as topoisomerase IIα inhibitors. Among the series of compounds, compounds **30** and **31**, containing the 6-aminobenzothiazole ring, exhibited significant cytotoxic activity against lung cancer (IC_{50}: 4.074 and 3.890 μM) and colon cancer (IC_{50}: 3.715 and 3.467 μM) cell lines when compared to the standard compound (amonafide **32**) (IC_{50}: 5.459 and 7.762 μM). They also investigated the DNA-binding properties of the active analogues using various techniques, such as DNA viscosity, CD, UV/Vis, fluorescence spectroscopy, and molecular docking, revealing a strong intercalation between the two DNA strands. Additionally, the most potent analogues, **30** and **31**, were successfully inhibited by DNA TOPO II (Figure 10) [76].

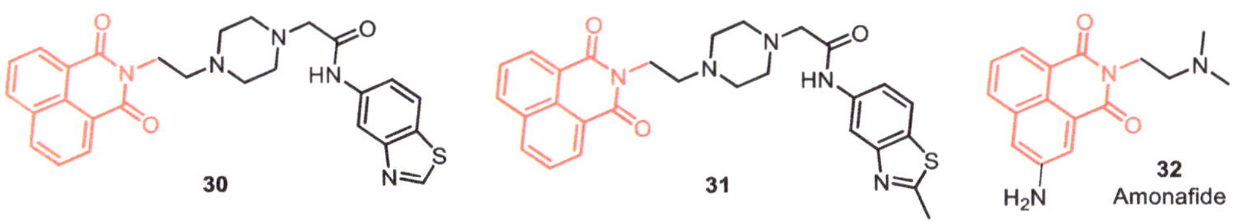

Figure 10. Naphthalimide–benzothiazole derivatives DNA–topoisomerase inhibitors.

4.2. β-Carboline Hybrids as Topoisomerase Inhibitors

In our previous studies, we have described the topoisomerase I inhibitory potential of β-carboline hybrids (**33–58**). A series of new β-carboline hybrids were prepared by introducing a phenyl group at the C1 position, along with chalcone/(N-acetyl)/pyrazole molecules at the position C3, and all the synthesized analogues were assessed for their anti-proliferative activity. Additionally, DNA photocleavage studies demonstrated that two of the analogues, compound **36** and compound **49**, successfully cleaved pBR322 plasmid DNA upon UV light irradiation. The potent hybrid compound **49** efficiently inhibited DNA TOPO I activity and maintained the DNA in the supercoiled form (Figure 11) [77].

In another study, dithiocarbamate-linked β-carboline analogues were reported as DNA TOPO II inhibitors. Among the derivatives, compounds **59** and **60** displayed a prominent anti-proliferative profile, with IC_{50} values of 1.34 μM and 0.79 μM on the DU-145 cell line, respectively. Both biophysical investigations and in silico studies indicated a complexion-type interaction between these analogues and DNA, distinguishing them from simple β-carbolines. To gain insight into their MOA, a DNA TOPO II inhibition assay was also performed (Figure 11) [78].

Similarly, C3-trans-cinnamide linked β-carboline analogues were described as significant anti-proliferative and DNA TOPO I poisons. These analogues were subjected to the cytotoxic profile against selected human cancer cell lines. The results indicated that the newly designed analogues displayed prominent activity against all the tested cell lines, having IC_{50} values in the range of 13–45 nM. Particularly, conjugates **61** and **62** demonstrated the highest activity against the breast cancer cell line (MCF-7 cells), with IC_{50} values of 14.05 nM and 13.84 nM, respectively. They also investigated the TOPO I inhibition assay, DNA-binding affinity, and in silico studies, disclosing that these novel analogues function as DNA-interactive TOPO I inhibitors (Figure 11) [79].

In related studies, β-carboline-combretastatin carboxamides were reported as DNA intercalation and TOPO II inhibitors. They were subjected to such studies for their potential DNA-binding affinity, cytotoxicity, and TOPO II inhibition activity. Among them, compounds **63** and **64** demonstrated potent cytotoxic effects against the A549 cell line, with

IC_{50} values of 1.01 μM and 1.17 μM, respectively. The most potent conjugate, **63**, was tested for DNA TOPO II inhibition activity (Figure 11) [80].

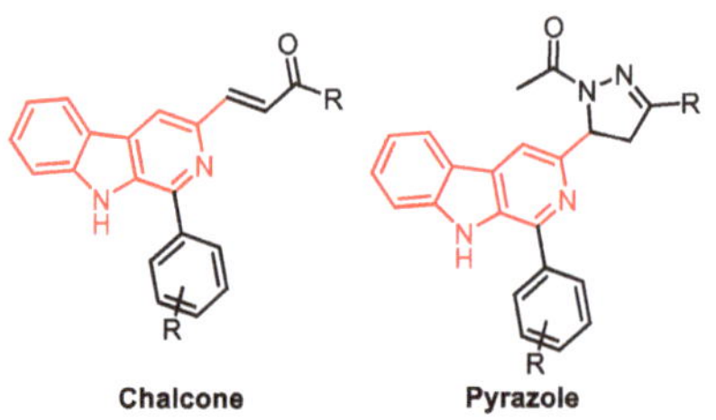

Chalcone	R	R2	Pyrazole
33	3-Cl	2-pyrrolyl	46
34	3-Cl	2-furanyl	47
35	4-CF3	2-pyrrolyl	48
36	4-CF3	2-furanyl	49
37	4-CF3	2-thiophenyl	50
38	4-CF3	4-amiophenyl	51
39	3,4,5-tri-OMe	2-thiophenyl	52
40	3,4,5-tri-OMe	2-furanyl	53
41	3,4,5-tri-OMe	4-amiophenyl	54
42	3,4-di-F	2-thiophenyl	55
43	3,4-di-F	4-amiophenyl	56
44	3,4-di-F	2-furanyl	57
45	3,4-di-F	3,4,5-trimethoxy phenyl	58

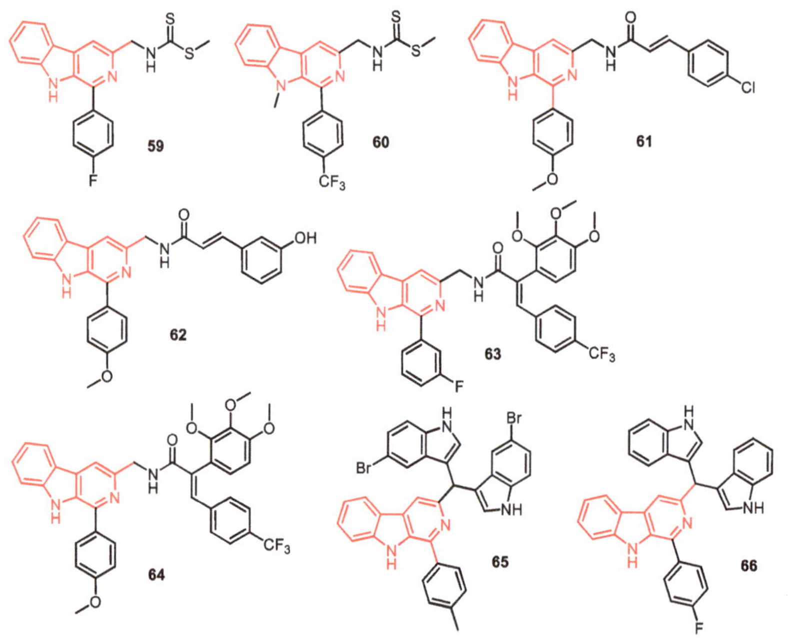

Figure 11. β-Carboline analogues as DNA–topoisomerase inhibitors.

In another example, β-carboline-*bis*indole analogues were described as DNA binding, photocleavage agents, and TOPO I inhibitors. A series of β-carboline-*bis*indole analogues were synthesized and subjected to studies for their anti-proliferative activity against various human cancer cell lines. All the analogues demonstrated significant anti-proliferative activity. Particularly, compounds **65** and **66** exhibited noteworthy activity against DU-145 cells, with IC_{50} values of 1.86 μM and 1.80 μM, respectively. The active conjugates significantly inhibited DNA TOPO I enzyme and were capable of cleaving the pBR322 plasmid under UV light irradiation. The results also reported a combilexin-type interaction between the compounds and DNA (Figure 11) [81].

4.3. Imidazopyridinyl-1,3,4-Oxadiazole Derivatives

A series of imidazopyridinyl-1,3,4-oxadiazole analogues were synthesized and subjected to an anti-proliferative assay, revealing promising results for some compounds. Notably, compound **67** (NSC 763639) exhibited significant growth inhibition with a single dose (10 μM) across all human cancer cell lines, meeting the threshold criteria. This compound was used in five different dose levels (0.01, 0.1, 1, 10, and 100 μM) and yielded

GI_{50} values ranging between 1.30 to 5.64 μM. Conjugate **67** exhibited significant inhibition of TOPO II activity, as observed in the TOPO II-mediated DNA relaxation assay (Figure 12) [82].

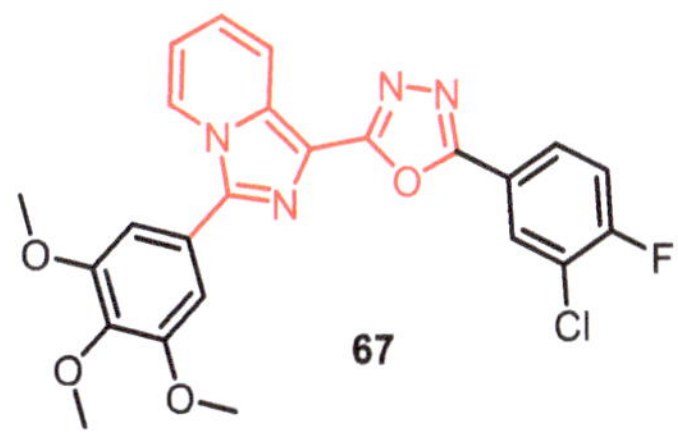

Figure 12. Imidazopyridinyl-1,3,4-oxadiazole derivatives as DNA–TOPO inhibitors.

4.4. *Pyrazole-Linked Benzothiazole-β-Naphthol Derivatives*

Pyrazole-linked benzothiazole-β-naphthol analogues were synthesized using environmentally friendly methods without the need for catalysts, yielding good to excellent yields. Notably, some derivatives, **68**, **69**, and **70**, exhibited prominent cytotoxicity against human cervical cancer cells (HeLa), with IC_{50} values ranging from 4.63–5.54 μM. Furthermore, these derivatives effectively inhibited TOPO I activity (Figure 13) [83].

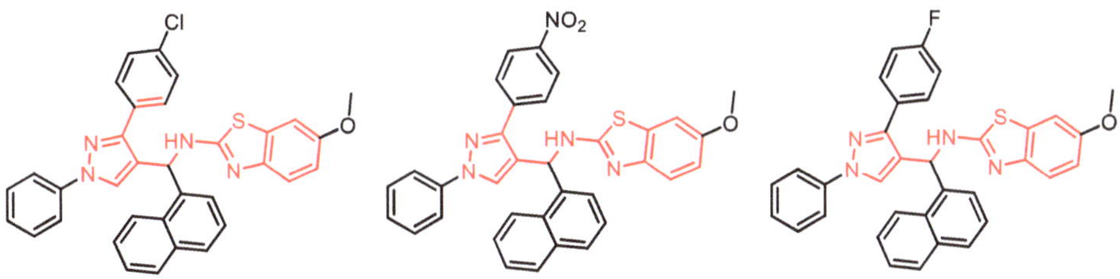

Figure 13. Pyrazole-linked benzothiazole-β-naphthol analogues as DNA–topoisomerase inhibitors.

4.5. *Podophyllotoxin Congeners*

Kamal and others reported a novel class of polyaromatic podophyllotoxin (**71–82**) congeners, specifically 4β-N-polyaromatic substituted podophyllotoxins, as DNA TOPO inhibitors (Figure 14) [84].

Similarly, novel 4β-sulphonamido and 4β-[(4-sulphonamido)benzamide] podophyllotoxin analogues demonstrated potential DNA topoisomerase IIα inhibitory potential. Among them, compounds **83**, **84**, and **85** exhibited greater potency than etoposide, a known anti-cancer drug. Additionally, compound **84** triggered both single-strand and double-strand DNA breaks, as noticed through a comet assay and c-H2AX analysis, respectively. Western blot analysis and related studies confirmed that compound **84** inhibited TOPO IIα activity. The compounds also activated ATM and Chk1 proteins, indicating effective DNA damage. Moreover, compound **84** induced apoptotic cell death, as evidenced by the triggering of caspase-3, the upregulation of p21 and p16, the downregulation of NF-kB, and the decreased expression of the Bcl-2 protein (Figure 14) [85].

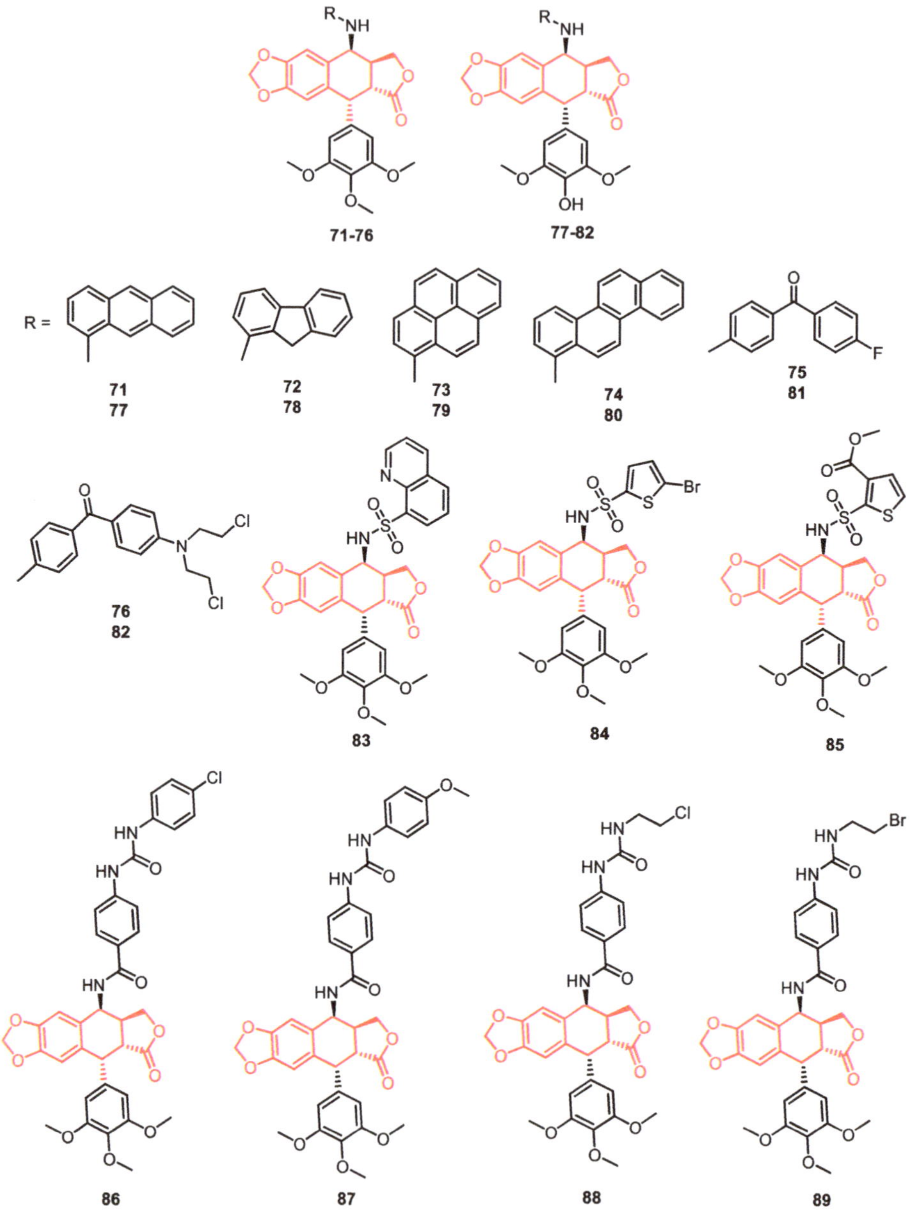

Figure 14. Podophyllotoxin congeners as DNA–topoisomerase inhibitors.

In another study, 4β-[4′-(1-(aryl)ureido)benzamide]podophyllotoxin congeners were reported as DNA TOPO I and IIα inhibitors. Some conjugates, **86, 87, 88**, and **89**, exhibited significant anti-proliferative activity in Colo 205 cells, surpassing the effectiveness of etoposide. Furthermore, enhanced inhibitory activities against DNA TOPO I and IIα enzymes were demonstrated. The active analogues also induced apoptosis by upregulating the caspase-3 protein, as observed through Western blotting and ELISA analysis (Figure 14) [86].

A highly efficient one-pot iodination method, utilizing zirconium tetrachloride and sodium iodide, has been successfully developed for synthesizing benzothiazolo-4β-anilino-podophyllotoxin (**90–97**) and benzothiazolo-4β-anilino-4-O-demethylepipodophyllotoxin (**98–105**) analogues. Selected representative conjugates were assessed with an anti-proliferat ive assay against specific human cancer cell lines and their capability to inhibit DNA TOPO II activity (Figure 15) [87].

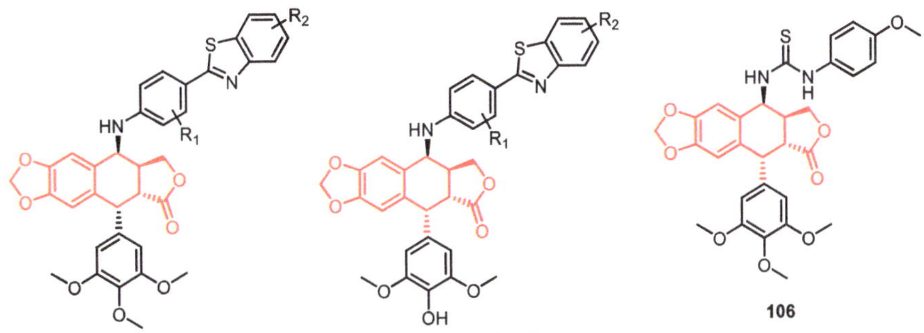

90-97 **98-105** **106**

Compound **90, 98**	R1 = H	R2 = H
Compound **91, 99**	R1 = 3-Cl	R2 = H
Compound **92, 100**	R1 = 2-Me	R2 = H
Compound **93, 101**	R1 = H	R2 = 6-OMe
Compound **94, 102**	R1 = H	R2 = 6-F
Compound **95, 103**	R1 = H	R2 = 4-Cl
Compound **96, 104**	R1 = H	R2 = 4,6 dichloro
Compound **97, 105**	R1 = 2-Br	R2 = H

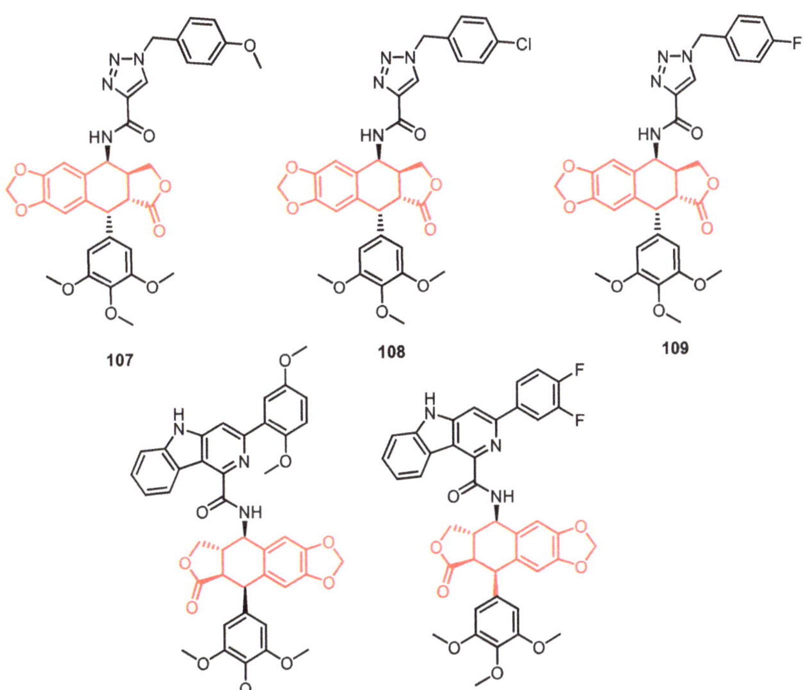

107 **108** **109** **110** **111**

Figure 15. Podophyllotoxin derivatives as DNA–topoisomerase inhibitors.

Similarly, podophyllotoxin–thiourea congeners were reported as DNA TOPO II inhibitors. Among them, selective cytotoxicity was observed on DU-145 (human prostate cancer) cells, with compound **106** displaying the most potent activity (IC_{50} of 0.50 ± 0.03 μM). Importantly, it demonstrated a favorable safety therapeutic window, exhibiting a 4-fold difference in potency compared to the non-cancerous human prostate cell line (RWPE-1, IC_{50} of 40.85 ± 0.78 μM). Flow cytometric analysis of this compound (**106**) revealed a significant G2/M-phase arrest and notable inhibition of TOPO II activity (Figure 15) [88].

In another study, a series of novel 4β-amidotriazole-linked podophyllotoxin congeners were designed and synthesized using click chemistry and assessed for their biological activities. Notably, compounds **107**, **108**, and **109** displayed remarkable cytotoxicity, with IC_{50} values of less than 1 μM against the tested cancer cell lines, surpassing the activity of the reference etoposide. Furthermore, these derivatives efficiently inhibited the activity of TOPO II, as demonstrated by topoisomerase-mediated DNA relaxation assays (Figure 15) [89].

Similarly, podophyllotoxin-linked β-carbolines analogous were reported as a significant anti-cancer agent and DNA TOPO II inhibitor. Among them, compounds **110** and **111** exhibited the highest cytotoxicity against the DU-145 cell line, with IC_{50} values of 1.07 ± 0.07 μM and 1.14 ± 0.16 μM, respectively. They also investigated cell cycle analysis, DNA-binding studies, a comet assay, TOPO II inhibition, and molecular modelling, which revealed that these derivatives interact with DNA and function as inhibitors of topoisomerase II (Figure 15) [90].

4.6. Benzimidazoles Congeners

Benzimidazoles have demonstrated their ability to disrupt DNA topoisomerases, as the significant class of enzymes involved in DNA manipulation. Notably, bibenzimidazole and terbenzimidazole compounds have emerged as distinctive Top1 poisons, constituting a class that stems from structural modifications of Hoechst 33342 (**112**), a blue fluorescent dye employed in DNA staining for molecular biology applications (Figure 16) [91].

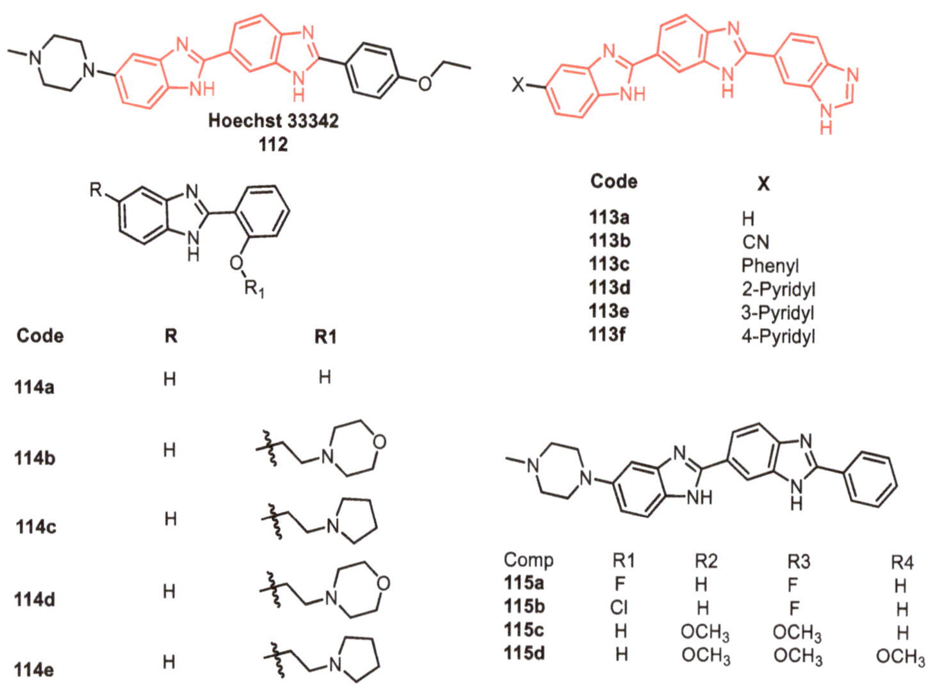

Figure 16. Benzimidazole derivatives as DNA–topoisomerase inhibitors.

The LaVoie Laboratory has been instrumental in advancing the understanding of terbenzimidazoles' impact on DNA. Multiple terbenzimidazoles were identified to induce DNA breakdowns in the presence of mammalian TOPO I. These conjugates were also subjected to analysis for their cytotoxic effects on various human cancer cell lines, supporting the notion of TOPO I-mediated DNA cleavage. Comparisons with the TOPO I poison compound **112,** based on Hoechst 33342, highlighted the comparable potency of several terbenzimidazoles. Compounds **113a** and **113b** displayed analogous TOPO I inhibition to that of compound **112** (Figure 15). The 5-phenyl-substituted terbenzimidazole **113c** exhibited approximately half the potency of compound **112** as a TOPO I poison. Meanwhile, the 3- and 4-pyridyl analogs **113e** and **113f** demonstrated greater activity than the 2-pyridyl derivative **113d,** both as TOPO I poisons and anti-proliferative agents (Figure 16) [92].

In 2009, Coban et al. [93] conducted research involving congeners of benzimidazole with alteration mainly at the second and fifth positions. Remarkably high activity was observed with analogs **114a** and **114c** (95.4% and 90.2% supercoiled vs. relaxed DNA bands), showing significant TOPO 1 inhibition and cytotoxicity (Figure 15). Additionally, in 2011, a series of benzimidazole congeners (2-aryl-5-substituted-2,5-bisbenzimidazole) were synthesized and subjected to analysis for their capacity to induce DNA cleavage in the presence of TOPO I (**115a–d**). These derivatives exhibited significant cancer growth inhibition against tested cancer cell lines, with IC_{50} values falling within the micromolar range (Figure 16) [94,95].

Ananda and others developed and tested a library of 11 dual inhibitors (DiPT-1 to DiPT-11) targeting PARP1 and TOP1. DiPT-4 (**116**) emerged as a standout, effectively inducing DNA damage, cell cycle arrest, and apoptosis, while sparing normal cells. DiPT-4 stabilizes the TOPO I-DNA complex, intensifying DNA breaks and impeding PARylation, countering resistance. It is a promising candidate for improved single-agent therapy, circumventing toxicities and enhancing clinical efficacy (Figure 17) [96].

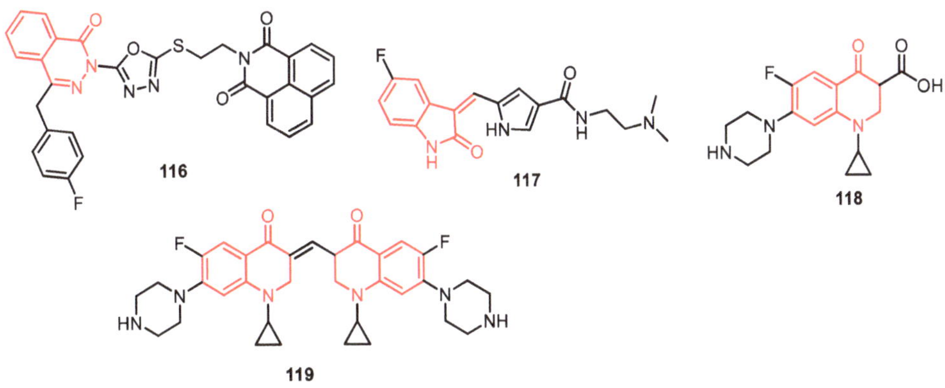

Figure 17. Miscellaneous derivatives as DNA–topoisomerase inhibitors.

Chauhan et al. reported on fluoroquinolones, known as bacterial topoisomerase inhibitors and vital antibacterial agents, which also exhibit anti-proliferative effects attributed to their impact on eukaryotic TOPO II, inducing DNA damage similar to TOPO II poisons [97]. Additional mechanisms, like tumor growth factor regulation, might contribute to their anti-proliferative actions. Leveraging the structures of tyrosine kinase inhibitor sunitinib (**117**) and bacterial topoisomerase inhibitor ciprofloxacin (**118**) [98], researchers designed hybrid molecules (e.g., HMNE3, **119**), integrating features of both drugs. HMNE3 displayed notable inhibitory activity against tyrosine kinases and TOPO II, demonstrating cytotoxicity across multiple cell lines (Panc-1, T24, BGC-823, PU145, HCG-27, Capan-1), with nanomolar IC_{50} values [99]. Since TOPO II and the epidermal growth factor receptor (EGFR) mutually influence expression, targeting them simultaneously emerges as a promising anti-cancer approach. Novel dual inhibitors of TOPO II and EGFR have been developed to enhance treatment efficacy (Figure 17) [100].

5. Conclusions

In conclusion, topoisomerase inhibitors have emerged as potential anti-cancer medications due to their ability to disrupt the normal function of these enzymes and induce DNA damage, leading to cell death. This review has discussed different types of TOPO inhibitors, including type I and II inhibitors, and their mechanisms of action in the treatment of cancer. TOPO I inhibitors stabilize the cleavage complex, while TOPO II inhibitors cleave the complex with DNA, resulting in DNA breaks. These inhibitors have shown therapeutic efficacy in the treatment of various cancers, such as breast, lung, and ovarian cancer. The development of targeted therapies that selectively inhibit overexpressed isoforms of topoisomerases in cancer cells holds great promise. By specifically targeting these isoforms, the inhibitors can disrupt DNA replication and repair processes in cancer cells, while minimizing damage to normal cells. Furthermore, combining TOPO inhibitors with other anti-cancer agents, such as chemotherapeutic drugs, immunotherapies, or targeted therapies, can enhance treatment outcomes and overcome drug resistance.

Researchers have also explored natural products and structural modifications to discover and optimize new topoisomerase inhibitors. Additionally, advances in structural biology, computational modelling, and virtual screening techniques have facilitated structure-based drug design, leading to the development of inhibitors with improved drug-like properties and the ability to overcome drug resistance mechanisms. These recent advances in the discovery and design of topoisomerase inhibitors offer promising avenues to address drug resistance, enhance treatment outcomes, and progress the overall efficacy and safety of cancer therapies. New methods to improve the therapeutic efficiency include nano-delivery systems and microneedle patches. Continued research and development in this field holds great potential for advancing cancer treatment approaches.

6. Recent Advances in the Discovery of New and Novel Topoisomerase Inhibitors

Researchers have been working on developing targeted therapies that specifically inhibit the isoforms of topoisomerases overexpressed in cancer cells. By selectively targeting these isoforms, the inhibitors can effectively disrupt DNA replication and repair processes in cancer cells, while minimizing damage to normal cells. Combining topoisomerase inhibitors with other anti-cancer agents has shown promise in overcoming drug resistance and enhancing treatment outcomes. Combinations with chemotherapy drugs, immunotherapies, or targeted therapies can synergistically enhance the cytotoxic effects and overcome resistance mechanisms. Dual inhibitors simultaneously target multiple isoforms of topoisomerases or combine topoisomerase inhibition with the inhibition of other essential cellular processes. This approach enhances therapeutic efficacy and minimizes the likelihood of resistance development. Natural products, such as marine-derived compounds and plant extracts, continue to be a valuable source for the discovery of new topoisomerase inhibitors. Structural modifications and synthesis of analogues derived from these natural compounds can enhance their potency, selectivity, and pharmacokinetic properties. Researchers have been exploring novel DNA binding modes for topoisomerase inhibitors to enhance their affinity and specificity. By targeting unique DNA-binding sites or inducing structural changes in DNA, these inhibitors can disrupt topoisomerase function more effectively. Advances in structural biology, computational modeling, and virtual screening techniques have enabled the rational design of topoisomerase inhibitors with improved drug-like properties. By utilizing the three-dimensional structures of topoisomerases, researchers can design inhibitors with optimized binding interactions and reduced off-target effects. Understanding the molecular mechanisms of drug resistance has paved the way for the development of inhibitors that can overcome these resistance mechanisms. Strategies include the development of inhibitors that are not susceptible to drug efflux pumps, bypass resistant mutations, or target alternative DNA repair pathways. These recent advances in the discovery of topoisomerase inhibitors offer promising avenues to address drug resistance, enhance treatment outcomes, and improve the overall efficacy and safety of cancer therapies.

Funding: This research was funded by the Department of Science and Technology (DST), the Government of India, grant number DST/IMRCD/BRICS/PilotCall2/CCT/2018-G, the Russian Foundation for Basic Research, grant number 18-515-80028, and the National Research Foundation (NRF) of South Africa, grant number 116014, under the BRICS STI cooperation program, and grant number BRICS2017-236.

Institutional Review Board Statement: Not applicable.

Informed Consent Statement: Not applicable.

Data Availability Statement: Data sharing is not applicable.

Conflicts of Interest: The authors declare no conflict of interest.

Abbreviations

ATP: adenosine triphosphate; ADP: adenosine diphosphate; TYR: tyrosine; AQD: anti-cancer quinolone derivative; DNA: deoxyribonucleic acid; NCI: National Cancer Institute; NCT: national clinical trial; PD: pharmacodynamics; TOPO I: topoisomerase 1; TOPO II: topoisomerase II; MTD: maximum tolerated dose; SAR: structure–activity relationships; PFS: progression free survival.

References

1. Delgado, J.L.; Hsieh, C.M.; Chan, N.L.; Hiasa, H. Topoisomerases as Anticancer Targets. *Biochem. J.* **2018**, *475*, 373–398. [CrossRef] [PubMed]
2. Jain, C.K.; Pradhan, B.S.; Banerjee, S.; Mondal, N.B.; Majumder, S.S.; Bhattacharyya, M.; Chakrabarti, S.; Roychoudhury, S.; Majumder, H.K. Sulfonoquinovosyl Diacylglyceride Selectively Targets Acute Lymphoblastic Leukemia Cells and Exerts Potent Anti-Leukemic Effects In Vivo. *Sci. Rep.* **2015**, *5*, 12082. [CrossRef] [PubMed]
3. Jang, J.Y.; Kim, D.; Kim, N.D. Recent Developments in Combination Chemotherapy for Colorectal and Breast Cancers with Topoisomerase Inhibitors. *Int. J. Mol. Sci.* **2023**, *24*, 8457. [CrossRef]
4. Capranico, G.; Marinello, J.; Chillemi, G. Type I DNA Topoisomerases. *J. Med. Chem.* **2017**, *60*, 2169–2192. [CrossRef] [PubMed]
5. Rajan, R.; Osterman, A.K.; Gast, A.T.; Mondragón, A. Biochemical Characterization of the Topoisomerase Domain of Methanopyrus Kandleri Topoisomerase V. *J. Biol. Chem.* **2014**, *289*, 28898–28909. [CrossRef]
6. Nitiss, J.L. DNA Topoisomerase II and Its Growing Repertoire of Biological Functions. *Nat. Rev. Cancer* **2009**, *9*, 327. [CrossRef]
7. Wang, Y.; Rakela, S.; Chambers, J.W.; Hua, Z.C.; Muller, M.T.; Nitiss, J.L.; Tse-Dinh, Y.C.; Leng, F. Kinetic Study of DNA Topoisomerases by Supercoiling-Dependent Fluorescence Quenching. *ACS Omega* **2019**, *4*, 18413–18422. [CrossRef]
8. Oppegard, L.M.; Delgado, J.L.; Kulkarni, C.A.; Towle, T.R.; Hart, D.E.; Williams, B.P.; Lentz, S.R.C.; Norris, B.J.; Flory, C.M.; Schumacher, R.J.; et al. Novel N-1 Substituted Fluoroquinolones Inhibit Human Topoisomerase I Activity and Exhibit Anti-Proliferative Activity. *Investig. New Drugs* **2019**, *37*, 378–383. [CrossRef]
9. Sooryakumar, D.; Dexheimer, T.S.; Teicher, B.A.; Pommier, Y. Molecular and Cellular Pharmacology of the Novel Noncamptothecin Topoisomerase I Inhibitor Genz-644282. *Mol. Cancer Ther.* **2011**, *10*, 1490–1499. [CrossRef]
10. McKie, S.J.; Neuman, K.C.; Maxwell, A. DNA Topoisomerases: Advances in Understanding of Cellular Roles and Multi-Protein Complexes via Structure-Function Analysis. *Bioessays* **2021**, *43*, e2000286. [CrossRef]
11. Buzun, K.; Bielawska, A.; Bielawski, K.; Gornowicz, A. DNA Topoisomerases as Molecular Targets for Anticancer Drugs. *J. Enzym. Inhib. Med. Chem.* **2020**, *35*, 1781–1799. [CrossRef] [PubMed]
12. Singh, B.N.; Mudgil, Y.; Sopory, S.K.; Reddy, M.K. Molecular Characterization of a Nuclear Topoisomerase II from Nicotiana Tabacum That Functionally Complements a Temperature-Sensitive Topoisomerase II Yeast Mutant. *Plant Mol. Biol.* **2003**, *52*, 1063–1076. [CrossRef] [PubMed]
13. Dasgupta, T.; Ferdous, S.; Tse-Dinh, Y.C. Mechanism of Type IA Topoisomerases. *Molecules* **2020**, *25*, 4769. [CrossRef] [PubMed]
14. Berney, D.M.; Shamash, J.; Gaffney, J.; Jordan, S.; Oliver, R.T.D. DNA Topoisomerase I and II Expression in Drug Resistant Germ Cell Tumours. *Br. J. Cancer* **2002**, *87*, 624–629. [CrossRef]
15. Marini, V.; Nikulenkov, F.; Samadder, P.; Juul, S.; Knudsen, B.R.; Krejci, L. MUS81 Cleaves TOP1-Derived Lesions and Other DNA–Protein Cross-Links. *BMC Biol.* **2023**, *21*, 110. [CrossRef]
16. Pommier, Y.; Barcelo, J.M.; Rao, V.A.; Sordet, O.; Jobson, A.G.; Thibaut, L.; Miao, Z.H.; Seiler, J.A.; Zhang, H.; Marchand, C.; et al. Repair of Topoisomerase I-Mediated DNA Damage. *Prog. Nucleic Acid. Res. Mol. Biol.* **2006**, *81*, 179. [CrossRef]

17. Yildizhan, H.; Barkan, N.P.; Turan, S.K.; Demiralp, Ö.; Demiralp, F.D.Ö.; Uslu, B.; Özkan, S.A. Treatment Strategies in Cancer from Past to Present. In *Drug Targeting and Stimuli Sensitive Drug Delivery Systems*; William Andrew Publishing: Norwich, NY, USA, 2018; pp. 1–37. [CrossRef]
18. Martin, S.A. The DNA Mismatch Repair Pathway. In *DNA Repair in Cancer Therapy: Molecular Targets and Clinical Applications: Second Edition*; Academic Press: Cambridge, MA, USA, 2016; pp. 151–177. [CrossRef]
19. Bjornsti, M.A.; Kaufmann, S.H. Topoisomerases and Cancer Chemotherapy: Recent Advances and Unanswered Questions. *F1000Research* **2019**, *8*, 1704. [CrossRef]
20. Radaeva, M.; Dong, X.; Cherkasov, A. The Use of Methods of Computer-Aided Drug Discovery in the Development of Topoisomerase II Inhibitors: Applications and Future Directions. *J. Chem. Inf. Model.* **2020**, *60*, 3703–3721. [CrossRef]
21. Pommier, Y. Topoisomerase I Inhibitors: Camptothecins and Beyond. *Nat. Rev. Cancer* **2006**, *6*, 789–802. [CrossRef]
22. Hu, W.; Huang, X.S.; Wu, J.F.; Yang, L.; Zheng, Y.T.; Shen, Y.M.; Li, Z.Y.; Li, X. Discovery of Novel Topoisomerase II Inhibitors by Medicinal Chemistry Approaches. *J. Med. Chem.* **2018**, *61*, 8947–8980. [CrossRef]
23. Bailly, C. Contemporary Challenges in the Design of Topoisomerase II Inhibitors for Cancer Chemotherapy. *Chem. Rev.* **2012**, *112*, 3611–3640. [CrossRef] [PubMed]
24. Deweese, J.E.; Osheroff, N. The DNA Cleavage Reaction of Topoisomerase II: Wolf in Sheep's Clothing. *Nucleic Acids Res.* **2009**, *37*, 738. [CrossRef] [PubMed]
25. Goffart, S.; Hangas, A.; Pohjoismäki, J.L.O. Twist and Turn—Topoisomerase Functions in Mitochondrial DNA Maintenance. *Int. J. Mol. Sci.* **2019**, *20*, 2041. [CrossRef] [PubMed]
26. McClendon, A.K.; Osheroff, N. DNA Topoisomerase II, Genotoxicity, and Cancer. *Mutat. Res.* **2007**, *623*, 83. [CrossRef] [PubMed]
27. El-Husseiny, W.M.; El-Sayed, M.A.-A.; El-Azab, A.S.; AlSaif, N.A.; Alanazi, M.M.; Abdel-Aziz, A.A.-M. Synthesis of New Polycyclic Systems with Potential Antitumor Activity and Angiogenic Biological Studies. *J. Enzym. Inhib. Med. Chem.* **2012**, *35*, 744–758. [CrossRef]
28. Ketron, A.C.; Osheroff, N. Phytochemicals as Anticancer and Chemopreventive Topoisomerase II Poisons. *Phytochem. Rev.* **2014**, *13*, 19–35. [CrossRef]
29. Ganga Reddy, V.; Srinivasa Reddy, T.; Privér, S.H.; Bai, Y.; Mishra, S.; Wlodkowic, D.; Mirzadeh, N.; Bhargava, S. Synthesis of Gold(I) Complexes Containing Cinnamide: In Vitro Evaluation of Anticancer Activity in 2D and 3D Spheroidal Models of Melanoma and In Vivo Angiogenesis. *Inorg. Chem.* **2019**, *58*, 5988–5999. [CrossRef]
30. Schoeffler, A.J.; Berger, J.M. DNA Topoisomerases: Harnessing and Constraining Energy to Govern Chromosome Topology. *Q. Rev. Biophys.* **2008**, *41*, 41–101. [CrossRef]
31. Jian, J.Y.; Osheroff, N. Telling Your Right Hand from Your Left: The Effects of DNA Supercoil Handedness on the Actions of Type II Topoisomerases. *Int. J. Mol. Sci.* **2023**, *24*, 11199. [CrossRef]
32. Hevener, K.E.; Verstak, T.A.; Lutat, K.E.; Riggsbee, D.L.; Mooney, J.W. Recent Developments in Topoisomerase-Targeted Cancer Chemotherapy. *Acta Pharm. Sin. B* **2018**, *8*, 844–861. [CrossRef]
33. Pommier, Y.; Leo, E.; Zhang, H.; Marchand, C. DNA Topoisomerases and Their Poisoning by Anticancer and Antibacterial Drugs. *Chem. Biol.* **2010**, *17*, 421–433. [CrossRef] [PubMed]
34. Johnson-Arbor, K.; Dubey, R. Doxorubicin. In *xPharm: The Comprehensive Pharmacology Reference*; Elsevier: Amsterdam, The Netherlands, 2022; pp. 1–5. [CrossRef]
35. Wang, H.; Xiao, X.; Xiao, Q.; Lu, Y.; Wu, Y. The Efficacy and Safety of Daunorubicin versus Idarubicin Combined with Cytarabine for Induction Therapy in Acute Myeloid Leukemia: A Meta-Analysis of Randomized Clinical Trials. *Medicine* **2020**, *99*, E20094. [CrossRef] [PubMed]
36. Ghorbani-Abdi-Saedabad, A.; Hanafi-Bojd, M.Y.; Parsamanesh, N.; Tayarani-Najaran, Z.; Mollaei, H.; Hoshyar, R. Anticancer and Apoptotic Activities of Parthenolide in Combination with Epirubicin in Mda-Mb-468 Breast Cancer Cells. *Mol. Biol. Rep.* **2020**, *47*, 5807–5815. [CrossRef] [PubMed]
37. Miyoshi, Y.; Kurosumi, M.; Kurebayashi, J.; Matsuura, N.; Takahashi, M.; Tokunaga, E.; Egawa, C.; Masuda, N.; Kim, S.J.; Okishiro, M.; et al. Topoisomerase IIalpha-Positive and BRCA1-Negative Phenotype: Association with Favorable Response to Epirubicin-Based Regimens for Human Breast Cancers. *Cancer Lett.* **2008**, *264*, 44–53. [CrossRef]
38. Kuznetsov, D.D.; Alsikafi, N.F.; O'Connor, R.C.; Steinberg, G.D. Intravesical Valrubicin in the Treatment of Carcinoma in Situ of the Bladder. *Expert. Opin. Pharmacother.* **2001**, *2*, 1009–1013. [CrossRef]
39. Nitiss, J.L.; Kiianitsa, K.; Sun, Y.; Nitiss, K.C.; Maizels, N. Topoisomerase Assays. *Curr. Protoc.* **2021**, *1*, e250. [CrossRef]
40. Chan, M.K.; Fadzil, N.A.; Chew, A.L.; Khoo, B.Y. New Molecular Biologist Perspective and Insight: DNA Topoisomerases Production by Recombinant DNA Technology for Medical Laboratory Application and Pharmaceutical Industry. *Electron. J. Biotechnol.* **2013**, *16*, 18. [CrossRef]
41. Sawyer, D.B.; Peng, X.; Chen, B.; Pentassuglia, L.; Lim, C.C. Mechanisms of Anthracycline Cardiac Injury: Can We Identify Strategies for Cardio-Protection? *Prog. Cardiovasc. Dis.* **2010**, *53*, 105. [CrossRef]
42. Dempke, W.C.M.; Zielinski, R.; Winkler, C.; Silberman, S.; Reuther, S.; Priebe, W. Anthracycline-Induced Cardiotoxicity—Are We about to Clear This Hurdle? *Eur. J. Cancer* **2023**, *185*, 94–104. [CrossRef]
43. Zinzani, P.L.; Corradini, P.; Martelli, M.; Minotti, G.; Oliva, S.; Spina, M.; Barosi, G.; Tura, S. Critical Concepts, Practice Recommendations, and Research Perspectives of Pixantrone Therapy in Non-Hodgkin Lymphoma: A SIE, SIES, and GITMO Consensus Paper. *Eur. J. Haematol.* **2016**, *97*, 554–561. [CrossRef]

44. Yakkala, P.A.; Panda, S.R.; Naidu, V.G.M.; Shafi, S.; Kamal, A. Pyridine-Based 1,2,4-Triazolo-Tethered Indole Conjugates Potentially Affecting TNKS and PI3K in Colorectal Cancer. *ACS Med. Chem. Lett.* **2023**, *14*, 260–269. [CrossRef] [PubMed]
45. Li, F.; Jiang, T.; Li, Q.; Ling, X. Camptothecin (CPT) and Its Derivatives Are Known to Target Topoisomerase I (Top1) as Their Mechanism of Action: Did We Miss Something in CPT Analogue Molecular Targets for Treating Human Disease Such as Cancer? *Am. J. Cancer Res.* **2017**, *7*, 2350. [PubMed]
46. Robati, M.; Holtz, D.; Dunton, C.J. A Review of Topotecan in Combination Chemotherapy for Advanced Cervical Cancer. *Ther. Clin. Risk Manag.* **2008**, *4*, 213. [CrossRef] [PubMed]
47. Goto, K.; Ohe, Y.; Shibata, T.; Seto, T.; Takahashi, T.; Nakagawa, K.; Tanaka, H.; Takeda, K.; Nishio, M.; Mori, K.; et al. Combined Chemotherapy with Cisplatin, Etoposide, and Irinotecan versus Topotecan Alone as Second-Line Treatment for Patients with Sensitive Relapsed Small-Cell Lung Cancer (JCOG0605): A Multicentre, Open-Label, Randomised Phase 3 Trial. *Lancet Oncol.* **2016**, *17*, 1147–1157. [CrossRef]
48. Fujita, K.I.; Kubota, Y.; Ishida, H.; Sasaki, Y. Irinotecan, a Key Chemotherapeutic Drug for Metastatic Colorectal Cancer. *World J. Gastroenterol.* **2015**, *21*, 12234. [CrossRef]
49. Kim, J.H.; Kim, H.S.; Han, A.R.; Moh, I.H.; Chung, D.C.; Choi, D.R.; Jang, H.J.; Kim, J.B.; Yang, D.H.; Lee, S.I.; et al. Irinotecan, Leucovorin and 5-Fluorouracil (Modified FOLFIRI) as Salvage Chemotherapy for Frail or Elderly Patients with Advanced Gastric Cancer. *Oncol. Lett.* **2012**, *4*, 751. [CrossRef]
50. Kim, Y.M.; Lee, S.W.; Kim, D.Y.; Kim, J.H.; Nam, J.H.; Kim, Y.T. The Efficacy and Toxicity of Belotecan (CKD-602), a Camptothericin Analogue Topoisomerase I Inhibitor, in Patients with Recurrent or Refractory Epithelial Ovarian Cancer. *J. Chemother.* **2010**, *22*, 197–200. [CrossRef]
51. Bethesda LiverTox. *LiverTox: Clinical and Research Information on Drug-Induced Liver Injury*; National Institute of Diabetes and Digestive and Kidney Diseases: Bethesda, MD, USA, 2012.
52. Agrawal, K. Teniposide. In *xPharm: The Comprehensive Pharmacology Reference*; Elsevier: Amsterdam, The Netherlands, 2007; pp. 1–4. [CrossRef]
53. Kollárová-Brázdová, P.; Jirkovská, A.; Karabanovich, G.; Pokorná, Z.; Piskáčková, H.B.; Jirkovský, E.; Kubeš, J.; Lenčová-Popelová, O.; Mazurová, Y.; Adamcová, M.; et al. Investigation of Structure-Activity Relationships of Dexrazoxane Analogs Reveals Topoisomerase IIβ Interaction as a Prerequisite for Effective Protection against Anthracycline Cardiotoxicity. *J. Pharmacol. Exp. Ther.* **2020**, *373*, 402–415. [CrossRef]
54. Deng, S.; Yan, T.; Jendrny, C.; Nemecek, A.; Vincetic, M.; Gödtel-Armbrust, U.; Wojnowski, L. Dexrazoxane May Prevent Doxorubicin-Induced DNA Damage via Depleting Both Topoisomerase II Isoforms. *BMC Cancer* **2014**, *14*, 842. [CrossRef]
55. Marinello, J.; Delcuratolo, M.; Capranico, G. Anthracyclines as Topoisomerase II Poisons: From Early Studies to New Perspectives. *Int. J. Mol. Sci.* **2018**, *19*, 3480. [CrossRef]
56. Saleem, T.; Kasi, A. Daunorubicin. In *xPharm: The Comprehensive Pharmacology Reference*; Elsevier: Amsterdam, The Netherlands, 2022; pp. 1–4. [CrossRef]
57. Volkova, M.; Raymond Russell, I. Anthracycline Cardiotoxicity: Prevalence, Pathogenesis and Treatment. *Curr. Cardiol. Rev.* **2011**, *7*, 214. [CrossRef] [PubMed]
58. Evison, B.J.; Sleebs, B.E.; Watson, K.G.; Phillips, D.R.; Cutts, S.M. Mitoxantrone, More than Just Another Topoisomerase II Poison. *Med. Res. Rev.* **2016**, *36*, 248–299. [CrossRef] [PubMed]
59. Rose, W.C.; Marathe, P.H.; Jang, G.R.; Monticello, T.M.; Balasubramanian, B.N.; Long, B.; Fairchild, C.R.; Wall, M.E.; Wani, M.C. Novel Fluoro-Substituted Camptothecins: In Vivo Antitumor Activity, Reduced Gastrointestinal Toxicity and Pharmacokinetic Characterization. *Cancer Chemother. Pharmacol.* **2006**, *58*, 73–85. [CrossRef] [PubMed]
60. Lv, C.; Liu, X.; Zheng, Q.; Chen, H.; Yang, X.; Zhong, J.; Wang, Y.; Duan, J.; Wang, Z.; Bai, H.; et al. Analysis of Topoisomerase I Expression and Identification of Predictive Markers for Efficacy of Topotecan Chemotherapy in Small Cell Lung Cancer. *Thorac. Cancer* **2018**, *9*, 1166–1173. [CrossRef] [PubMed]
61. Vennepureddy, A.; Atallah, J.-P.; Terjanian, T. Role of Topotecan in Non-Small Cell Lung Cancer: A Review of Literature. *World J. Oncol.* **2015**, *6*, 429. [CrossRef]
62. Tsavaris, N.; Ziras, N.; Kosmas, C.; Giannakakis, T.; Gouveris, P.; Vadiaka, M.; Dimitrakopoulos, A.; Karadima, D.; Rokana, S.; Papalambros, E.; et al. Two Different Schedules of Irinotecan (CPT-11) in Patients with Advanced Colorectal Carcinoma Relapsing after a 5-Fluorouracil and Leucovorin Combination. A Randomized Study. *Cancer Chemother. Pharmacol.* **2003**, *52*, 514–519. [CrossRef]
63. Smith, N.A.; Byl, J.A.W.; Mercer, S.L.; Deweese, J.E.; Osheroff, N. Etoposide Quinone Is a Covalent Poison of Human Topoisomerase IIβ. *Biochemistry* **2014**, *53*, 3229–3236. [CrossRef]
64. Fraser, J.; Wills, L.; Fardus-Reid, F.; Irvine, L.; Elliss-Brookes, L.; Fern, L.; Cameron, A.L.; Pritchard-Jones, K.; Feltbower, R.G.; Shelton, J.; et al. Oral Etoposide as a Single Agent in Childhood and Young Adult Cancer in England: Still a Poorly Evaluated Palliative Treatment. *Pediatr. Blood Cancer* **2021**, *68*, e29204. [CrossRef]
65. Cushman, M. Design and Synthesis of Indenoisoquinolines Targeting Topoisomerase I and Other Biological Macromolecules for Cancer Chemotherapy. *J. Med. Chem.* **2021**, *64*, 17572–17600. [CrossRef]
66. Pommier, Y.; Cushman, M. The Indenoisoquinolines Non-Camptothecin Topoisomerase I Inhibitors: Update and Perspectives. *Mol. Cancer Ther.* **2009**, *8*, 1008. [CrossRef]

67. Zuco, V.; Supino, R.; Favini, E.; Tortoreto, M.; Cincinelli, R.; Croce, A.C.; Bucci, F.; Pisano, C.; Zunino, F. Efficacy of ST1968 (Namitecan) on a Topotecan-Resistant Squamous Cell Carcinoma. *Biochem. Pharmacol.* **2010**, *79*, 535–541. [CrossRef] [PubMed]
68. Benton, C.B.; Ravandi, F. Targeting Acute Myeloid Leukemia with TP53-Independent Vosaroxin. *Future Oncol.* **2017**, *13*, 125–133. [CrossRef]
69. Faruqi, A.; Tadi, P. Cytarabine. In *xPharm: The Comprehensive Pharmacology Reference*; Elsevier: Amsterdam, The Netherlands, 2022; pp. 1–5. [CrossRef]
70. Gadade, D.D.; Pekamwar, S.S. Cyclodextrin Based Nanoparticles for Drug Delivery and Theranostics. *Adv. Pharm. Bull.* **2020**, *10*, 166. [CrossRef]
71. Pettengell, R.; Długosz-Danecka, M.; Andorsky, D.; Belada, D.; Georgiev, P.; Quick, D.; Singer, J.W.; Singh, S.B.; Pallis, A.; Egorov, A.; et al. Pixantrone plus Rituximab versus Gemcitabine plus Rituximab in Patients with Relapsed Aggressive B-Cell Non-Hodgkin Lymphoma Not Eligible for Stem Cell Transplantation: A Phase 3, Randomized, Multicentre Trial (PIX306). *Br. J. Haematol.* **2020**, *188*, 240–248. [CrossRef] [PubMed]
72. Chamberlain, F.E.; Jones, R.L.; Chawla, S.P. Aldoxorubicin in Soft Tissue Sarcomas. *Future Oncol.* **2019**, *15*, 1429–1435. [CrossRef] [PubMed]
73. Lazareva, N.F.; Baryshok, V.P.; Lazarev, I.M. Silicon-Containing Analogs of Camptothecin as Anticancer Agents. *Arch. Pharm.* **2018**, *351*, 1700297. [CrossRef]
74. Van Arnam, E.B.; Ruzzini, A.C.; Sit, C.S.; Currie, C.R.; Clardy, J. A Rebeccamycin Analog Provides Plasmid-Encoded Niche Defense. *J. Am. Chem. Soc.* **2015**, *137*, 14272–14274. [CrossRef]
75. Robey, R.W.; Obrzut, T.; Shukla, S.; Polgar, O.; MacAlou, S.; Bahr, J.C.; Di Pietro, A.; Ambudkar, S.V.; Bates, S.E. Becatecarin (Rebeccamycin Analog, NSC 655649) Is a Transport Substrate and Induces Expression of the ATP-Binding Cassette Transporter, ABCG2, in Lung Carcinoma Cells. *Cancer Chemother. Pharmacol.* **2009**, *64*, 575. [CrossRef]
76. Sankara Rao, N.; Nagesh, N.; Lakshma Nayak, V.; Sunkari, S.; Tokala, R.; Kiranmai, G.; Regur, P.; Shankaraiah, N.; Kamal, A. Design and Synthesis of DNA-Intercalative Naphthalimide-Benzothiazole/Cinnamide Derivatives: Cytotoxicity Evaluation and Topoisomerase-IIα Inhibition. *Medchemcomm* **2019**, *10*, 72–79. [CrossRef]
77. Kamal, A.; Srinivasulu, V.; Nayak, V.L.; Sathish, M.; Shankaraiah, N.; Bagul, C.; Reddy, N.V.S.; Rangaraj, N.; Nagesh, N. Design and Synthesis of C3-Pyrazole/Chalcone-Linked Beta-Carboline Hybrids: Antitopoisomerase I, DNA-Interactive, and Apoptosis-Inducing Anticancer Agents. *ChemMedChem* **2014**, *9*, 2084–2098. [CrossRef]
78. Kamal, A.; Sathish, M.; Nayak, V.L.; Srinivasulu, V.; Kavitha, B.; Tangella, Y.; Thummuri, D.; Bagul, C.; Shankaraiah, N.; Nagesh, N. Design and Synthesis of Dithiocarbamate Linked β-Carboline Derivatives: DNA Topoisomerase II Inhibition with DNA Binding and Apoptosis Inducing Ability. *Bioorg. Med. Chem.* **2015**, *23*, 5511–5526. [CrossRef] [PubMed]
79. Sathish, M.; Chetan Dushantrao, S.; Nekkanti, S.; Tokala, R.; Thatikonda, S.; Tangella, Y.; Srinivas, G.; Cherukommu, S.; Hari Krishna, N.; Shankaraiah, N.; et al. Synthesis of DNA Interactive C3-Trans-Cinnamide Linked β-Carboline Conjugates as Potential Cytotoxic and DNA Topoisomerase I Inhibitors. *Bioorg. Med. Chem.* **2018**, *26*, 4916–4929. [CrossRef] [PubMed]
80. Jadala, C.; Sathish, M.; Reddy, T.S.; Reddy, V.G.; Tokala, R.; Bhargava, S.K.; Shankaraiah, N.; Nagesh, N.; Kamal, A. Synthesis and in Vitro Cytotoxicity Evaluation of β-Carboline-Combretastatin Carboxamides as Apoptosis Inducing Agents: DNA Intercalation and Topoisomerase-II Inhibition. *Bioorg. Med. Chem.* **2019**, *27*, 3285–3298. [CrossRef]
81. Kovvuri, J.; Nagaraju, B.; Nayak, V.L.; Akunuri, R.; Rao, M.P.N.; Ajitha, A.; Nagesh, N.; Kamal, A. Design, Synthesis and Biological Evaluation of New β-Carboline-Bisindole Compounds as DNA Binding, Photocleavage Agents and Topoisomerase I Inhibitors. *Eur. J. Med. Chem.* **2018**, *143*, 1563–1577. [CrossRef] [PubMed]
82. Subba Rao, A.V.; Vishnu Vardhan, M.V.P.S.; Subba Reddy, N.V.; Srinivasa Reddy, T.; Shaik, S.P.; Bagul, C.; Kamal, A. Synthesis and Biological Evaluation of Imidazopyridinyl-1,3,4-Oxadiazole Conjugates as Apoptosis Inducers and Topoisomerase IIα Inhibitors. *Bioorg. Chem.* **2016**, *69*, 7–19. [CrossRef] [PubMed]
83. Nagaraju, B.; Kovvuri, J.; Kumar, C.G.; Routhu, S.R.; Shareef, M.A.; Kadagathur, M.; Adiyala, P.R.; Alavala, S.; Nagesh, N.; Kamal, A. Synthesis and Biological Evaluation of Pyrazole Linked Benzothiazole-β-Naphthol Derivatives as Topoisomerase I Inhibitors with DNA Binding Ability. *Bioorg. Med. Chem.* **2019**, *27*, 708–720. [CrossRef]
84. Kamal, A.; Kumar, B.A.; Suresh, P.; Agrawal, S.K.; Chashoo, G.; Singh, S.K.; Saxena, A.K. Synthesis of 4β-N-Polyaromatic Substituted Podophyllotoxins: DNA Topoisomerase Inhibition, Anticancer and Apoptosis-Inducing Activities. *Bioorg. Med. Chem.* **2010**, *18*, 8493–8500. [CrossRef]
85. Kamal, A.; Suresh, P.; Ramaiah, M.J.; Mallareddy, A.; Imthiajali, S.; Pushpavalli, S.N.C.V.L.; Lavanya, A.; Pal-Bhadra, M. Synthesis and Biological Evaluation of 4β-Sulphonamido and 4β-[(4′-Sulphonamido)Benzamide]Podophyllotoxins as DNA Topoisomerase-IIα and Apoptosis Inducing Agents. *Bioorg. Med. Chem.* **2012**, *20*, 2054–2066. [CrossRef]
86. Kamal, A.; Suresh, P.; Ramaiah, M.J.; Srinivasa Reddy, T.; Kapavarapu, R.K.; Rao, B.N.; Imthiajali, S.; Lakshminarayan Reddy, T.; Pushpavalli, S.N.C.V.L.; Shankaraiah, N.; et al. 4β-[4′-(1-(Aryl)Ureido)Benzamide]Podophyllotoxins as DNA Topoisomerase i and IIα Inhibitors and Apoptosis Inducing Agents. *Bioorg. Med. Chem.* **2013**, *21*, 5198–5208. [CrossRef]
87. Kamal, A.; Kumar, B.A.; Suresh, P.; Shankaraiah, N.; Kumar, M.S. An Efficient One-Pot Synthesis of Benzothiazolo-4β-Anilino-Podophyllotoxin Congeners: DNA Topoisomerase-II Inhibition and Anticancer Activity. *Bioorg. Med. Chem. Lett.* **2011**, *21*, 350–353. [CrossRef]

88. Shankaraiah, N.; Kumar, N.P.; Amula, S.B.; Nekkanti, S.; Jeengar, M.K.; Naidu, V.G.M.; Reddy, T.S.; Kamal, A. One-Pot Synthesis of Podophyllotoxin-Thiourea Congeners by Employing NH2SO3H/NaI: Anticancer Activity, DNA Topoisomerase-II Inhibition, and Apoptosis Inducing Agents. *Bioorg. Med. Chem. Lett.* **2015**, *25*, 4239–4244. [CrossRef] [PubMed]
89. Reddy, V.G.; Bonam, S.R.; Reddy, T.S.; Akunuri, R.; Naidu, V.G.M.; Nayak, V.L.; Bhargava, S.K.; Kumar, H.M.S.; Srihari, P.; Kamal, A. 4β-Amidotriazole Linked Podophyllotoxin Congeners: DNA Topoisomerase-IIα Inhibition and Potential Anticancer Agents for Prostate Cancer. *Eur. J. Med. Chem.* **2018**, *144*, 595–611. [CrossRef] [PubMed]
90. Sathish, M.; Kavitha, B.; Nayak, V.L.; Tangella, Y.; Ajitha, A.; Nekkanti, S.; Alarifi, A.; Shankaraiah, N.; Nagesh, N.; Kamal, A. Synthesis of Podophyllotoxin Linked β-Carboline Congeners as Potential Anticancer Agents and DNA Topoisomerase II Inhibitors. *Eur. J. Med. Chem.* **2018**, *144*, 557–571. [CrossRef] [PubMed]
91. Kim, J.S.; Gatto, B.; Yu, C.; Liu, A.; Liu, L.F.; LaVoie, E.J. Substituted 2,5′-Bi-1H-Benzimidazoles: Topoisoraerase I Inhibition and Cytotoxicity. *J. Med. Chem.* **1996**, *39*, 992–998. [CrossRef] [PubMed]
92. Kim, J.S.; Yu, C.; Liu, A.; Liu, L.F.; LaVoie, E.J. Terbenzimidazoles: Influence of 2″-, 4-, and 5-Substituents on Cytotoxicity and Relative Potency as Topoisomerase I Poisons. *J. Med. Chem.* **1997**, *40*, 2818–2824. [CrossRef]
93. Coban, G.; Zencir, S.; Zupkó, I.; Réthy, B.; Gunes, H.S.; Topcu, Z. Synthesis and Biological Activity Evaluation of 1H-Benzimidazoles via Mammalian DNA Topoisomerase I and Cytostaticity Assays. *Eur. J. Med. Chem.* **2009**, *44*, 2280–2285. [CrossRef]
94. Singh, M.; Tandon, V. Synthesis and Biological Activity of Novel Inhibitors of Topoisomerase I: 2-Aryl-Substituted 2-Bis-1H-Benzimidazoles. *Eur. J. Med. Chem.* **2011**, *46*, 659–669. [CrossRef]
95. Talukdar, A.; Kundu, B.; Sarkar, D.; Goon, S.; Mondal, M.A. Topoisomerase I Inhibitors: Challenges, Progress and the Road Ahead. *Eur. J. Med. Chem.* **2022**, *236*, 114304. [CrossRef]
96. Majumdar, A.G.; Shree, S.; Das, A.; Kumar, B.K.; Dey, P.; Subramanian, M.; Patro, B.S. Design, Synthesis and Development of a Dual Inhibitor of Topoisomerase 1 and Poly (ADP-Ribose) Polymerase 1 for Efficient Killing of Cancer Cells. *Eur. J. Med. Chem.* **2023**, *258*, 115598. [CrossRef]
97. Idowu, T.; Schweizer, F. Ubiquitous Nature of Fluoroquinolones: The Oscillation between Antibacterial and Anticancer Activities. *Antibiotics* **2017**, *6*, 26. [CrossRef]
98. Beberok, A.; Wrześniok, D.; Rok, J.; Rzepka, Z.; Respondek, M.; Buszman, E. Ciprofloxacin Triggers the Apoptosis of Human Triple-Negative Breast Cancer MDA-MB-231 Cells via the P53/Bax/Bcl-2 Signaling Pathway. *Int. J. Oncol.* **2018**, *52*, 1727–1737. [CrossRef]
99. Ma, Y.C.; Wang, Z.X.; Jin, S.J.; Zhang, Y.X.; Hu, G.Q.; Cui, D.T.; Wang, J.S.; Wang, M.; Wang, F.Q.; Zhao, Z.J. Dual Inhibition of Topoisomerase II and Tyrosine Kinases by the Novel Bis-Fluoroquinolone Chalcone-Like Derivative HMNE3 in Human Pancreatic Cancer Cells. *PLoS ONE* **2016**, *11*, e0162821. [CrossRef]
100. Skok, Ž.; Zidar, N.; Kikelj, D.; Ilaš, J. Dual Inhibitors of Human DNA Topoisomerase II and Other Cancer-Related Targets. *J. Med. Chem.* **2020**, *63*, 884–904. [CrossRef]

pharmaceuticals

MDPI

Review

The Implication of Topoisomerase II Inhibitors in Synthetic Lethality for Cancer Therapy

Victor M. Matias-Barrios [1,2,*] and Xuesen Dong [1]

1 The Vancouver Prostate Centre, Department of Urologic Sciences, University of British Columbia, 2660 Oak Street, Vancouver, BC V6H 3Z6, Canada

2 School of Medicine and Health Sciences, Tecnologico de Monterrey, Avenida Eugenio Garza Sada 2501, Monterrey 64849, Mexico

* Correspondence: vmatiasbarrios@gmail.com

Abstract: DNA topoisomerase II (Top2) is essential for all eukaryotic cells in the regulation of DNA topology through the generation of temporary double-strand breaks. Cancer cells acquire enhanced Top2 functions to cope with the stress generated by transcription and DNA replication during rapid cell division since cancer driver genes such as Myc and EZH2 hijack Top2 in order to realize their oncogenic transcriptomes for cell growth and tumor progression. Inhibitors of Top2 are therefore designed to target Top2 to trap it on DNA, subsequently causing protein-linked DNA breaks, a halt to the cell cycle, and ultimately cell death. Despite the effectiveness of these inhibitors, cancer cells can develop resistance to them, thereby limiting their therapeutic utility. To maximize the therapeutic potential of Top2 inhibitors, combination therapies to co-target Top2 with DNA damage repair (DDR) machinery and oncogenic pathways have been proposed to induce synthetic lethality for more thorough tumor suppression. In this review, we will discuss the mode of action of Top2 inhibitors and their potential applications in cancer treatments.

Keywords: DNA topoisomerase II; topoisomerase inhibitors; synthetic lethality; DNA damage repair; DNA repair inhibitors; Myc; EZH2

1. Introduction

DNA topoisomerase II (Top2) is necessary for DNA replication and transcription, as well as for chromosome condensation and segregation. Top2 inhibitors such as doxorubicin and etoposide are medications commonly used to treat breast, lung, and testicular cancer, as well as lymphomas, sarcomas, and other neoplasms, even though they present dose-limiting toxicity and side effects [1,2]. Despite the emergence of revolutionary medications in the context of personalized approaches, these therapies are still regarded as essential in cancer treatment. However, cancer cells generate resistance to these drugs by using driver genes such as Myc and EZH2 to increase Top2 catalytic function and DNA damage repair (DDR) machinery, thus overcoming replication and transcription stress and avoiding cell death. In this regard, synthetic lethality techniques may provide a broader range of druggable targets to deal with therapy resistance developed toward Top2 inhibitors, which should thus increase the susceptibility of cancer cells to genotoxic therapy. "Synthetic lethality" occurs when numerous defects in two or more related genes occur concurrently, causing cell death or apoptosis; a single loss in one of these genes that is tolerated results in cell survival [3]. With advances in genomic investigations of gene mutations and expression patterns, it is now possible to analyze the effect of single gene deletion on tumor cell survival on a large scale, allowing for the discovery of novel synthetic lethal targets. Initially, synthetic lethality strategies included combining a mutation with a drug targeting a specific tumor pathway. PARP inhibitors were the first targeted treatments to employ synthetic lethality and were used to destroying cancers through DNA repair failure (e.g., BRCA1/2 mutation) [4]. As an alternative to mutations, a second drug that functionally duplicates the effect of the mutation may be employed to generate a combination of synthetic lethality. Combination treatments based on synthetic lethality have the potential to be effective at controlling tumor growth because they permit the administration of drugs at lower dosages, reduce cytotoxicity in tumor cells, and minimize drug resistance selection. Drugs that change or impede DNA repair pathways or enhance Top2's genotoxic effects have the potential to sensitize cells to commonly used Top2 inhibitors. They may lead to more robust tumor suppression, longer periods of remaining disease-free, increased survival, and even enhanced quality of life.

In addition, targeting genes that potentiate the epigenetic and chromosome activity of Top2, such as Myc and EZH2, can also be used to introduce synthetic lethality. Myc has been recently used as a target for new chemotherapeutics [5,6]. Since its primary mechanism is to induce cancer survival genes [7], dual inhibition of Myc and Top2 can be beneficial. In addition, Top2 has been demonstrated to be essential in efforts to relax DNA and expose the Myc promoter region for transcription [8]. Thus, the combination of Myc–Top2 inhibition can help treat tumors. On the other hand, EZH2, which serves as an epigenetic modifier that inhibits the expression of tumor suppressors, has been shown as a potential drug target both alone and in the combination with Top2 inhibition [9]. Here, we explore the mechanism of action of Top2 inhibitors and propose the application of DDR, Myc, and EZH2 inhibitors in combination with Top2 inhibitors to induce synthetic lethality.

2. DNA Topoisomerases

DNA topoisomerases are well-studied proteins that are required for transcription, DNA replication, and chromosome segregation through cell cycling and act by temporarily introducing single- or double-strand breaks in DNA and then resealing them [10]. As a result, topoisomerases are essential for maintaining DNA integrity during transcription and replication when cells are proliferating.

Based on the number of DNA strands they cleave, topoisomerases can be divided into types I and II. Type I enzymes cleave just one strand of DNA, whereas type II enzymes cleave both strands to prevent supercoiling or entanglements [10,11]. Furthermore, according to the kinds of covalent phosphotyrosyl intermediate they generate (5′ or 3′ linkage), the structures and reaction mechanisms of topoisomerases can be classified into five types: IA, IB, IC, IIA, and IIB (Table 1). These enzymes function by temporarily rupturing and reuniting DNA strands. This process involves a nucleophilic assault on a DNA phosphodiester bond that is carried out by a topoisomerase tyrosine residue [12].

Table 1. Type and function of topoisomerases.

Type	Subfamily	Mechanism	Human Proteins	Function	Co-Factor	Substrate
I	IA	Strand passage	Top3a	Relaxed DNA Sc, decatenation activity	Mg^{2+}	DNA Sc, hemicatenanes, double Holliday junctions, and D loops.
			Top3b	Relaxed DNA Sc, RNA helicase activity		DNA HSc, RNA knots, and R loops.
	IB	Controlled rotation	Top1	Relaxed negative and positive DNA Sc	None	DNA positive and negative Sc.
			Top1mt			
	IC		Not found in humans			
II	IIA	Strand passage	Top2a	Relaxed negative and positive DNA Sc, potent decatenation activity	Mg^{2+}, ATP	DNA positive and negative Sc, DNA knots, and DNA catenanes.
			Top2b			
	IIB		Not found in humans			

Sc: supercoiled; HSc: hyper-supercoiled; mt: mitochondrial.

Eukaryotic type I topoisomerases include two monomeric enzymes. Type IA cleaves a single-strand segment and lets the intact strand pass through the split [13]. Topoisomerase IIIa (Top3a) and IIIb (Top3b) form the IA subfamily in humans. They are known to relax hyper-supercoiled DNA segments [14]. On the other hand, type IB allows the broken strand to revolve around the intact strand, does not need divalent metal ions, and covalently binds to the 3′-terminal phosphate of the DNA. In humans, topoisomerase I (Top1) and topoisomerase I mitochondrial (Top1mt) have been observed. They relax negative and positive supercoiled DNA. Finally, type IC works using a controlling rotation mechanism similar to the IB type; however, it is only found in *Methanopyrus genus* [14]. Although these enzymes can modulate the under- and over-winding of DNA, they cannot remove knots or tangles from duplex DNA. They also do not require a high-energy co-factor to function [15].

Eukaryotic type II topoisomerases, also known as Top2, are the enzymes responsible for cleaving both DNA strands and allowing the DNA duplex to continue past the breakage. These enzymes carry out their functions as homodimers and require the presence of divalent metal ions, in addition to adenosine triphosphate (ATP), for total catalytic activity [16]. The type IIB subfamily is found in the archaea and the bacteria domain of life [17], while the type IIA subfamily is found in bacteria and mammals. The topoisomerase IIa (Top2a) and topoisomerase IIb (Top2b) isoforms are found in humans. Despite their close relationship, they are encoded by two separate genes and have distinct molecular masses: Top2a is 170 kDa in size, while Top2b is 180 kDa [18]. The two enzymes have different expression patterns in vertebrate cells and thus different physiological activities. Top2a is essential to actively dividing cells, and its protein levels are regulated throughout the cell cycle, with concentrations peaking at the G2/M phase. During the process of mitosis, Top2a stays firmly attached to the chromosomes. On the other hand, Top2b demonstrates an autonomous position regarding proliferation and distances itself from chromosomes in the process of mitosis [19]. Top2b is incapable of making up for the absence of Top2a in mammalian cells, which suggests that the two enzymes serve distinct and independent functions in the body.

The functions of DNA topoisomerases are different in normal cells and cancer cells. In normal cells, they maintain genome integrity by modifying the chromatin structure and promoting safe chromatid segregation during mitosis. In cancer cells, their protein levels or activities are enhanced to cooperate with increased cell replication and to enhance cell survival by serving as a co-factor protein in biomolecular processes [20,21]. Top2a protein levels have been correlated with proliferation rate and the carcinogenesis of neoplasms in multiple types of cancers [22], and Top2a expression is associated with poor prognoses [23]. On the other hand, Top2b activity is reported to be upregulated in steroid hormone cellular signaling in breast and prostate cancers, thus aiding cancer cell survival. This survival mechanism is accomplished by an increase in double-strand breaks in androgen receptor (AR)- or estrogen receptor (ER)-positive cells, which relaxes DNA and thus allows these receptors to bind to their target promoters to induce their transcriptional activity [24,25]. These findings rationalize DNA topoisomerase inhibition as a therapeutic method. Decades of research have led to DNA topoisomerase inhibitors becoming one of the leading cancer therapeutics in the clinical setting.

3. DNA Topoisomerase Inhibitors

Topoisomerase inhibitors are known to be among the most powerful and most frequently prescribed anticancer treatments [26]. They target the enzymatic activities of topoisomerase in DNA cleavage and ligation processes. Inhibition of these processes will result in either increased cleavage complexes or replication stall during DNA synthesis and lead to DNA damage, cell cycling arrest, or even cell death [1]. In addition, these medications exhibit substantial selectivity and non-ambiguity by not cross-reacting to each topoisomerase [27].

3.1. DNA Topoisomerase II Inhibitors

Type II topoisomerases inhibitors block the activity of Top2a/Top2b at different stages of their catalytic cycle. The enzymes consist of three domains, namely the N-terminal domain, the cleavage/DNA binding domain, and the C-terminal domain [28]. These domains work collaboratively to create double-strand breaks. There are six steps in the Top2 catalytic cycle involving DNA structure dynamics (Figure 1): (1) Top2 binds to two DNA segments; (2) in a Mg^{2+}-dependent process, one DNA strand is bent and a double-strand break is induced by initiating a nucleophilic attack on the phosphodiester DNA backbone (Tyr805 and Tyr821 in human Top2a and Top2b, respectively); (3) Top2 conformational changes following ATP binding induce segment trapping; (4) the hydrolysis of ATP facilitates the passage of the other DNA strand through the break; (5) DNA strand breakages are re-ligated; and (6) Top2 dissociates from DNA after ATP hydrolysis [29,30].

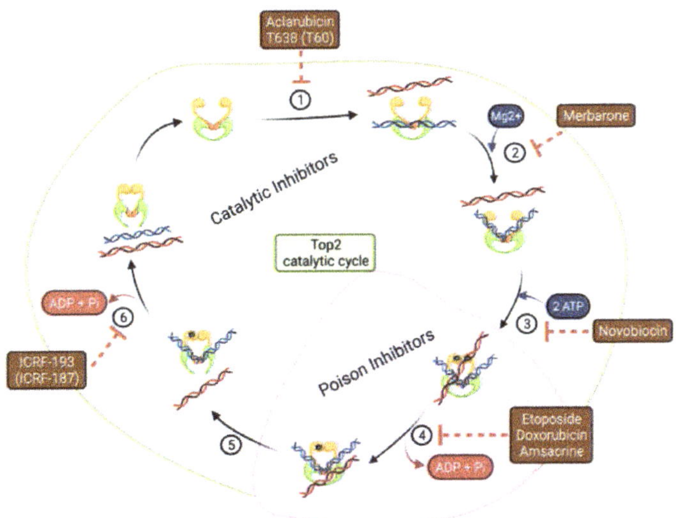

Figure 1. Catalytic cycle of Top2. Relaxed/tangled DNA dynamics process: (1) the enzyme binds to double-strand DNA; (2) the DNA cleavage reaction requires Mg^{2+}; (3) two molecules of ATP bind to the N-terminal domain; (4) ATP hydrolysis provides the energy for DNA passage; (5) DNA repair/religation; and (6) dissociation of DNA–Top2 complex after the second ATP hydrolysis. Top2 is then ready to start a new cycle of enzymatic activity. The figure was created on biorender.com.

Top2 inhibitors contain a broad range of compounds that block Top2 activity through various modes of action. However, they can be roughly divided into two groups based on their destructive effects on cells, namely Top2 poison inhibitors and Top2 catalytic inhibitors (Table 2).

Table 2. Topoisomerase II inhibitors.

Name	Mechanism of Action	Mode of Inhibition	Application	Refs.
Aclarubicin	Prevents binding of Top2 to DNA	Catalytic	Acute myeloid leukemia	[31]
T638 (T60)			Inhibition compared to etoposide with less cytotoxicity in K562 cancer cells and xenografts	[32–34]

Table 2. *Cont.*

Name	Mechanism of Action	Mode of Inhibition	Application	Refs.
Novobiocin	Binds to ATP binding site		BRCA-deficient tumors with acquired PARP inhibitor resistance	[35]
Merbarone	Blocks DNA Cleavage		Limitations due to nephrotoxicity	[36]
ICRF-193 (ICRF-187)	Blocks ATP hydrolysis and traps enzymes in the closed clamp		Addresses cardiotoxicity caused by Top2 poison	[37]
Etoposide	Stabilizes covalent cleavage complexes	Poison	Small cell lung cancer, lymphomas, refractory testicular tumors	[38]
Doxorubicin			Leukemia, ovarian and breast carcinomas	[39]
Amsacrine			Acute myeloid leukemia	[40]

3.1.1. DNA Topoisomerase II Poisons

Poison inhibitors halt the Top2 catalytic cycle following DNA cleavage. They increase the amount of Top2-DNA cleavage complexes (Top2cc), resulting in genotoxic entities within cells due to induced DNA damage caused by a deficient DNA repair system. Consequently, the buildup of these DNA breaks will eventually result in programmed cell death [41]. Etoposide, an exemplary Top2 poison, is frequently used in oncology (typically in conjunction with other chemotherapeutics) to treat a range of malignancies (including ovarian, testicular, and small cell lung cancer, as well as leukemia and lymphoma) [38]. Because of its low affinity toward free DNA, weak DNA intercalating activity, strong selectivity toward the Top2–DNA cleavage complex, and its ability to trap cleavage complexes with high frequency, etoposide is an appealing Top2 poison [42]. However, etoposide usage is associated with long-term toxicities, secondary malignancies, and dose-limiting cardiotoxicity [43]. Patients can also develop tolerance to Top2 poisons because they activate DDR machinery (phosphorylation of ATM and activation of downstream targets in both the HR and NHEJ pathways) [44]. However, a recent genome-wide investigation revealed genes that may predict resistance to Top2 poisons, increasing their potential as precision medicine therapies [45].

3.1.2. DNA Topoisomerase II Catalytic Inhibitors

Catalytic Top2 inhibitors are chemicals that inhibit Top2 enzymatic activity. They block the enzyme before DNA breaks are carried out or after DNA re-ligation is completed. Therefore, they do not induce Top2cc accumulation. Due to the inability of Top2 to relax DNA supercoils or decatenate sister chromatids during mitosis, failed cell division and subsequent cell death can be established [46,47].

Catalytic Top2 inhibitors do not have side effects that are caused by Top2 poisons, but their therapeutic use is restricted. Despite the limited success of Top2 catalytic inhibitors in drug development, several compounds have been identified, such as novobiocin, merbarone, suramin, aclarubicin, ICRF-187, ICRF-193, T60, and T638 [32–34,46]. However, novobiocin, merbarone, suramin, and aclarubicin have multiple targets in addition to Top2, limiting their future characterization and development as anticancer drugs [46]. These off-target effects might be explained by poor drug development processes before clinical trials. As for T60 and its derivates, compounds designed using a novel pocket of the enzyme with computer-aided drug development methods seem to have fewer off-target effects than the known catalytic inhibitors [32]. Regarding its clinical use, ICRF-187 is authorized for human use as a calcium chelating reagent and is used to treat cardiotoxicity induced by Top2 poisons [48]. ICRF-193, T60, and T638 have demonstrated anticancer activity in animal models [32,33,49]. While several Top2 catalytic inhibitors are undergoing preclinical development and have advanced to varying stages of research [32,50,51], Top2 catalytic inhibitors offer promise as potential anticancer medicines.

4. Double-Strand Break (DSB) Repair

DNA DSBs can be lethal to cells. They require immediate repair to avert potentially harmful chromosomal disruption. On the DNA lesion site, several DSB sensors are needed to detect DNA damage and activate downstream effectors. PARP1 and the Ku70/Ku80 complex are the first proteins recruited to DNA DSBs [52]. They target proteins to promote chromatin de-condensation and recruit successive second-cell messengers.

Once the lesion has been identified, ATM is recruited by the MRN complex (RAD50, NBS1, and MRE11) to create a checkpoint arrest amplification. Additionally, ATM activates the histone H2AX, which results in the production of gamma-H2AX (H2AX) [53]. This modification is necessary for the recruitment of several factors, such as the DNA damage checkpoint 1 mediator (MDC1), and it is accompanied by a concomitant increase in the E3 ubiquitin–protein ligase RNF168, which enhances chromatin relaxation even further and makes it possible to recruit additional DNA repair factors [54].

During this DSB recognition stage, the decision of applying homologous recombination (HR) or a non-homologous end-joining (NHEJ) repair mechanism depends mainly on the cell cycle phase during DNA resection (Figure 2). DSB end resection stimulators favor HR at the S and G2 phases, primarily through CtIP and BRCA1 and MRN complex [55]. On the other hand, during the G1 phase of the cell cycle, the DSB end resection process is inhibited by the Ku70/Ku80 heterodimer and TP53BP1 [56].

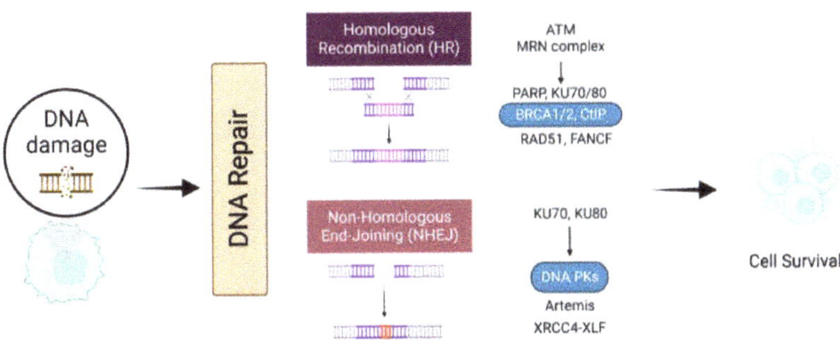

Figure 2. DNA damage response and repair pathways. Several DSB repair mechanisms are available based on the phase of the cell cycle and the existence of homology sequences in the DNA strands. The DNA repair mechanism could undergo the HR or the NHEJ pathway according to the cell cycle stage at the moment of the lesion and the acquired mutations in the cell. The figure was created on biorender.com.

4.1. *DNA Repair Mechanism*

DNA repair is regulated by a complex network of sensors and effectors and begins with the identification of DNA damage and the selection of the most viable repair route, which is determined by the cell cycle stage and the type of DNA damage. During the G0/G1 phase, small lesions are resolved by nucleotide excision repair and base excision repair. During the S phase, DNA is repaired using the Fanconi anemia. mismatch repair DSBs, Alt-NHEJ (alternative non-homologous end-joining), and single-strand annealing. Each of these repair mechanisms is unique [57]. However, they are not mutually exclusive, which creates a network of proteins that share DNA repair functions. Mutations in these systems may cause an accumulation of poorly repaired DNA, predisposing cells to cancer [58].

Additionally, abnormalities in DNA repair systems that preserve genome integrity allow cancer cells to become aggressive and resistant to cancer therapy [59]. These abnormalities in cancer cells include overexpression, downregulation, mutations, and the polymorphism of proteins in different DNA repair pathways. In the NHEJ pathway, overexpression of the Ku70/80 complex has been identified in gastric and breast cancers [60,61]. DNA-PK was reported to be highly expressed in oral, lung, and esophageal carcinoma [62]. As for the HR pathway, BRCA1/BRCA2-deficient cells are found in breast cancers [63]. However, overexpression of RAD51 genes has also been demonstrated in pancreatic cancer and leukemia [64]. Thus, cancer cells can optimize cell survival by enhancing their intact DNA repair mechanism in environments where the other pathway is defective. Thus, this generates an opportunity to treat cancer cells with synthetic lethality by inhibiting the remaining intact DNA repair pathway and other DNA damage agents.

4.1.1. Homologous Recombination

MRE11's endonuclease activity cleaves the DNA close to the DSB in HR and then removes the DNA edges [65]. The replication protein (RPA) will then bind the DNA after forming a single strand to prevent it from being degraded [66]. RPA is then displaced by BRCA2 and PALB2 to increase the binding of RAD52 to ssDNA [67]. This mechanism is believed to displace the Ku70/Ku80 complex in order to allow resection factors, including DNA2, EXO1, BLM, WRN, and RPA, to approach the target sites [68].

Subsequently, RAD41 invades the complementary strands containing homologous sequences, while RAD51 mediates complementary annealing of the DNA strands by mediating the DNA–DNA interaction [69]. Subsequently, a displacement loop (D-loop) is generated by lengthening the DNA strand through the use of replicative or translation DNA polymerases [70]. At this point, three different pathways can be used to resolve D-loops. Double-strand break repair (DSBR) induces the production of crossover and non-crossover products [71]. Synthesis-dependent strand annealing (SDSA) then increases the non-crossover effects [72]. Finally, break-induced replication (BIR) generates half-crossover products and frequently promotes mutagenesis [73]. It has been shown that TP53BP1 promotes a fork cleavage-free pathway, while BRCA1 facilitates the BIR pathway coupled with SLX–MUS complex-mediated fork cleavage [56]. This TP53BP1 function is performed by reducing the activity of helicases, which limits D-loop stability and makes it more likely for crossover and BIR events to occur.

A systematic review of 33 types of cancer revealed that multiple mutations in the HR pathway create a defected DNA repair mechanism in ovarian cancers. The most common mutations in the HR pathway occur in the BRCA1, BRCA2, RAD51, BLM, and RAD50 genes [74]. RAD51 overexpression has been linked to worse prognosis in patients with sporadic gastric cancer and neuroblastoma who present modified MRE11 [75,76]. In contrast, RAD50 and NBS1 mutations are associated with carcinogenesis during endometrial and prostate cancer development [77,78].

4.1.2. Non-Homologous End-Joining

NHEJ is the principal mechanism by which DSBs are repaired in human cells. During the classical NHEJ (c-NHEJ) pathway, Ku70/Ku80 heterodimer binds to the DSB ends before resection, allowing for the recruitment of DNA-PK. The endonuclease ARTEMIS then processes the fragmented ends until cohesion is restored. In the final step, DSB ligation is accelerated by DNA ligase 4 interacting with NHEJ factor 1 (XLF) [79]. Alternatively, DSBs may be repaired by SSA or Alt-NHEJ, which leads to DNA deletion depending on the extent of end resection. These pathways require ATM signaling [80]. In the SSA pathway, RAD52 anneals homologous sequences during resection, while DNA polymerase fills gaps [81]. On the other hand, during MMEJ, PARP1 works together with helicases to displace RPA from ssDNA. This process reveals microhomologies and promotes the stabilization of the ssDNA ends [82]. To finalize the ligation process, DNA ligase 3 (LIG3) forms a complex with X-ray repair cross-complementing protein 1 (XRCC1) to catalyze the reaction [83]. The decision between c-NHEJ and Alt-NHEJ is influenced by the WRN complex, which prevents MRE11 and CtIP from being recruited on the DNA lesion, thus favoring c-NHEJ. Another critical step in Alt-NHEJ and SSA is the elimination of non-homologous DNA strands, which is mainly mediated by nucleases [84].

One of the mechanisms of PARP1i resistance during this DNA repair pathway is the decreased expression of proteins in the ATM–CHK2 pathway. This process has been linked to the downregulation of the late-phase HR factor TP53BP1 [85]. In general, abnormal expression of the TP53BP1 protein has been shown to contribute to tumor growth [86,87].

In addition, relatively few malignancies have been linked to the downregulation or modification of genes involved in c-NHEJ. In colon and endometrial cancers, only uncommon Ku70/Ku80, LIG4, ARTEMIS, and XLF mutations were identified [88].

Alt-NHEJ and SSA are innately mutagenic due to the production of deletions and the creation of genomic instability, which is associated with many neoplasms. Other genes have been postulated to be potentially carcinogenic through routes such as CtIP [89]. CtIP inactivation reduced carcinogenesis in a mammary mouse model lacking TP53 [90]. In addition, Alt-NHEJ is dependent on Polθ, and enhanced POLQ protein expression has been reported in various cancers, including breast and ovarian cancer [91].

5. Topoisomerase II inhibition Synergizes with DNA Repair Inhibitors

The key role of DNA repair pathways in cancer cells when encountering DNA damage created by Top2 poisons inhibitors rationalizes therapies targeting the remaining functioning pathways that are essential for cancer cell survival and proliferation. This is the foundation of the notion of synthetic lethality, which is the genetic interaction of two genes. Cell viability is unaffected if only one of the two genes is altered. However, cell death ensues when both genes are mutated simultaneously. Identifying several synthetic lethal interactions between proteins involved in DNA repair enabled the creation of tailored therapeutic combinations that target DNA repair enzymes and topoisomerase II inhibitors with the aim of eradicating cancer cells. This dual inhibition ensures that malignant cells are forced into apoptosis by mitosis catastrophe without trying to repair the damage caused by cell cycle arrest (Figure 3). Numerous DDR-targeting compounds are being investigated in clinical trials, with encouraging outcomes for cancer therapy expected (Table 3).

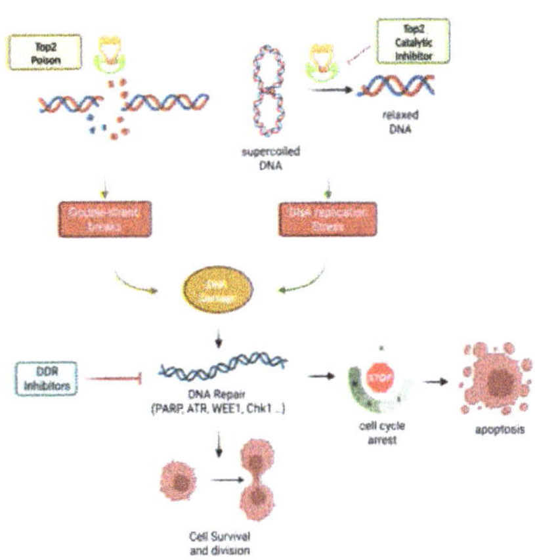

Figure 3. Synthetic lethality of Top2 inhibition in combination with DDR inhibitors. Top2 inhibition generates DNA damage during double-strand breaks or DNA replication stress. These perturbations to the cell can be abolished by DNA damage repair pathways, thus allowing the cell to survive. However, inhibiting the proteins involved in the repair mechanism can increase the efficacy of Top2 inhibitors by producing synthetic lethality. The figure was created on biorender.com.

Table 3. DNA damage repair inhibitors.

DNA Repair Protein	Inhibitor	Application	Clinical Trial	Refs.
PARP	Olaparib	BRCA or HR+ ovarian cancer mutations	NCT02476968 NCT03286842	[92,93]
ATM	AZD0156	Advanced solid tumors	NCT02588105	[94]
	AZD1390	Non-small cell lung cancer and brain cancer	NCT03423628 NCT04550104	[95,96]
ATR	M6620	Metastatic urothelial cancer and solid tumors	NCT02589522 NCT02567409	[97,98]
CHK1	AZD7762	Refractory solid tumors	NCT00937664	[99]
WEE1	AZD1775	Glioblastoma and refractory solid tumors	NCT02095132 NCT01849146	[100,101]
DNA-PK	VX-984	Advanced solid tumors	NCT02644278	[102]
	M3814	Advanced rectal cancer	NCT03770689	[103]

5.1. PARP Inhibitors

Poly (ADP-ribose) polymerase-1 inhibitors (PARP1i) are the first effective instance of targeted medicines that employ synthetic lethality to eradicate malignancies through DNA repair failure [4]. They were first utilized in cancers with BRCA1/BRCA2 mutations. Olaparib was the first FDA approved inhibitor of PARP1, and several derivates, along with new molecules, are now being investigated for treating BRCA-mutant cancer [4,104,105]. Due to PARP1's functions in DNA repair, replication stress responses, and chromatin remodeling, other applications for PARP1i in combination treatment are still being investigated [106]. Overexpression of PARP1 and other DNA damage repair proteins may aid in repairing DNA lesions caused by Top2 inhibitors, allowing tumor cells to withstand treatment involving Top2 inhibitors [107,108]. Thus, Top2 and PARP1 dual inhibitions are considered an alternative for achieving a synergistic impact on tumor cells. BRCA1 maintains genome integrity by preventing the formation of estrogen-induced pathogenic Top2–DNA complexes [109]. BRCA1-deficient cells exhibited hypersensitivity to etoposide and daunomycin that was comparable to Olaparib [110]. A recent study shows that in ovarian cancer cells, the combination treatment of the topoisomerase inhibitor and PARP inhibitor (PARPi) is superior to PARP inhibition or topoisomerase inhibition alone [107].

5.2. ATM/ATR Inhibitors

ATM and ATR activate DDR repair pathways for cell survival in response to DNA damage. Thus, they play a crucial role in cancer treatment. The ATR protein is localized to the DSB site by the ATR-interacting protein (ATRIP) after RPA coats the ssDNA. Consequently, the CHK1 signaling cascade will induce G2-M phase cell cycle arrest, allowing DNA damage to be fixed. On the other hand, ATM uses the MRN complex to respond to DNA damage [44]. By inhibiting these major proteins (ATR and ATM), DNA damage response pathways can be interfered with, causing unresolved DNA damage in growing cancer cells and, eventually, cell death. Non-tumor cancer cells may endure DNA damage because they possess additional DDR components, such as MMR and NER, that detect and fix DNA replication mistakes. As a result, this inhibition may be a successful method in cancer therapy. Some small molecules of ATM and ATR inhibitors have been developed and are being investigated in preclinical trials (Table 3).

Since ATM and ATR are the first molecules activated after DNA damage, their dual inhibition with Top2 inhibitors made them a potential combination therapy for cancer proliferation arrest and apoptosis. In a recent study, ATM inhibition increased the susceptibility of tumor cells to Top1/Top2 inhibitors (e.g., camptothecin, doxorubicin) [111]. In addition, ATR downregulation also increased the efficacy of topoisomerase poisons [112].

5.3. WEE1/CHK1 Inhibitors

CHK1 and WEE1 are overexpressed in various tumors [113–115]. The WEE1 kinase halts the cell cycle at the G2/M phase by inhibiting CDK2/CDK1, which is a process that allows DNA repair mechanisms to be carried out [116]. Inhibiting WEE1 will cause tumor cells to become more susceptible to DNA damage treatments. The CHK1 kinase has several roles in DNA damage-activated signal pathways. CHEK1 inhibition can cause cell cycle arrest at the G2 and S checkpoints, promote DNA damage accumulation, and eventually lead to tumor cell death. Through suppressing CHK1, CHK2, or WEE1, cells suffering DNA damage would lack the ability to activate the checkpoint mechanism and therefore fail to arrest the cell cycle resulting in mitosis aberration with the proportional accumulation of lethal mutations to the cells [117]. This uninterrupted cell replication eventually leads to cell death.

Different approaches have been created to test synergism between Top2 and WEE1/CHK1/CHK2 inhibitors. Interestingly, CHK1 and Plk1 regulate mitotic entrance after Top2 catalytic inhibition, which opens the possibility that its inhibition will lead to a mitotic catastrophe instead of cell cycle arrest [118]. WEE1 inhibition makes ovarian, colon, cervical, osteosarcoma, glioblastoma, and lung cancer cells more vulnerable to DNA damage caused by irradiation and topoisomerase inhibition [119]. In addition, the apoptotic effects of doxorubicin were greatly enhanced in all cell lines by suppressing the expression of WEE1 [120].

5.4. DNA-PK Inhibitors

DNA-PK is a crucial response to DSBs in the c-NHEJ pathway. c-NHEJ is initiated by the Ku complex. DNA-PK then binds to the DSB and begins DNA repair [121]. It has been discovered that overexpression of DNA-PK in tumor tissues after radiation therapy and DNA-PK inhibition decreases DNA repair, indicating that DNA-PK plays a role in overcoming the drug resistance of tumor cells and that DNA-PK inhibition has a superior therapeutic impact. As a result, DNA-PK has been proposed as a viable pharmacological target for anticancer therapies [122–124].

Current studies that combine peposertib (a DNA-PK inhibitor) with Top2 inhibitors (e.g., doxorubicin and etoposide) showed the improved effectiveness of these drugs in ovarian cancer xenografts [125]. Compared to monotherapy controls, the combination of peposertib and Top2 inhibitors reduced tumor development in murine grafts. A clinical study of peposertib in conjunction with cytotoxic treatment is now underway for ovarian cancer [126]. In addition, new DNA-PK inhibitors such as ZL-2201 have demonstrated notable synergy with Top2 inhibitors regardless of ATM status [127]. Moreover, the discovery of a need for TP53 in Top2a-dependent G2 arrest and the activation of DNA-PK pathways opens the possibility of inducing selective synthetic lethality by blocking either PK or the TP53 protein [128].

6. Chromatin Remodeling and Gene Transcription

Restoring damaged DNA requires chromatin remodeling, which involves epigenetic modifications to histones (e.g., methylation or acetylation) [129]. These modifications may activate or inhibit gene clusters responsible for cancer cell survival [130]. In addition to histone modifiers, DNA topoisomerases are eukaryotes' most essential chromatin dynamic regulators. Top2 recruitment affects the epigenetic state of target promoters regulated by histone modifiers, thus resulting in chromatin accessibility [131]. Indeed, the loss of Top2a activity is associated with histone modifications, as seen in decreased H3K27me3 enrichment [132,133]. Besides Top2's functions as a chromatin modulator, epigenetic modifier, and DNA damage inducer, it also plays a role in inducing gene transcription [12]. This transcriptional mechanism is induced by DSBs in gene promoter sites [12,134,135].

During neuronal differentiation, Top2b induces housekeeping genes while proliferation genes are silenced. Additionally, Top2a influences development-specific genes in embryonic stem cells targeting bivalent histone modifications, and these genes are expressed and occupied by Top2b during differentiation [136]. This provides a rationale for combining Top2 inhibitors with inhibitors of histone modifiers or cancer cell survival genes to create a novel synthetic lethality approach for cancer therapy.

6.1. EZH2 and Topoisomerase II

Enhancer of zeste homolog 2 (EZH2) is a member of the polycomb group of genes, which is a family of essential transcription-repressing epigenetic regulators. EZH2 is a component of polycomb repressive complex 2 (PRC2) that methylates the Lys-27 residue in histone 3 (H3K27me3) to repress gene transcription [9]. EZH2 controls the trimethylation of H3K27 and is overexpressed in kidney, breast, and lung cancers, increasing cell motility, colony formation, and genomic instability [137]. The significance of its involvement in cancer pathophysiology is generally recognized by its functions in cell proliferation, apoptosis, and senescence. Therefore, targeting EZH2 for cancer treatment is now a popular area of research, and several EZH2 inhibitors have been developed [9]. Tazemetostat is an EZH2 inhibitor that prevents H3K27 methylation. By promoting the buildup of excessive DNA damage, tazemetostat sensitizes cells to genotoxic treatments (such as the use of Top2 inhibitors), resulting in their death [138,139].

Recent investigations have demonstrated EZH2 overexpression in metastatic prostate cancers [140–143]. Its overexpression has been especially linked to neuroendocrine prostate cancer (NEPC) progression. EZH2 was reported to suppress AR signaling to induce the development of neuroendocrine prostate cancer by forming the N-Myc/AR/EZH2-PRC2 complex. This gene suppressor mechanism is particularly reliant on the presence of EZH2 [142,144]. In prostate cancers, EZH2 has dual functions in that it acts as a transcriptional activator mediated by the induction of AR-regulated genes and as a gene silencer mediated by the epigenetic modification suppression of AR-regulated genes [145]. In AR-negative NEPC cells, enzymatic EZH2 inhibitors are substantially more effective than their counterparts since their actions in AR function in prostate adenocarcinoma are independent of their catalytic activity. Compounds capable of degrading EZH2 protein, analogous to EZH2 knockdown, could outperform enzymatic EZH2 inhibitors significantly and would have superior specificity in inhibiting both of EZH2's activities [146].

Top2 inhibitors can play a critical role in the synergism of EZH2 inhibitors. By promoting chromatin relaxation and accessibility, EZH2 inhibitors create the optimal scenario for Top2 poisons to be able to access DNA double strands and generate breaks [147]. Increased accessibility to the H3K27me3-marked chromatin in leukemia cells after treatment with EZH2 inhibitors suggests that Top2 inhibitors induce DNA damage and cell death [148]. Additionally, the use of doxorubicin with EZH2 inhibitors increases the production of proapoptotic genes, which may have contributed to the death of AML cells [9]. In preclinical prostate cancer mouse cell line models, the simultaneous elevation of Top2a and EZH2 results in hypersensitivity to combined treatment with etoposide, a Top2-targeting toxin, and inhibitors of EZH2 [149]. In addition, combined EZH2 and Top2 inhibition is an alternative treatment for EGFR-mutant tumors, including those that continuously acquire resistance to EGFR tyrosine kinase inhibitors [150].

Furthermore, Top2 catalytic inhibitors can be combined with EZH2 degraders to abrogate AR transcriptional activity, which is methylation independent. Top2 increases androgen signaling and promotes tumor cell growth by relaxing chromatin, exposing promoter regions, and enhancing the transcription of androgen- and estrogen-responsive genes [24]. Catalytic Top2 inhibitors disrupt AR signaling and reduce tumor formation in castration-resistant xenografts, suggesting their potential application in the treatment of castration-resistant cancers [49]. Furthermore, the use of catalytic Top2 inhibitors such as ICRF-193 or T60/T633 in combination with enzalutamide has demonstrated a synergistic effect that blocks AR signaling and cancer cell replication [32–34,49]. In addition to the AR activity abolishment created by Top2 catalytic inhibitors in PCa genesis, they might also be used to reduce stem gene transcription by inhibiting Top2, as previously demonstrated by ICRF-193 in neuronal ES cells reducing essential genes such as LIN28 and NANOG with the extra function of Top2 inhibitors, thus inducing cell cycle arrest [136]. This stemness gene suppression suggests the possibility of using a continuum of Top2 catalytic inhibitors in the progression of PCa disease with the synergism of an EZH2 degrader in AR-dependent PCa cells and EZH2 catalytic inhibitors in NEPC tumor cells.

6.2. *Myc and Topoisomerase II*

The Myc protein family regulates oncogenic gene networks inside cancer cells through transcriptional programming to promote cell proliferation, metabolism, anti-apoptosis, and the inhibition of differentiation [7,151]. Dysregulation of Myc, particularly N-Myc, is correlated with the development of therapy-resistant tumors [7,152].

L-Myc, c-Myc, and N-Myc are involved in the development and progression of the whole spectrum of prostate cancer, from localized adenocarcinoma to the most advanced and treatment-resistant forms, including castration-resistant prostate cancer (CRPC) and neuroendocrine prostate cancer (NEPC) [153,154]. The most detected genomic changes linked to clinical phases and subtypes of PCa are amplifications of Myc family members. L-Myc amplification is mutually exclusive with c-Myc in 27% of localized PCa, while c-Myc amplification is prevalent in all PCa stages and subtypes [155]. Notably, c-Myc overexpression inhibits the transcriptional activity of the androgen receptor (AR), which is the driving force in PCa and the principal therapeutic target in the advanced tumor stage [156]. c-Myc overexpression increases AR-V7, the constitutively active and ligand-independent AR splice variant that promotes CRPC [157]. N-Myc amplification is significantly upregulated in NEPC [158]. N-Myc increases the stability of AURKA by blocking its association with the E3 ubiquitin ligase FBXW7 and binds to the promoters of target genes such as NSE, Syn, and AR to control their expression, ultimately resulting in NEPC tumor progression [159]. In addition, the development of NEPC is facilitated by the synergistic stimulation of the Wnt/β-catenin signaling pathway, which is enabled by amplifying the ALK and N-Myc genes [153].

Myc transcriptional activity is associated with Top2, whereby Myc puts together a complex called the "topoisome" to open chromatin and access promoter regions [8]. This complex is made up of Top1/Top2 proteins [8]. c-Myc joins with Top1 and Top2a in proliferating cells, while N-Myc only joins with Top1 and Top2b in terminal differentiating cells. This differentiation in topoisome proteins is in line with the fact that N-Myc is highly expressed in post-mitotic cells such as neurons. The topoisome controls chromatin morphology to maintain high transcriptional output and makes it possible for Myc to transcriptionally regulate cancer cell proliferation [8].

The discovery of new compounds that target N-Myc offers considerable therapeutic promise for NEPC. Recent research has discovered new ways of targeting Myc proteins by breaking up the Myc–Max heterodimer or stopping the Myc–Max–DNA complex from forming [5,6,160]. Top2 catalytic activity, which is required for chromatin remodeling and accessibility, should also be considered a significant molecule in Myc function. It can also be targeted and provides a proof-of-principal for the development of synthetic lethality strategies that use both N-Myc and Top2 catalytic inhibitors (Figure 4).

Dual Myc–Top2 catalytic inhibitors can be used in different phases of PCa development with a particular mechanism of action that is different for each survival pathway at different tumor stages. In CRPC, where c-Myc is most abundant, AR-V7 or AR splicing variant signals can be abolished by the dual inhibition of Top2 and Myc [32,49]. On the other hand, during NEPC development, Top2 catalytic inhibitors will prevent DNA accessibility for N-Myc, leading to survival transcription gene deficiency and cell death. Besides direct gene transcription arrest, the Top2 catalytic inhibitor abrogates the Wnt-β-catenin epithelial to mesenchymal transition pathway, which is also regulated by N-Myc [161]. Top2a physically interacts with β-catenin and enhances β-catenin's entrance into glioma cell nuclei [162]. Top2a substantially increases the metastatic behaviors of glioma cells in a β-catenin-dependent way. Top2 inhibitors block the transcriptional activity of the TCF/lymphoid enhancer factor mediated by β-catenin [163]. Together, these findings underscore the potential application of therapies that co-target Top2 and Myc to achieve complete tumor suppression.

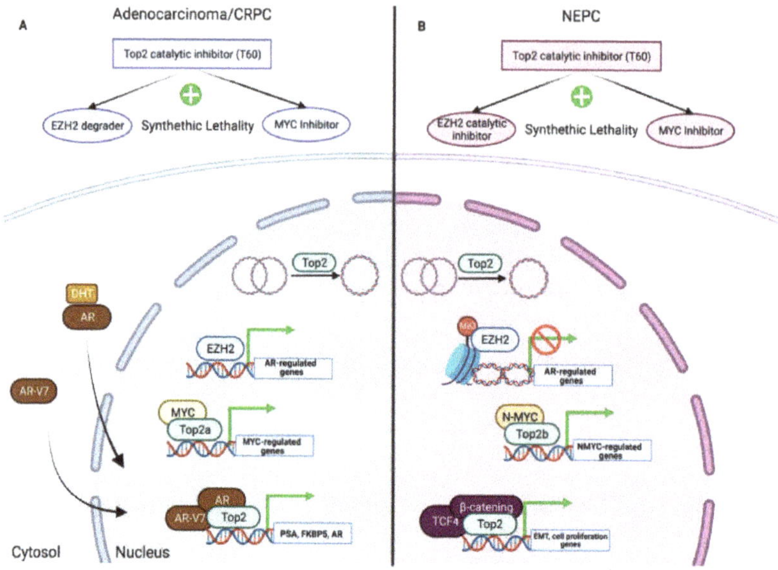

Figure 4. Top2 inhibition synthetic lethality with Myc/EZH2. (**A**) Top2–EZH2–Myc relationship with AR-dependent and CRPC tumor cell survival signaling pathways. (**B**) Top2–EZH2–Myc relationship with NEPC survival signaling pathways. The figure was created on biorender.com.

7. Top2 Catalytic Inhibitors and Synthetic Lethality

DNA repair is a typical process that occurs in cells after DNA has been damaged. Therefore, activity can be boosted by blocking DNA repair proteins such as PARP1 or ATM/ATR. This creates the opportunity to co-target the DNA repair mechanism and Top2 activity. Therapies with catalytic Top2 inhibitors will significantly slow the pace of cancer cell replication, decrease Top2 mobility, and prevent Top2 recruitment to DSBs and gene promoters [46]. Thus, Top2 catalytic inhibition is an underestimated approach in the treatment of tumor cells. Several Top2 catalytic inhibitors have been used for decades to stop cell replication without their mechanism of action on chromatin dynamics and transcription regulation having been explored. Although ICRF-193 is a compound under clinical trials for adjuvant treatment for some cancers, it is far from becoming a key regulator in cancer treatment. Deficiencies in drug development techniques and the use of previous toxic non-selective skeleton molecules make this kind of drug unsuitable for further development. DNA–protein interactions (such as those found in Myc–Max inhibitors) and protein degradation (such as that caused by EZH2 inhibitors) have been proposed as novel therapeutic means to target critical components in cancer survival pathways [5,6,9]. Although some previous drugs have emerged to prevent Top2–DNA interaction pathways, such as aclarubicin, their mechanism of action is not entirely elucidated [46]. This prevents them from advancing to clinical trials.

Our earlier research used a computer-aided drug design approach to discover a druggable site at the Top2–DNA interaction interface. We also designed a series of small molecule inhibitors to prevent Top2 from binding its DNA substrate. Both methods were successful. The lead compound, known as T60, has demonstrated highly effective inhibitory effects against both Top2a and Top2b. Despite its low genotoxicity, it suppresses the multiplication of cancer cells. Importantly, T60 and its derivative T638 have also been shown to block the activity of the AR and the proliferation of AR-positive prostate cancer cells. These findings demonstrate that it is a viable pharmacological candidate that warrants further investigation [32–34]. Because it has low cytotoxicity toward cells, T638 can be combined with other cancer drugs to increase the level of tumor suppression with minimum doses, making it a suitable candidate for combination therapy with DNA repair inhibitors without the side effects associated with high doses of Top2 poisons.

8. Conclusions

Several benefits are associated with synthetic lethality-based therapeutic techniques, such as the ability to overcome resistance to current targeted cancer therapies and the capacity to provide synergism when paired with other chemotherapies. Several DDR-targeting drugs are now being tested in clinical trials and can be further investigated in combination with Top2 inhibitors to increase tumor treatment sensitivity by preventing cancer cell DNA repair. Although Top2 inhibitors are among the oldest target treatments, there are still unknown molecular mechanisms involved in cellular reactions to Top2 inhibitors that might enable novel synthetic lethality medicines, such as the synergism between chromatin remodeling and transcription pathways that can be observed in Myc and EZH2 signaling. Catalytic Top2 inhibitors may play a crucial role in combination therapy since large doses of these drugs may be employed to inhibit Top2 activity without causing significant genotoxicity. In addition, Top2–DNA interaction disruptors can be used to target cell proliferation in ES cells and inhibit the transcription of neurological factors in neuroendocrine or post-mitotic cells. This form of Top2 inhibition is poorly investigated but has the potential to be used in co-targeting therapies with other tumor-surviving genes.

Author Contributions: Conceptualization—V.M.M.-B. and X.D.; original draft—V.M.M.-B.; table and figure preparation—V.M.M.-B.; editing—V.M.M.-B. and X.D.; supervision—X.D. All authors have read and agreed to the published version of the manuscript.

Funding: This work was funded by project grants given to XD by the Canadian Institute of Health Research (CIHR) (PJT156150, PTJ178063, and MOP-137007).

Institutional Review Board Statement: Not applicable.

Informed Consent Statement: Not applicable.

Data Availability Statement: Not applicable.

Conflicts of Interest: The authors declare no conflict of interest.

References

1. Delgado, J.L.; Hsieh, C.M.; Chan, N.L.; Hiasa, H. Topoisomerases as Anticancer Targets. *Biochem. J.* **2018**, *475*, 373–398. [CrossRef] [PubMed]
2. Bjornsti, M.A.; Kaufmann, S.H. Topoisomerases and Cancer Chemotherapy: Recent Advances and Unanswered Questions. *F1000Research* **2019**, *8*, F1000 Faculty Rev-1704. [CrossRef] [PubMed]
3. Huang, A.; Garraway, L.A.; Ashworth, A.; Weber, B. Synthetic Lethality as an Engine for Cancer Drug Target Discovery. *Nat. Rev. Drug Discov.* **2020**, *19*, 23–38. [CrossRef] [PubMed]
4. Dziadkowiec, K.N.; Gasiorowska, E.; Nowak-Markwitz, E.; Jankowska, A. PARP Inhibitors: Review of Mechanisms of Action and BRCA1/2 Mutation Targeting. *Prz. Menopauzalny* **2016**, *15*, 215–219. [CrossRef] [PubMed]
5. Bayliss, R.; Burgess, S.G.; Leen, E.; Richards, M.W. A Moving Target: Structure and Disorder in Pursuit of Myc Inhibitors. *Biochem. Soc. Trans.* **2017**, *45*, 709–717. [CrossRef]
6. Carabet, L.A.; Lallous, N.; Leblanc, E.; Ban, F.; Morin, H.; Lawn, S.; Ghaidi, F.; Lee, J.; Mills, I.G.; Gleave, M.E.; et al. Computer-Aided Drug Discovery of Myc-Max Inhibitors as Potential Therapeutics for Prostate Cancer. *Eur. J. Med. Chem.* **2018**, *160*, 108–119. [CrossRef]
7. Dang, C.V. MYC on the Path to Cancer. *Cell* **2012**, *149*, 22–35. [CrossRef]
8. Das, S.K.; Kuzin, V.; Cameron, D.P.; Sanford, S.; Jha, R.K.; Nie, Z.; Rosello, M.T.; Holewinski, R.; Andresson, T.; Wisniewski, J.; et al. MYC Assembles and Stimulates Topoisomerases 1 and 2 in a "Topoisome". *Mol. Cell* **2022**, *82*, 140–158.e12. [CrossRef]
9. Adema, V.; Colla, S. EZH2 Inhibitors: The Unpacking Revolution. *Cancer Res.* **2022**, *82*, 359–361. [CrossRef]
10. Champoux, J.J. DNA Topoisomerases: Structure, Function, and Mechanism. *Annu. Rev. Biochem.* **2001**, *70*, 369–413. [CrossRef]
11. Corbett, K.D.; Berger, J.M. Structure, Molecular Mechanisms, and Evolutionary Relationships in DNA Topoisomerases. *Annu. Rev. Biophys. Biomol. Struct.* **2004**, *33*, 95–118. [CrossRef] [PubMed]

12. Pommier, Y.; Sun, Y.; Huang, S.Y.N.; Nitiss, J.L. Roles of Eukaryotic Topoisomerases in Transcription, Replication and Genomic Stability. *Nat. Rev. Mol. Cell Biol.* **2016**, *17*, 703–721. [CrossRef] [PubMed]
13. Spakman, D.; Bakx, J.A.M.; Biebricher, A.S.; Peterman, E.J.G.; Wuite, G.J.L.; King, G.A. Unravelling the Mechanisms of Type 1A Topoisomerases Using Single-Molecule Approaches. *Nucleic Acids Res.* **2021**, *49*, 5470–5492. [CrossRef]
14. Type I DNA Topoisomerases | Journal of Medicinal Chemistry. Available online: https://pubs.acs.org/doi/10.1021/acs.jmedchem.6b00966 (accessed on 27 December 2022).
15. McKie, S.J.; Neuman, K.C.; Maxwell, A. DNA Topoisomerases: Advances in Understanding of Cellular Roles and Multi-Protein Complexes via Structure-Function Analysis. *BioEssays* **2021**, *43*, e2000286. [CrossRef] [PubMed]
16. Nitiss, J.L. DNA Topoisomerase II and Its Growing Repertoire of Biological Functions. *Nat. Rev. Cancer* **2009**, *9*, 327–337. [CrossRef] [PubMed]
17. Dalvie, E.D.; Osheroff, N. DNA Recognition/Processing | DNA Topoisomerases: Type II ☆. In *Encyclopedia of Biological Chemistry III*, 3rd ed.; Jez, J., Ed.; Elsevier: Oxford, UK, 2021; pp. 479–486. ISBN 978-0-12-822040-5.
18. Lang, A.J.; Mirski, S.E.L.; Cummings, H.J.; Yu, Q.; Gerlach, J.H.; Cole, S.P.C. Structural Organization of the Human TOP2A and TOP2B Genes. *Gene* **1998**, *221*, 255–266. [CrossRef]
19. Lee, J.H.; Berger, J.M. Cell Cycle-Dependent Control and Roles of DNA Topoisomerase II. *Genes* **2019**, *10*, 859. [CrossRef]
20. Bredel, M.; Slavc, I.; Birner, P.; Czech, T.; Haberler, C.; Ströbel, T.; Wolfsberger, S.; Budka, H.; Hainfellner, J.A. DNA Topoisomerase IIalpha Expression in Optic Pathway Gliomas of Childhood. *Eur. J. Cancer* **2002**, *38*, 393–400. [CrossRef]
21. Feng, Y.; Zhang, H.; Gao, W.; Wen, S.; Huangfu, H.; Sun, R.; Bai, W.; Wang, B. Expression of DNA Topoisomerase II-α: Clinical Significance in Laryngeal Carcinoma. *Oncol. Lett.* **2014**, *8*, 1575–1580. [CrossRef]
22. Zhao, Q.; Li, H.; Zhu, L.; Hu, S.; Xi, X.; Liu, Y.; Liu, J.; Zhong, T. Bioinformatics Analysis Shows That TOP2A Functions as a Key Candidate Gene in the Progression of Cervical Cancer. *Biomed. Rep.* **2020**, *13*, 21. [CrossRef]
23. Wang, X.; Wang, J.; Lyu, L.; Gao, X.; Cai, Y.; Tang, B. Oncogenic Role and Potential Regulatory Mechanism of Topoisomerase IIα in a Pan-Cancer Analysis. *Sci. Rep.* **2022**, *12*, 1–16. [CrossRef]
24. Haffner, M.C.; Aryee, M.J.; Toubaji, A.; Esopi, D.M.; Albadine, R.; Gurel, B.; Isaacs, W.B.; Bova, G.S.; Liu, W.; Xu, J.; et al. Androgen-Induced TOP2B-Mediated Double-Strand Breaks and Prostate Cancer Gene Rearrangements. *Nat. Genet.* **2010**, *42*, 668–675. [CrossRef] [PubMed]
25. Manville, C.M.; Smith, K.; Sondka, Z.; Rance, H.; Cockell, S.; Cowell, I.G.; Lee, K.C.; Morris, N.J.; Padget, K.; Jackson, G.H.; et al. Genome-Wide ChIP-Seq Analysis of Human TOP2B Occupancy in MCF7 Breast Cancer Epithelial Cells. *Biol. Open* **2015**, *4*, 1436–1447. [CrossRef] [PubMed]
26. Liang, X.; Wu, Q.; Luan, S.; Yin, Z.; He, C.; Yin, L.; Zou, Y.; Yuan, Z.; Li, L.; Song, X.; et al. A Comprehensive Review of Topoisomerase Inhibitors as Anticancer Agents in the Past Decade. *Eur. J. Med. Chem.* **2019**, *171*, 129–168. [CrossRef]
27. Pommier, Y. Drugging Topoisomerases: Lessons and Challenges. *ACS Chem. Biol.* **2013**, *8*, 82–95. [CrossRef]
28. Berger, J.M.; Gamblin, S.J.; Harrison, S.C.; Wang, J.C. Structure and Mechanism of DNA Topoisomerase II. *Nature* **1996**, *379*, 225–232. [CrossRef] [PubMed]
29. Burden, D.A.; Osheroff, N. Mechanism of Action of Eukaryotic Topoisomerase II and Drugs Targeted to the Enzyme. *Biochim. Biophys. Acta (BBA)-Gene Struct. Expr.* **1998**, *1400*, 139–154. [CrossRef]
30. Bromberg, K.D.; Osheroff, N. Mechanism of Action of Topoisomerase II-Targeted Anticancer Drugs. In *DNA Topoisomerases in Cancer Therapy: Present and Future*; Andoh, T., Ed.; Springer: Boston, MA, USA, 2003; pp. 53–78. ISBN 978-1-4615-0141-1.
31. Wang, L.; Luo, J.; Chen, G.; Fang, M.; Wei, X.; Li, Y.; Liu, Z.; Zhang, Y.; Gao, S.; Shen, J.; et al. Chidamide, Decitabine, Cytarabine, Aclarubicin, and Granulocyte Colony-Stimulating Factor (CDCAG) in Patients with Relapsed/Refractory Acute Myeloid Leukemia: A Single-Arm, Phase 1/2 Study. *Clin. Epigenetics* **2020**, *12*, 132. [CrossRef] [PubMed]
32. Matias-Barrios, V.M.; Radaeva, M.; Song, Y.; Alperstein, Z.; Lee, A.R.; Schmitt, V.; Lee, J.; Ban, F.; Xie, N.; Qi, J.; et al. Discovery of New Catalytic Topoisomerase II Inhibitors for Anticancer Therapeutics. *Front. Oncol.* **2021**, *10*, 633142. [CrossRef]
33. Matias-Barrios, V.M.; Radaeva, M.; Ho, C.-H.; Lee, J.; Adomat, H.; Lallous, N.; Cherkasov, A.; Dong, X. Optimization of New Catalytic Topoisomerase II Inhibitors as an Anti-Cancer Therapy. *Cancers* **2021**, *13*, 3675. [CrossRef]
34. Barrios, M.; Manuel, V. New Catalytic Topoisomerase II Inhibitors Discovered for Anticancer Therapeutics, University of British Columbia. *Front. Oncol.* **2021**, *10*, 633142. [CrossRef] [PubMed]
35. Patterson-Fortin, J.; Jadhav, H.; Phan, T.; D'Andrea, A.; Shapiro, G.I. Abstract 1133: Novobiocin-Mediated Polymerase Theta Inhibition Induces CGAS/STING Pathway Activation and T-Cell Infiltration in BRCA-Associated Cancers. *Cancer Res.* **2022**, *82*, 1133. [CrossRef]
36. Kraut, E.H.; Bendetti, J.; Balcerzak, S.P.; Doroshow, J.H. Phase II Trial of Merbarone in Soft Tissue Sarcoma. A Southwest Oncology Group Study. *Invest New Drugs* **1992**, *10*, 347–349. [CrossRef] [PubMed]
37. Yi, L.L.; Kerrigan, J.E.; Lin, C.P.; Azarova, A.M.; Tsai, Y.C.; Ban, Y.; Liu, L.F. Topoisomerase IIbeta Mediated DNA Double-Strand Breaks: Implications in Doxorubicin Cardiotoxicity and Prevention by Dexrazoxane. *Cancer Res.* **2007**, *67*, 8839–8846. [CrossRef]
38. Reyhanoglu, G.; Tadi, P. *Etoposide*; StatPearls Publishing: Treasure Island, FL, USA, 2022.
39. Johnson-Arbor, K.; Dubey, R. Doxorubicin. In *StatPearls*; StatPearls Publishing: Treasure Island, FL, USA, 2022.
40. Kessler, T.; Mohr, M.; Müller-Tidow, C.; Krug, U.; Brunnberg, U.; Mohr, B.; Schliemann, C.; Sauerland, C.; Serve, H.; Büchner, T.; et al. Amsacrine Containing Induction Therapy in Elderly AML Patients: Comparison to Standard Induction Regimens in a Matched-Pair Analysis. *Leuk Res.* **2008**, *32*, 491–494. [CrossRef]

41. Liu, L.F. DNA Topoisomerase Poisons as Antitumor Drugs. *Annu. Rev. Biochem.* **1989**, *58*, 351–375. [CrossRef]
42. Meresse, P.; Dechaux, E.; Monneret, C.; Bertounesque, E. Etoposide: Discovery and Medicinal Chemistry. *Curr. Med. Chem.* **2012**, *11*, 2443–2466. [CrossRef]
43. Montgomery, B.; Lin, D.W. Chapter 10-Toxicities of Chemotherapy for Genitourinary Malignancies. In *Complications of Urologic Surgery*, 4th ed.; Taneja, S.S., Ed.; W.B. Saunders: Philadelphia, PA, USA, 2010; pp. 117–123. ISBN 978-1-4160-4572-4.
44. Maréchal, A.; Zou, L. DNA Damage Sensing by the ATM and ATR Kinases. *Cold Spring Harb. Perspect Biol.* **2013**, *5*, a012716. [CrossRef]
45. Wijdeven, R.H.; Pang, B.; van der Zanden, S.Y.; Qiao, X.; Blomen, V.; Hoogstraat, M.; Lips, E.H.; Janssen, L.; Wessels, L.; Brummelkamp, T.R.; et al. Genome-Wide Identification and Characterization of Novel Factors Conferring Resistance to Topoisomerase II Poisons in Cancer. *Cancer Res.* **2015**, *75*, 4176–4187. [CrossRef]
46. Larsen, A.K.; Escargueil, A.E.; Skladanowski, A. Catalytic Topoisomerase II Inhibitors in Cancer Therapy. *Pharmacol. Ther.* **2003**, *99*, 167–181. [CrossRef]
47. Portugal, J.; Mansilla, S.; Bataller, M. Mechanisms of Drug-Induced Mitotic Catastrophe in Cancer Cells. *Curr. Pharm. Des.* **2010**, *16*, 69–78. [CrossRef]
48. Weiss, G.; Loyevsky, M.; Gordeuk, V.R. Dexrazoxane (ICRF-187). *Gen. Pharmacol.* **1999**, *32*, 155–158. [CrossRef] [PubMed]
49. Li, H.; Xie, N.; Gleave, M.E.; Dong, X.; Li, H.; Xie, N.; Gleave, M.E.; Dong, X. Catalytic Inhibitors of DNA Topoisomerase II Suppress the Androgen Receptor Signaling and Prostate Cancer Progression. *Oncotarget* **2015**, *6*, 20474–20484. [CrossRef] [PubMed]
50. Kawatani, M.; Takayama, H.; Muroi, M.; Kimura, S.; Maekawa, T.; Osada, H. Identification of a Small-Molecule Inhibitor of DNA Topoisomerase II by Proteomic Profiling. *Chem. Biol.* **2011**, *18*, 743–751. [CrossRef] [PubMed]
51. Dong, G.; Wu, Y.; Sun, Y.; Liu, N.; Wu, S.; Zhang, W.; Sheng, C. Identification of Potent Catalytic Inhibitors of Human DNA Topoisomerase II by Structure-Based Virtual Screening. *Med. Chem. Commun.* **2018**, *9*, 1142–1146. [CrossRef]
52. Yang, G.; Liu, C.; Chen, S.H.; Kassab, M.A.; Hoff, J.D.; Walter, N.G.; Yu, X. Super-Resolution Imaging Identifies PARP1 and the Ku Complex Acting as DNA Double-Strand Break Sensors. *Nucleic Acids Res.* **2018**, *46*, 3446–3457. [CrossRef]
53. Jazayeri, A.; Balestrini, A.; Garner, E.; Haber, J.E.; Costanzo, V. Mre11–Rad50–Nbs1-Dependent Processing of DNA Breaks Generates Oligonucleotides That Stimulate ATM Activity. *EMBO J.* **2008**, *27*, 1953–1962. [CrossRef]
54. Bartocci, C.; Denchi, E.L. Put a RING on It: Regulation and Inhibition of RNF8 and RNF168 RING Finger E3 Ligases at DNA Damage Sites. *Front Genet* **2013**, *4*, 128. [CrossRef]
55. Scully, R.; Panday, A.; Elango, R.; Willis, N.A. DNA Double Strand Break Repair Pathway Choice in Somatic Mammalian Cells. *Nat. Rev. Mol. Cell Biol.* **2019**, *20*, 698–714. [CrossRef]
56. Chapman, J.R.; Taylor, M.R.G.; Boulton, S.J. Playing the End Game: DNA Double-Strand Break Repair Pathway Choice. *Mol. Cell* **2012**, *47*, 497–510. [CrossRef]
57. Branzei, D.; Foiani, M. Regulation of DNA Repair throughout the Cell Cycle. *Nat. Rev. Mol. Cell Biol.* **2008**, *9*, 297–308. [CrossRef] [PubMed]
58. Torgovnick, A.; Schumacher, B. DNA Repair Mechanisms in Cancer Development and Therapy. *Front Genet* **2015**, *6*, 157. [CrossRef] [PubMed]
59. Hsieh, P.; Yamane, K. DNA Mismatch Repair: Molecular Mechanism, Cancer, and Ageing. *Mech. Ageing Dev.* **2008**, *129*, 391–407. [CrossRef]
60. Li, W.; Xie, C.; Yang, Z.; Chen, J.; Lu, N.-H. Abnormal DNA-PKcs and Ku 70/80 Expression May Promote Malignant Pathological Processes in Gastric Carcinoma. *World J. Gastroenterol.* **2013**, *19*, 6894–6901. [CrossRef] [PubMed]
61. Alshareeda, A.T.; Negm, O.H.; Albarakati, N.; Green, A.R.; Nolan, C.; Sultana, R.; Madhusudan, S.; Benhasouna, A.; Tighe, P.; Ellis, I.O.; et al. Clinicopathological Significance of KU70/KU80, a Key DNA Damage Repair Protein in Breast Cancer. *Breast Cancer Res. Treat* **2013**, *139*, 301–310. [CrossRef] [PubMed]
62. Goodwin, J.F.; Knudsen, K.E. Beyond DNA Repair: DNA-PK Function in Cancer. *Cancer Discov.* **2014**, *4*, 1126–1139. [CrossRef] [PubMed]
63. Nicolas, E.; Bertucci, F.; Sabatier, R.; Gonçalves, A. Targeting BRCA Deficiency in Breast Cancer: What Are the Clinical Evidences and the Next Perspectives? *Cancers* **2018**, *10*, 506. [CrossRef] [PubMed]
64. Liu, H.; Weng, J. A Pan-Cancer Bioinformatic Analysis of RAD51 Regarding the Values for Diagnosis, Prognosis, and Therapeutic Prediction. *Front. Oncol.* **2022**, *12m*, 858756. [CrossRef]
65. Reginato, G.; Cejka, P. The MRE11 Complex: A Versatile Toolkit for the Repair of Broken DNA. *DNA Repair* **2020**, *91–92*, 102869. [CrossRef]
66. Chen, R.; Wold, M.S. Replication Protein A: Single-Stranded DNA's First Responder: Dynamic DNA-Interactions Allow Replication Protein A to Direct Single-Strand DNA Intermediates into Different Pathways for Synthesis or Repair. *Bioessays* **2014**, *36*, 1156–1161. [CrossRef]
67. Ma, C.J.; Kwon, Y.; Sung, P.; Greene, E.C. Human RAD52 Interactions with Replication Protein A and the RAD51 Presynaptic Complex. *J. Biol. Chem.* **2017**, *292*, 11702–11713. [CrossRef] [PubMed]
68. Ranjha, L.; Howard, S.M.; Cejka, P. Main Steps in DNA Double-Strand Break Repair: An Introduction to Homologous Recombination and Related Processes. *Chromosoma* **2018**, *127*, 187–214. [CrossRef] [PubMed]

69. Wu, Y.; Kantake, N.; Sugiyama, T.; Kowalczykowski, S.C. Rad51 Protein Controls Rad52-Mediated DNA Annealing. *J. Biol. Chem.* **2008**, *283*, 14883–14892. [CrossRef]
70. Kwon, Y.; Sung, P. Biochemical Analysis of D-Loop Extension and DNA Strand Displacement Synthesis. *Methods Mol. Biol.* **2021**, *2153*, 87–99. [CrossRef] [PubMed]
71. Wright, W.D.; Shah, S.S.; Heyer, W.D. Homologous Recombination and the Repair of DNA Double-Strand Breaks. *J. Biol. Chem.* **2018**, *293*, 10524–10535. [CrossRef] [PubMed]
72. Multiple Pathways of Recombination Induced by Double-Strand Breaks in Saccharomyces Cerevisiae-PMC. Available online: https://www.ncbi.nlm.nih.gov/pmc/articles/PMC98970/ (accessed on 27 December 2022).
73. Malkova, A.; Ira, G. Break-Induced Replication: Functions and Molecular Mechanism. *Curr. Opin. Genet Dev.* **2013**, *23*, 271–279. [CrossRef]
74. Knijnenburg, T.A.; Wang, L.; Zimmermann, M.T.; Chambwe, N.; Gao, G.F.; Cherniack, A.D.; Fan, H.; Shen, H.; Way, G.P.; Greene, C.S.; et al. Genomic and Molecular Landscape of DNA Damage Repair Deficiency across The Cancer Genome Atlas. *Cell Rep.* **2018**, *23*, 239–254. [CrossRef]
75. Wu, Y.; Zhao, H. CTBP1 Strengthens the Cisplatin Resistance of Gastric Cancer Cells by Upregulating RAD51 Expression. *Oncol Lett* **2021**, *22*, 810. [CrossRef]
76. Xu, Y.; Chen, K.; Cai, Y.; Cheng, C.; Zhang, Z.; Xu, G. Overexpression of Rad51 Predicts Poor Prognosis and Silencing of Rad51 Increases Chemo-Sensitivity to Doxorubicin in Neuroblastoma. *Am. J. Transl. Res.* **2019**, *11*, 5788–5799.
77. García-Sanz, P.; Triviño, J.C.; Mota, A.; Pérez López, M.; Colás, E.; Rojo-Sebastián, A.; García, Á.; Gatius, S.; Ruiz, M.; Prat, J.; et al. Chromatin Remodelling and DNA Repair Genes Are Frequently Mutated in Endometrioid Endometrial Carcinoma. *Int. J. Cancer* **2017**, *140*, 1551–1563. [CrossRef]
78. Cybulski, C.; Górski, B.; Debniak, T.; Gliniewicz, B.; Mierzejewski, M.; Masojć, B.; Jakubowska, A.; Matyjasik, J.; Złowocka, E.; Sikorski, A.; et al. NBS1 Is a Prostate Cancer Susceptibility Gene. *Cancer Res.* **2004**, *64*, 1215–1219. [CrossRef] [PubMed]
79. Davis, A.J.; Chen, D.J. DNA Double Strand Break Repair via Non-Homologous End-Joining. *Transl. Cancer Res.* **2013**, *2*, 130–143. [CrossRef]
80. Valikhani, M.; Rahimian, E.; Ahmadi, S.E.; Chegeni, R.; Safa, M. Involvement of Classic and Alternative Non-Homologous End Joining Pathways in Hematologic Malignancies: Targeting Strategies for Treatment. *Exp. Hematol. Oncol.* **2021**, *10*, 51. [CrossRef]
81. Blasiak, J. Single-Strand Annealing in Cancer. *Int. J. Mol. Sci.* **2021**, *22*, 2167. [CrossRef]
82. McVey, M.; Lee, S.E. MMEJ Repair of Double-Strand Breaks (Director's Cut): Deleted Sequences and Alternative Endings. *Trends Genet.* **2008**, *24*, 529–538. [CrossRef]
83. Lu, G.; Duan, J.; Shu, S.; Wang, X.; Gao, L.; Guo, J.; Zhang, Y. Ligase I and Ligase III Mediate the DNA Double-Strand Break Ligation in Alternative End-Joining. *Proc. Natl. Acad. Sci. USA* **2016**, *113*, 1256–1260. [CrossRef]
84. Shamanna, R.A.; Lu, H.; De Freitas, J.K.; Tian, J.; Croteau, D.L.; Bohr, V.A. WRN Regulates Pathway Choice between Classical and Alternative Non-Homologous End Joining. *Nat. Commun.* **2016**, *7*, 13785. [CrossRef]
85. Yao, J.; Huang, A.; Zheng, X.; Liu, T.; Lin, Z.; Zhang, S.; Yang, Q.; Zhang, T.; Ma, H. 53BP1 Loss Induces Chemoresistance of Colorectal Cancer Cells to 5-Fluorouracil by Inhibiting the ATM–CHK2–P53 Pathway. *J. Cancer Res. Clin. Oncol.* **2017**, *143*, 419–431. [CrossRef] [PubMed]
86. Expression of DNA Damage Checkpoint 53BP1 Is Correlated with Prognosis, Cell Proliferation and Apoptosis in Colorectal Cancer-PMC. Available online: https://www.ncbi.nlm.nih.gov/pmc/articles/PMC4525819/ (accessed on 27 December 2022).
87. Matsuda, K.; Kawasaki, T.; Akazawa, Y.; Hasegawa, Y.; Kondo, H.; Suzuki, K.; Iseki, M.; Nakashima, M. Expression Pattern of P53-Binding Protein 1 as a New Molecular Indicator of Genomic Instability in Bladder Urothelial Carcinoma. *Sci. Rep.* **2018**, *8*, 15477. [CrossRef] [PubMed]
88. Sishc, B.J.; Davis, A.J. The Role of the Core Non-Homologous End Joining Factors in Carcinogenesis and Cancer. *Cancers* **2017**, *9*, 81. [CrossRef]
89. CtIP Contributes to Non-Homologous End Joining Formation through Interacting with Ligase IV and Promotion of TMZ Resistance in Glioma Cells-PubMed. Available online: https://pubmed.ncbi.nlm.nih.gov/30915754/ (accessed on 27 December 2022).
90. Reczek, C.R.; Shakya, R.; Miteva, Y.; Szabolcs, M.; Ludwig, T.; Baer, R. The DNA Resection Protein CtIP Promotes Mammary Tumorigenesis. *Oncotarget* **2016**, *7*, 32172–32183. [CrossRef] [PubMed]
91. Expression of POLQ in Cancer-Summary-The Human Protein Atlas. Available online: https://www.proteinatlas.org/ENSG00000051341-POLQ/pathology (accessed on 27 December 2022).
92. To Assess the Efficacy and Safety of Olaparib Maintenance Monotherapy in the Treatment of Ovarian Cancer-Full Text online. Available online: https://clinicaltrials.gov/ct2/show/NCT02476968 (accessed on 27 December 2022).
93. AstraZeneca. An Open Label, Single Arm, Multicentre Study to Assess the Clinical Effectiveness and Safety of Lynparza (Olaparib) Capsules Maintenance Monotherapy in Platinum Sensitive Relapsed Somatic or Germline BRCA Mutated Ovarian Cancer Patients Who Are in Complete or Partial Response Following Platinum Based Chemotherapy (ORZORA). clinicaltrials.gov. 2022. Available online: https://clinicaltrials.gov/ct2/show/NCT03286842 (accessed on 27 December 2022).
94. AstraZeneca. A Phase I, Open-Label Study to Assess the Safety, Tolerability, Pharmacokinetics and Preliminary Efficacy of Ascending Doses of AZD0156 Monotherapy or in Combination With Either Cytotoxic Chemotherapies or Novel Anti-Cancer Agents in Patients With Advanced Malignancies; clinicaltrials.gov. 2022. Available online: https://clinicaltrials.gov/ct2/show/NCT02588105 (accessed on 27 December 2022).

95. A Study to Assess the Safety and Tolerability of AZD1390 Given With Radiation Therapy in Patients With Brain Cancer-Full Text View-ClinicalTrials.Gov. Available online: https://clinicaltrials.gov/ct2/show/NCT03423628 (accessed on 27 December 2022).
96. Walls, G.M.; Oughton, J.B.; Chalmers, A.J.; Brown, S.; Collinson, F.; Forster, M.D.; Franks, K.N.; Gilbert, A.; Hanna, G.G.; Hannaway, N.; et al. CONCORDE: A Phase I Platform Study of Novel Agents in Combination with Conventional Radiotherapy in Non-Small-Cell Lung Cancer. *Clin. Transl. Radiat. Oncol.* **2020**, *25*, 61–66. Available online: https://clinicaltrials.gov/ct2/show/NCT04550104 (accessed on 27 December 2022). [CrossRef]
97. Testing the Safety of M6620 (VX-970) When Given With Standard Whole Brain Radiation Therapy for the Treatment of Brain Metastases From Non-Small Cell Lung Cancer, Small Cell Lung Cancer, or Neuroendocrine Tumors-Full Text View-ClinicalTrials.Gov. Available online: https://clinicaltrials.gov/ct2/show/NCT02589522 (accessed on 27 December 2022).
98. Pal, S.K.; Frankel, P.H.; Mortazavi, A.; Milowsky, M.; Vaishampayan, U.; Parikh, M.; Lyou, Y.; Weng, P.; Parikh, R.; Teply, B.; et al. Effect of Cisplatin and Gemcitabine with or without Berzosertib in Patients with Advanced Urothelial Carcinoma: A Phase 2 Randomized Clinical Trial. *JAMA Oncol.* **2021**. Available online: https://clinicaltrials.gov/ct2/show/NCT02567409 (accessed on 27 December 2022). [CrossRef]
99. Safety and Tolerability Study of AZD7762 in Combination with Gemcitabine-Full Text View-ClinicalTrials.Gov. Available online: https://clinicaltrials.gov/ct2/show/NCT00937664 (accessed on 27 December 2022).
100. Adavosertib and Irinotecan Hydrochloride in Treating Younger Patients With Relapsed or Refractory Solid Tumors-Full Text View-ClinicalTrials.Gov. Available online: https://clinicaltrials.gov/ct2/show/NCT02095132 (accessed on 27 December 2022).
101. Adavosertib, Radiation Therapy, and Temozolomide in Treating Patients With Newly Diagnosed or Recurrent Glioblastoma-Full Text View-ClinicalTrials.Gov. Available online: https://clinicaltrials.gov/ct2/show/NCT01849146 (accessed on 27 December 2022).
102. First-in-Human Study of the Safety, Tolerability, and Pharmacokinetic/Pharmacodynamic Profile of VX-984 in Combination With Chemotherapy-Full Text View-ClinicalTrials.Gov. Available online: https://clinicaltrials.gov/ct2/show/NCT02644278 (accessed on 27 December 2022).
103. Study of Peposertib in Combination With Capecitabine and RT in Rectal Cancer-Full Text View-ClinicalTrials.Gov. Available online: https://clinicaltrials.gov/ct2/show/NCT03770689 (accessed on 27 December 2022).
104. Gelmon, K.A.; Fasching, P.A.; Couch, F.J.; Balmaña, J.; Delaloge, S.; Labidi-Galy, I.; Bennett, J.; McCutcheon, S.; Walker, G.; O'Shaughnessy, J.; et al. Clinical Effectiveness of Olaparib Monotherapy in Germline BRCA-Mutated, HER2-Negative Metastatic Breast Cancer in a Real-World Setting: Phase IIIb LUCY Interim Analysis. *Eur. J. Cancer* **2021**, *152*, 68–77. [CrossRef]
105. Bochum, S.; Berger, S.; Martens, U.M. Olaparib. *Recent Results Cancer Res.* **2018**, *211*, 217–233. [CrossRef]
106. Gupta, N.; Huang, T.T.; Horibata, S.; Lee, J.M. Cell Cycle Checkpoints and beyond: Exploiting the ATR/CHK1/WEE1 Pathway for the Treatment of PARP Inhibitor–Resistant Cancer-ScienceDirect. *Pharmacol. Res.* **2022**, *178*, 106162. [CrossRef]
107. Cal, M.; Szakonyi, Z.; Flörkemeier, I.; Hillmann, J.S.; Weimer, J.P.; Hildebrandt, J.; Hedemann, N.; Rogmans, C.; Dempfle, A.; Arnold, N.; et al. Combined PARP and Dual Topoisomerase Inhibition Potentiates Genome Instability and Cell Death in Ovarian Cancer. *Int. J. Mol. Sci.* **2022**, *23*, 10503. [CrossRef]
108. Boerner, J.L.; Nechiporchik, N.; Mueller, K.L.; Polin, L.; Heilbrun, L.; Boerner, S.A.; Zoratti, G.L.; Stark, K.; Lorusso, P.M.; Burger, A. Protein Expression of DNA Damage Repair Proteins Dictates Response to Topoisomerase and PARP Inhibitors in Triple-Negative Breast Cancer. *PLoS ONE* **2015**, *10*, e0119614. [CrossRef]
109. Sasanuma, H.; Tsuda, M.; Morimoto, S.; Saha, L.K.; Rahman, M.M.; Kiyooka, Y.; Fujiike, H.; Cherniack, A.D.; Itou, J.; Moreu, E.C.; et al. BRCA1 Ensures Genome Integrity by Eliminating Estrogen-Induced Pathological Topoisomerase II-DNA Complexes. *Proc. Natl. Acad. Sci. USA* **2018**, *115*, E10642–E10651. [CrossRef]
110. Treszezamsky, A.D.; Kachnic, L.A.; Feng, Z.; Zhang, J.; Tokadjian, C.; Powell, S.N. BRCA1- and BRCA2-Deficient Cells Are Sensitive to Etoposide-Induced DNA Double-Strand Breaks via Topoisomerase II. *Cancer Res.* **2007**, *67*, 7078–7081. [CrossRef]
111. ATM Orchestrates the DNA-Damage Response to Counter Toxic Non-Homologous End-Joining at Broken Replication Forks-PubMed. Available online: https://pubmed.ncbi.nlm.nih.gov/30622252/ (accessed on 27 December 2022).
112. Wagner, J.M.; Kaufmann, S.H. Prospects for the Use of ATR Inhibitors to Treat Cancer. *Pharmaceuticals* **2010**, *3*, 1311–1334. [CrossRef] [PubMed]
113. Davies, K.D.; Cable, P.L.; Garrus, J.E.; Sullivan, F.X.; von Carlowitz, I.; Huerou, Y.L.; Wallace, E.; Woessner, R.D.; Gross, S. Chk1 Inhibition and Wee1 Inhibition Combine Synergistically to Impede Cellular Proliferation. *Cancer Biol. Ther.* **2011**, *12*, 788–796. [CrossRef] [PubMed]
114. Sørensen, C.S.; Syljuåsen, R.G. Safeguarding Genome Integrity: The Checkpoint Kinases ATR, CHK1 and WEE1 Restrain CDK Activity during Normal DNA Replication. *Nucleic Acids Res.* **2012**, *40*, 477–486. [CrossRef]
115. Mir, S.E.; De Witt Hamer, P.C.; Krawczyk, P.M.; Balaj, L.; Claes, A.; Niers, J.M.; Van Tilborg, A.A.G.; Zwinderman, A.H.; Geerts, D.; Kaspers, G.J.L.; et al. In Silico Analysis of Kinase Expression Identifies WEE1 as a Gatekeeper against Mitotic Catastrophe in Glioblastoma. *Cancer Cell* **2010**, *18*, 244–257. [CrossRef]
116. Regulation of G2/M Transition by Inhibition of WEE1 and PKMYT1 Kinases-PMC. Available online: https://www.ncbi.nlm.nih.gov/pmc/articles/PMC6149964/ (accessed on 27 December 2022).
117. ATM, ATR, CHK1, CHK2 and WEE1 Inhibitors in Cancer and Cancer Stem Cells-PMC. Available online: https://www.ncbi.nlm.nih.gov/pmc/articles/PMC6072143/ (accessed on 27 December 2022).

118. Arroyo, M.; Cañuelo, A.; Calahorra, J.; Hastert, F.D.; Sánchez, A.; Clarke, D.J.; Marchal, J.A. Mitotic Entry upon Topo II Catalytic Inhibition Is Controlled by Chk1 and Plk1. *FEBS J.* **2020**, *287*, 4933–4951. [CrossRef] [PubMed]
119. WEE1 Inhibition Enhances the Antitumor Immune Response to PD-L1 Blockade by the Concomitant Activation of STING and STAT1 Pathways in SCLC-ScienceDirect. Available online: https://www.sciencedirect.com/science/article/pii/S221112472200585X (accessed on 27 December 2022).
120. Ghelli Luserna Di Rorà, A.; Beeharry, N.; Imbrogno, E.; Ferrari, A.; Robustelli, V.; Righi, S.; Sabattini, E.; Verga Falzacappa, M.V.; Ronchini, C.; Testoni, N.; et al. Targeting WEE1 to Enhance Conventional Therapies for Acute Lymphoblastic Leukemia. *J Hematol. Oncol.* **2018**, *11*, 99. [CrossRef] [PubMed]
121. Blackford, A.N.; Jackson, S.P. ATM, ATR, and DNA-PK: The Trinity at the Heart of the DNA Damage Response. *Mol. Cell* **2017**, *66*, 801–817. [CrossRef]
122. Hu, S.; Hui, Z.; Lirussi, F.; Garrido, C.; Ye, X.Y.; Xie, T. Small Molecule DNA-PK Inhibitors as Potential Cancer Therapy: A Patent Review (2010–Present). *Expert Opin. Ther. Pat.* **2021**, *31*, 435–452. [CrossRef]
123. Damia, G. Targeting DNA-PK in Cancer. *Mutat. Res.-Fundam. Mol. Mech. Mutagen.* **2020**, *821*, 111692. [CrossRef]
124. Timme, C.R.; Rath, B.H.; O'Neill, J.W.; Camphausen, K.; Tofilon, P.J. The DNA-PK Inhibitor VX-984 Enhances the Radiosensitivity of Glioblastoma Cells Grown in Vitro and as Orthotopic Xenografts. *Mol. Cancer Ther.* **2018**, *17*, 1207–1216. [CrossRef]
125. Zenke, F.T.; Zimmermann, A.; Sirrenberg, C.; Dahmen, H.; Kirkin, V.; Pehl, U.; Grombacher, T.; Wilm, C.; Fuchss, T.; Amendt, C.; et al. Pharmacologic Inhibitor of DNA-PK, M3814, Potentiates Radiotherapy and Regresses Human Tumors in Mouse Models. *Mol. Cancer Ther.* **2020**, *19*, 1091–1101. [CrossRef]
126. A Study Combining the Peposertib (M3814) Pill With Standard Chemotherapy in Patients With Ovarian Cancer With an Expansion in High Grade Serous Ovarian Cancer and Low Grade Serous Ovarian Cancer-Full Text View-ClinicalTrials.Gov. Available online: https://clinicaltrials.gov/ct2/show/NCT04092270 (accessed on 27 December 2022).
127. Lal, S.; Sun, B.-C.; Chen, Y.; Huang, T.; Bhola, N.; Morton, V.; Chen, K.; Xia, S.; Zhang, H.; Ye, Q.; et al. Abstract 2594: Discovery and Characterization of ZL-2201, a Potent, Highly-Selective, and Orally Bioavailable Small-Molecule DNA-PK Inhibitor. *Cancer Res.* **2022**, *82*, 2594. [CrossRef]
128. Lockwood, N.; Martini, S.; Lopez-Pardo, A.; Deiss, K.; Segeren, H.A.; Semple, R.K.; Collins, I.; Repana, D.; Cobbaut, M.; Soliman, T.; et al. Genome-Protective Topoisomerase 2a-Dependent G2 Arrest Requires P53 in HTERT-Positive Cancer Cells. *Cancer Res.* **2022**, *82*, 1762–1773. [CrossRef] [PubMed]
129. Song, H.; Shen, R.; Liu, X.; Yang, X.; Xie, K.; Guo, Z.; Wang, D. Histone Post-Translational Modification and the DNA Damage Response. *Genes Dis.* **2022**, *128*, 28–36. [CrossRef]
130. Lu, Y.; Chan, Y.-T.; Tan, H.-Y.; Li, S.; Wang, N.; Feng, Y. Epigenetic Regulation in Human Cancer: The Potential Role of Epi-Drug in Cancer Therapy. *Mol. Cancer* **2020**, *19*, 79. [CrossRef]
131. Lu, L.-Y.; Kuang, H.; Korakavi, G.; Yu, X. Topoisomerase II Regulates the Maintenance of DNA Methylation. *J. Biol. Chem.* **2015**, *290*, 851–860. [CrossRef]
132. Li, L.; Wang, Y. Cross-Talk between the H3K36me3 and H4K16ac Histone Epigenetic Marks in DNA Double-Strand Break Repair. *J. Biol. Chem.* **2017**, *292*, 11951–11959. [CrossRef] [PubMed]
133. Top2a Promotes the Development of Social Behavior via PRC2 and H3K27me3. Science Advances. Available online: https://www.science.org/doi/10.1126/sciadv.abm7069 (accessed on 27 December 2022).
134. Bunch, H.; Lawney, B.P.; Lin, Y.F.; Asaithamby, A.; Murshid, A.; Wang, Y.E.; Chen, B.P.C.; Calderwood, S.K. Transcriptional Elongation Requires DNA Break-Induced Signalling. *Nat. Commun.* **2015**, *6*, 10191. [CrossRef] [PubMed]
135. Ju, B.G.; Lunyak, V.V.; Perissi, V.; Garcia-Bassets, I.; Rose, D.W.; Glass, C.K.; Rosenfeld, M.G. A Topoisomerase IIβ-Mediated DsDNA Break Required for Regulated Transcription. *Science* **2006**, *312*, 1798–1802. [CrossRef]
136. Thakurela, S.; Garding, A.; Jung, J.; Schübeler, D.; Burger, L.; Tiwari, V.K. Gene Regulation and Priming by Topoisomerase IIα in Embryonic Stem Cells. *Nat. Commun. 2013 4:1* **2013**, *4*, 1–13. [CrossRef]
137. Eich, M.L.; Athar, M.; Ferguson, J.E.; Varambally, S. EZH2-Targeted Therapies in Cancer: Hype or a Reality. *Cancer Res.* **2020**, *80*, 5449–5458. [CrossRef] [PubMed]
138. Yuan, H.; Nishikori, M.; Otsuka, Y.; Arima, H.; Kitawaki, T.; Takaori-Kondo, A. The EZH2 Inhibitor Tazemetostat Upregulates the Expression of CCL17/TARC in B-Cell Lymphoma and Enhances T-Cell Recruitment. *Cancer Sci.* **2021**, *112*, 4604–4616. [CrossRef] [PubMed]
139. Morschhauser, F.; Tilly, H.; Chaidos, A.; McKay, P.; Phillips, T.; Assouline, S.; Batlevi, C.L.; Campbell, P.; Ribrag, V.; Damaj, G.L.; et al. Tazemetostat for Patients with Relapsed or Refractory Follicular Lymphoma: An Open-Label, Single-Arm, Multicentre, Phase 2 Trial. *Lancet Oncol.* **2020**, *21*, 1433–1442. [CrossRef]
140. Bryant, R.J.; Cross, N.A.; Eaton, C.L.; Hamdy, F.C.; Cunliffe, V.T. EZH2 Promotes Proliferation and Invasiveness of Prostate Cancer Cells. *Prostate* **2007**, *67*, 547–556. [CrossRef] [PubMed]
141. Bryant, R.J.; Winder, S.J.; Cross, S.S.; Hamdy, F.C.; Cunliffe, V.T. The Polycomb Group Protein EZH2 Regulates Actin Polymerization in Human Prostate Cancer Cells. *Prostate* **2008**, *68*, 255–263. [CrossRef]
142. Dardenne, E.; Beltran, H.; Benelli, M.; Gayvert, K.; Berger, A.; Puca, L.; Cyrta, J.; Sboner, A.; Noorzad, Z.; MacDonald, T.; et al. N-Myc Induces an EZH2-Mediated Transcriptional Program Driving Neuroendocrine Prostate Cancer. *Cancer Cell* **2016**, *30*, 563–577. [CrossRef]

143. Varambally, S.; Dhanasekaran, S.M.; Zhou, M.; Barrette, T.R.; Kumar-Sinha, C.; Sanda, M.G.; Ghosh, D.; Pienta, K.J.; Sewalt, R.G.A.B.; Otte, A.P.; et al. The Polycomb Group Protein EZH2 Is Involved in Progression of Prostate Cancer. *Nature* **2002**, *419*, 624–629. [CrossRef]
144. Zhang, Y.; Zheng, D.; Zhou, T.; Song, H.; Hulsurkar, M.; Su, N.; Liu, Y.; Wang, Z.; Shao, L.; Ittmann, M.; et al. Androgen Deprivation Promotes Neuroendocrine Differentiation and Angiogenesis through CREB-EZH2-TSP1 Pathway in Prostate Cancers. *Nat. Commun.* **2018**, *9*, 4080. [CrossRef]
145. Park, S.H.; Fong, K.-W.; Mong, E.; Martin, M.C.; Schiltz, G.E.; Yu, J. Going beyond Polycomb: EZH2 Functions in Prostate Cancer. *Oncogene* **2021**, *40*, 5788–5798. [CrossRef]
146. Discovery of a First-in-Class EZH2 Selective Degrader | Nature Chemical Biology. Available online: https://www.nature.com/articles/s41589-019-0421-4 (accessed on 27 December 2022).
147. Targeting EZH2 in Cancer-PMC. Available online: https://www.ncbi.nlm.nih.gov/pmc/articles/PMC4918227/ (accessed on 27 December 2022).
148. Porazzi, P.; Petruk, S.; Pagliaroli, L.; De Dominici, M.; Deming, D.; Puccetti, M.V.; Kushinsky, S.; Kumar, G.; Minieri, V.; Barbieri, E.; et al. Targeting Chemotherapy to De-Condensed H3K27me3-Marked Chromatin of AML Cells Enhances Leukemia Suppression. *Cancer Res.* **2022**, *82*, 458–471. [CrossRef]
149. Labbé, D.P.; Sweeney, C.J.; Brown, M.; Galbo, P.; Rosario, S.; Wadosky, K.M.; Ku, S.-Y.; Sjöström, M.; Alshalalfa, M.; Erho, N.; et al. TOP2A and EZH2 Provide Early Detection of an Aggressive Prostate Cancer Subgroup. *Clin. Cancer Res.* **2017**, *23*, 7072–7083. [CrossRef] [PubMed]
150. Fillmore, C.M.; Xu, C.; Desai, P.T.; Berry, J.M.; Rowbotham, S.P.; Lin, Y.J.; Zhang, H.; Marquez, V.E.; Hammerman, P.S.; Wong, K.K.; et al. EZH2 Inhibition Sensitizes BRG1 and EGFR Mutant Lung Tumours to TopoII Inhibitors. *Nature* **2015**, *520*, 239–242. [CrossRef] [PubMed]
151. Dang, C.V. C-Myc Target Genes Involved in Cell Growth, Apoptosis, and Metabolism. *Mol. Cell Biol.* **1999**, *19*, 1–11. [CrossRef] [PubMed]
152. Yin, Y.; Xu, L.; Chang, Y.; Zeng, T.; Chen, X.; Wang, A.; Groth, J.; Foo, W.C.; Liang, C.; Hu, H.; et al. N-Myc Promotes Therapeutic Resistance Development of Neuroendocrine Prostate Cancer by Differentially Regulating MiR-421/ATM Pathway. *Mol. Cancer* **2019**, *18*. [CrossRef]
153. Xie, Y.; Ning, S.; Hu, J. Jianpeng Molecular Mechanisms of Neuroendocrine Differentiation in Prostate Cancer Progression. *J. Cancer Res. Clin. Oncol.* **2022**, *148*, 1813–1823. [CrossRef]
154. Yamada, Y.; Beltran, H. Clinical and Biological Features of Neuroendocrine Prostate Cancer. *Curr. Oncol. Rep.* **2021**, *23*. [CrossRef]
155. Zeng, W.; Sun, H.; Meng, F.; Liu, Z.; Xiong, J.; Zhou, S.; Li, F.; Hu, J.; Hu, Z.; Liu, Z. Nuclear C-MYC Expression Level Is Associated with Disease Progression and Potentially Predictive of Two Year Overall Survival in Prostate Cancer. *Int. J. Clin. Exp. Pathol.* **2015**, *8*, 1878–1888.
156. C-Myc Antagonises the Transcriptional Activity of the Androgen Receptor in Prostate Cancer Affecting Key Gene Networks-PMC. Available online: https://www.ncbi.nlm.nih.gov/pmc/articles/PMC5405195/ (accessed on 27 December 2022).
157. Kallio, H.M.L.; Hieta, R.; Latonen, L.; Brofeldt, A.; Annala, M.; Kivinummi, K.; Tammela, T.L.; Nykter, M.; Isaacs, W.B.; Lilja, H.G.; et al. Constitutively Active Androgen Receptor Splice Variants AR-V3, AR-V7 and AR-V9 Are Co-Expressed in Castration-Resistant Prostate Cancer Metastases. *Br. J. Cancer* **2018**, *119*, 347–356. [CrossRef]
158. Lee, J.K.; Phillips, J.W.; Smith, B.A.; Park, J.W.; Stoyanova, T.; McCaffrey, E.F.; Baertsch, R.; Sokolov, A.; Meyerowitz, J.G.; Mathis, C.; et al. N-Myc Drives Neuroendocrine Prostate Cancer Initiated from Human Prostate Epithelial Cells. *Cancer Cell* **2016**, *29*, 536–547. [CrossRef]
159. Choi, S.H.; Wright, J.B.; Gerber, S.A.; Cole, M.D. Myc Protein Is Stabilized by Suppression of a Novel E3 Ligase Complex in Cancer Cells. *Genes Dev.* **2010**, *24*, 1236–1241. [CrossRef]
160. Fletcher, S.; Prochownik, E.V. Small-Molecule Inhibitors of the Myc Oncoprotein. *Biochim. Biophys. Acta-Gene Regul. Mech.* **2015**, *1849*, 525–543. [CrossRef] [PubMed]
161. TOP2A Induces Malignant Character of Pancreatic Cancer through Activating β-Catenin Signaling Pathway-PubMed. Available online: https://pubmed.ncbi.nlm.nih.gov/29045811/ (accessed on 27 December 2022).
162. Liu, Y.; Ma, J.; Song, J.S.; Zhou, H.Y.; Li, J.H.; Luo, C.; Geng, X.; Zhao, H.X. DNA Topoisomerase II Alpha Promotes the Metastatic Characteristics of Glioma Cells by Transcriptionally Activating β-Catenin. *Bioengineered* **2022**, *13*, 2207–2216. [CrossRef] [PubMed]
163. Huang, L.; Shitashige, M.; Satow, R.; Honda, K.; Ono, M.; Yun, J.; Tomida, A.; Tsuruo, T.; Hirohashi, S.; Yamada, T. Functional Interaction of DNA Topoisomerase IIα With the β-Catenin and T-Cell Factor-4 Complex. *Gastroenterology* **2007**, *133*, 1569–1578. [CrossRef] [PubMed]

pharmaceuticals

MDPI

Review

Developments in Non-Intercalating Bacterial Topoisomerase Inhibitors: Allosteric and ATPase Inhibitors of DNA Gyrase and Topoisomerase IV

Scott Grossman *, Colin W. G. Fishwick and Martin J. McPhillie *

School of Chemistry, University of Leeds, Leeds LS2 9JT, UK
* Correspondence: s.grossman@leeds.ac.uk (S.G.); m.j.mcphillie@leeds.ac.uk (M.J.M.)

Abstract: Increases in antibiotic usage and antimicrobial resistance occurrence have caused a dramatic reduction in the effectiveness of many frontline antimicrobial treatments. Topoisomerase inhibitors including fluoroquinolones are broad-spectrum antibiotics used to treat a range of infections, which stabilise a topoisomerase-DNA cleavage complex via intercalation of the bound DNA. However, these are subject to bacterial resistance, predominantly in the form of single-nucleotide polymorphisms in the active site. Significant research has been undertaken searching for novel bioactive molecules capable of inhibiting bacterial topoisomerases at sites distal to the fluoroquinolone binding site. Notably, researchers have undertaken searches for anti-infective agents that can inhibit topoisomerases through alternate mechanisms. This review summarises work looking at the inhibition of topoisomerases predominantly through non-intercalating agents, including those acting at a novel allosteric site, ATPase domain inhibitors, and those offering unique binding modes and mechanisms of action.

Keywords: topoisomerase; DNA gyrase; topo IV; antibiotic; allosteric; ATPase

1. Introduction

Infectious disease contributes significantly to worldwide mortality rates, with over 10 million deaths attributed to this class of disease in 2019 [1]. Following the publication of the O'Neill report in 2016, it was shown that rising trends in antimicrobial resistance (AMR) will have profound effects on both mortality rates as well as the global economy [2]. Data from the GRAM report suggested that in 2019, AMR infections were a factor in nearly 5 million deaths, directly causing mortality in 1.2 million cases [3]. As such, there is an urgent need to curb deaths due to infectious diseases, particularly through the development of novel antibiotics.

Resistant infections are increasing due to a multitude of reasons, including overprescribing antibiotics, their use in agriculture, and poor public understanding of the causes and implications of antibiotic resistance [4–6]. Current estimates predict that should these trends remain on course, by 2050, deaths due to AMR could exceed 10 million, more than currently die worldwide from all cancers combined [2]. Therefore, one approach to reducing AMR is the production of new treatments, particularly drugs with new mechanisms of action, in order to evade pre-existing resistance methods utilised by bacteria.

Broad-spectrum antibiotic activity can be achieved through the inhibition of essential proteins and enzymes found to be relatively well conserved across bacteria. For example, topoisomerases are found in all prokaryotes and eukaryotes, and some viruses [7]. Topoisomerases are responsible for topological changes in DNA, notably changes in supercoiling, decatenation, and unknotting, and are heavily involved in the replication and transcription processes by regulating supercoiling and by decatenating intertwined DNA strands [8]. Without these topological adjustments, DNA would remain unusable, leading to cell death.

Topoisomerases are divided into classes I and II dependent on whether they act via a single-strand DNA break (type I) or a double-strand DNA break (type II), with bacteria generally containing up to four examples [9]. For example, *Escherichia coli* contains topoisomerase I and topoisomerase III (type I), both of which are dimeric and reduce negative supercoiling in DNA [10]. Within type II, *E. coli* produces topoisomerase II (DNA gyrase) and topoisomerase IV (topo IV). DNA gyrase is found to introduce negative supercoils into DNA, whilst topo IV decatenates DNA and relaxes positive supercoiling. Whilst type I topoisomerases are, in theory, suitable drug targets, to date, the only clinically used drugs targeting bacterial topoisomerases are type II inhibitors.

DNA gyrase and topo IV are tetrameric proteins formed from two copies each of two proteins. DNA gyrase consists of two subunits of GyrA and two subunits of GyrB, with the protein found in the form A_2B_2 (Figure 1). Likewise, with topo IV, two subunits each of ParC and ParE form a homologous A_2B_2 tetramer [11]. In DNA gyrase, the enzyme functions through a complex mechanism, detailed studies of which have been described elsewhere [12]. Briefly, a segment of DNA (G segment) associates with DNA gyrase at the interface of the GyrA and GyrB subunits, and is wrapped around the enzyme (Figure 2). Upon the winding of approximately 130 bp around the enzyme, the conformation allows for a different segment of the same DNA strand (T segment) to enter the complex through the N-terminal gate of the GyrB subunits, positioning it above the G segment of DNA already associated.

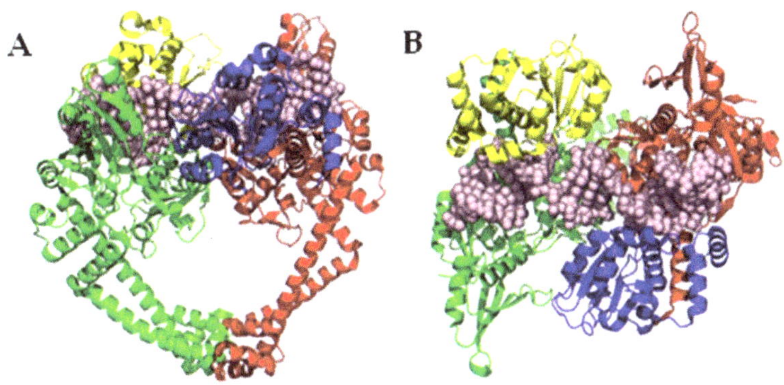

Figure 1. (**A**) Front view of *Staphylococcus aureus* DNA gyrase (PDB ID 6QX2 [13]), representative of the structure of type II topoisomerases. (**B**) Top view of *S. aureus* DNA gyrase, rotated 90° forwards from (**A**). The GyrA/ParC units of topoisomerase (green and red cartoons) are known to be critical to the actions of DNA strand breakage and covalent linkage of the phosphate backbone (whole DNA segment in pink spheres) to a conserved tyrosine residue in the active site, whilst the GyrB/ParE units (yellow and blue cartoons) are required for interactions with ATP. Both of these interactions are putative targets for topoisomerase inhibitors.

The GyrB subunit contains an ATPase domain, and the binding of ATP to this protein causes the N gate to shut, trapping the T segment, followed by double strand cleavage of the G segment and subsequent covalent linkage of the phosphate backbone to a conserved tyrosine located on the GyrA subunit [14]. The strand break allows the T segment to pass through the broken G segment, thereby changing the link number by two, and performing its function to decatenate or alter supercoiling. The T segment leaves the protein through an exit gate, formed through a conformational change of the GyrA subunits. Religation of the broken strand, promoted through ATP hydrolysis, reforms the G segment, which exits via the N gate, reopened through the conversion of ATP to ADP and its subsequent dissociation from the ATPase domain. Topo IV acts through a similar mechanism, though with homologous subunits ParC and ParE forming the complex with DNA in place of

GyrA and GyrB, respectively. Whilst ATP provides the energy required for DNA gyrase to introduce negative supercoils into DNA, in the absence of ATP, it relaxes negative supercoiling in DNA.

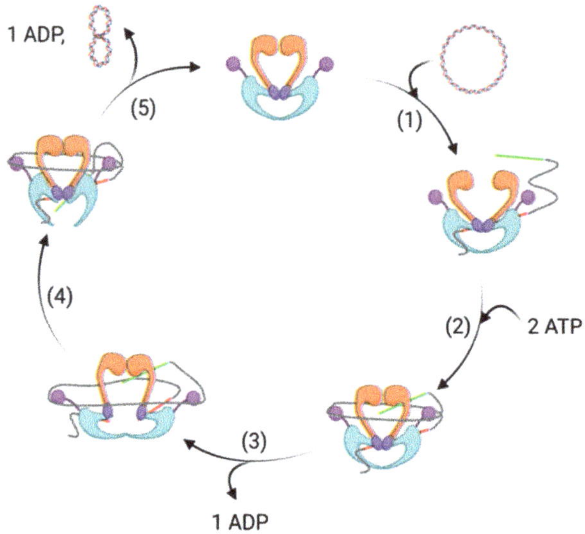

Figure 2. A generalised catalytic cycle of a topoisomerase. This figure depicts a cycle increasing supercoiling in circular DNA, though a similar cycle is used to reduce supercoiling, and decatenate and unknot DNA; GyrA units in blue; GyrB units in orange; G segment of DNA in red; T segment of DNA in green; non-interacting DNA in grey. (1) DNA interacts with the topoisomerase, and the G segment associates at the interface between the GyrA and GyrB subunits. (2) The non-interacting DNA loops around the topoisomerase and the T segment enters the topoisomerase, positioning itself above the G segment. The binding of ATP to the ATPase domain of GyrB causes the N-terminal gate to shut. (3) The G segment is cleaved, creating a change in conformation, enabling strand passage. (4) The T segment passes through the cleavage complex, altering the link number by two, and leaves via an exit gate. (5) Religation of the G segment, promoted by ATP hydrolysis, reopens the N-terminal gate, through which it leaves in its supercoiled form. The release of ADP reverts the conformation of topoisomerase back to its apo form.

The crucial interaction between the GyrA/ParC unit and DNA is the target for the fluoroquinolone (FQ) class of drugs [15]. Following DNA strand breakage, the FQs are able to stabilise the cleavage complex by stacking between DNA base pairs at the site of cleavage, thus preventing further advancement of the strand passage and religation process. The FQs are often referred to as being 'gyrase poisons'. Many FQs are used as frontline drugs, such as ciprofloxacin and moxifloxacin (**1** and **2**, Figure 3), in part due to their broad spectrum of activity across both Gram-positive and Gram-negative bacteria [16].

FQs have been in use since the 1970s, though the precursor drug nalidixic acid (**3**, Figure 3), a 1,8-naphthyridinone, was discovered in 1962 [17]. Since then, they have been increasingly used as frontline drugs to treat a wide range of infections, particularly urinary tract infections [18]. They do, however, have several adverse side-effects, which can affect patients significantly. Of note, treatment with FQs has been shown to increase the risk of tendon disorders, including tendon ruptures and tendinitis [19]. The severity of these risks led the FDA to introduce a black box warning on many FQ antibiotics [20]. Moreover, FQ use has been shown to increase the bacterial SOS response, causing increased mutation and antibiotic resistance in persister cells [21,22].

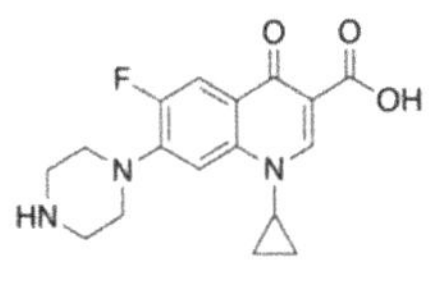

Ciprofloxacin, **1**

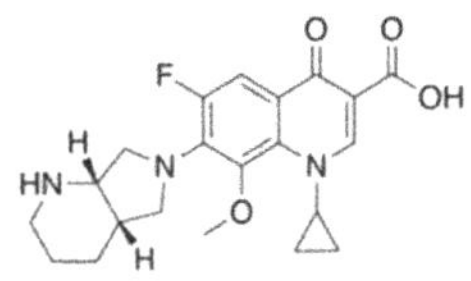

Moxifloxacin, **2**

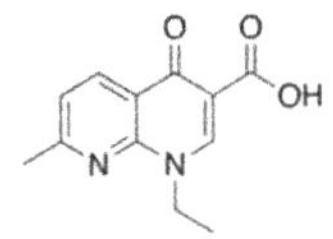

Nalidixic acid, **3**

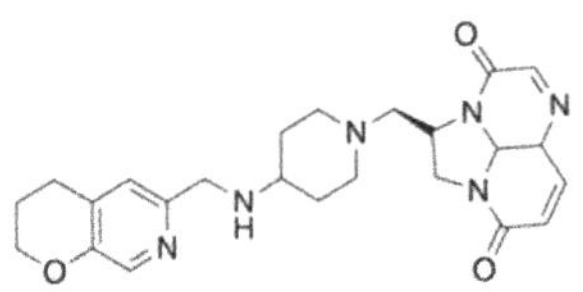

Example NBTI, gepotidacin, **4**

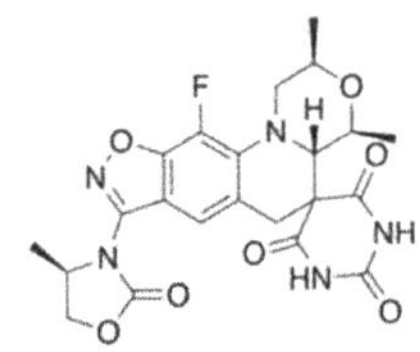

Zoliflodacin, **5**
(PDB ID 8BP2)

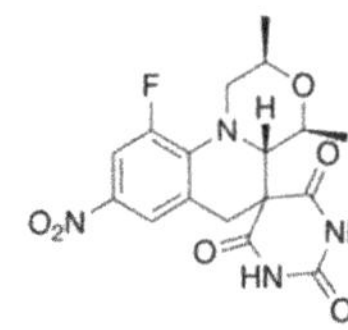

QPT-1, **6**
(PDB ID 5CDM)

Figure 3. Examples of DNA intercalating drugs. Ciprofloxacin **1** and moxifloxacin **2** are examples of fluoroquinolones (FQ), which are readily used as frontline drugs currently. They are both derivatives of an earlier class of drug, 1,8-naphthyridinones, of which nalidixic acid **3** was the first-in-class. Gepotidacin **4** is an example of a class of topoisomerase inhibitor, which intercalates DNA at an alternative site, leading to a different mechanism of action. Zoliflodacin **5** and QPT-1 **6** are examples of spiropyrimidinetriones, found to overlap the FQ site, but act with an alternate mode of inhibition.

Although usage in high-income countries has stabilised since the year 2000, overall usage is increasing steadily worldwide, with findings by Browne et al. suggesting that global use per capita has approximately doubled since 2000 [23]. With this trend, an increase in FQ resistance has emerged in pathogens affecting humans, wildlife, farmed animals, and aquatic environments, with multiple mutations frequently observed in clinical isolates, notably single-nucleotide polymorphisms (SNP) at sites $GyrA^{83}$, $GyrA^{87}$, $ParC^{80}$, and $ParC^{84}$ (*E. coli* numbering).

Whilst plasmid-mediated quinolone resistance does occur, SNP occurrence is the most common form of FQ resistance. Hamed et al. found that, in their study of 169 cancer patients suffering infection due to ciprofloxacin-resistant pathogens, all isolates contained at least one mutation in GyrA, and 93.3% had a further mutation in ParC, thereby inhibiting the crucial interaction between ciprofloxacin and the GyrA/ParC subunit [24]. Variations in the FQ core have subsequently been developed, which evade these forms of resistance, including using a 2-pyridone, quinazolinedione, isothiazoloquinolone, and oxazoloquinolin-2-one core [25–28].

As FQ resistance poses a significant threat to global healthcare, drugs targeting topoisomerases that further differentiate from FQs through distinct binding positions and mechanisms of inhibition are highly desirable. One area where researchers have made progress in is the development of Novel Bacterial Topoisomerase Inhibitors (NBTIs), such as gepotidacin (**4**, Figure 3). These compounds, similarly to FQs, act as DNA intercalating agents, though they occupy a different site within the cleavage complex to evade some FQ resistance mechanisms. Successes in NBTI developments have been thoroughly reviewed elsewhere [29–31].

One further interesting scaffold which has been developed in recent years is that of the spiropyrimidinetrione, the most studied example of which is zoliflodacin, also known as AZD0914 and ETX0914 (**5**, Figure 3) [32,33]. Initially, this compound was used to target *Neisseria gonorrhoeae*, and is currently in Phase III trials (NCT number NCT03959527), but it shows further strong activity against a range of other bacteria [34,35]. Work by Basarab et al. showed that zoliflodacin displays a novel mode of inhibition with respect to ciprofloxacin, and interacts with GyrB more strongly than GyrA [32,36], whilst still intercalating DNA at a site overlapping that of FQs, as evidenced by crystallographic data on both zoliflodacin (PDB ID 8BP2) as well as a related compound QPT-1 (**6**, Figure 3, PDB ID 5CDM) [37,38]. Advances in DNA intercalating topoisomerase inhibitors including spiropyrimidinetriones have been reviewed elsewhere [39,40].

In addition to NBTIs and spiropyrimidinetriones, significant work has been produced studying inhibition of alternative binding sites of topoisomerases, including that of the second catalytic domain within DNA gyrase and topo IV, the ATPase domain found in GyrB/ParE, as well as a novel allosteric site identified in the vicinity of the GyrA active site. Through the targeting of distal sites within topoisomerases, it may be possible to reduce the current usage of FQs, thereby extending their lifespan and giving potential for combined-therapeutic treatments targeting DNA gyrase and topo IV. This review aims to summarise the research to date primarily focussing on non-intercalating inhibitors targeting alternative sites in bacterial topoisomerases.

2. Allosteric Inhibitors of GyrA

Researchers at GSK identified an allosteric site in DNA gyrase in 2017, providing a potential route to novel alternatives to FQs. A high-throughput screen by Chan et al. identified a thiophene-bearing compound, which was subsequently shown to bind to DNA gyrase in a previously unidentified pocket, dubbed the allosteric site, at the interface between GyrA and GyrB in the vicinity of the active site of the enzyme (Figure 4) [41]. Optimisation of the structure from **7** to **8** (Figure 5) improved activity across a range of bacterial species, including both Gram-positive bacteria (*S. aureus, Streptococcus pneumoniae*) and Gram-negative bacteria (*Acinetobacter baumannii, Pseudomonas aeruginosa, E. coli, Klebsiella pneumoniae*). Efflux of both compounds was proposed, with inhibition in efflux pump mutants of *E. coli, P. aeruginosa,* and *A. baumannii* improving efficacy by up to 64-fold.

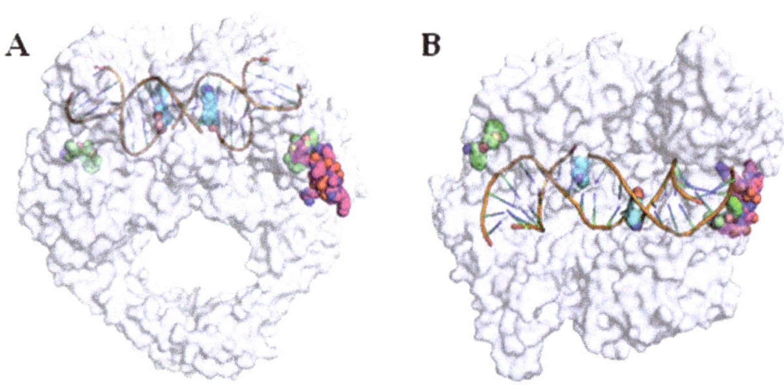

Figure 4. (**A**) Front view of superimposed DNA gyrase crystal structures highlighting the primary active site of ciprofloxacin (cyan spheres, PDB ID 2XCT [42]), and a novel allosteric site recently identified (green spheres, PDB ID 5NPP [41]). A further natural product, evybactin, was found to occupy the same allosteric site in *Mycobacterium tuberculosis* (magenta spheres, PDB ID 7UGW [43]). (**B**) Top view of the above-listed superimposed DNA gyrase crystal structures, rotated 90° forwards from (**A**).

Importantly, **8** was found be unaffected by mutations to key residues in GyrA, ParC, and ParE, with potent inhibition observed in FQ-resistant strains. However, one alternate mutation in GyrB-Glu^{793} (*E. coli* numbering, Glu^{634} in *S. aureus*) was identified which reduced inhibition caused by **8**. Further studies confirmed that the mechanism of action was different to that of FQs, as single- and double-strand DNA cleavage complexes were formed in equal quantities, as opposed to the entirely double-strand cleavage caused by the actions of ciprofloxacin. Although compound **8** showed promise due to its novel mode of action, this particular series was discontinued due to in vivo toxicity issues [41].

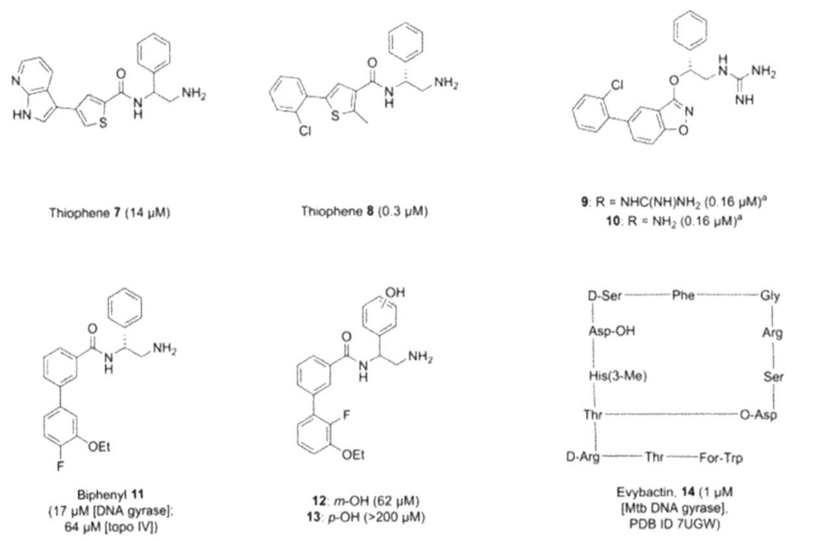

Figure 5. Inhibitors of bacterial DNA gyrase's allosteric pocket, with IC_{50} values against *Escherichia coli* DNA gyrase in brackets unless stated otherwise. Thiophene compounds **7** and **8** were the basis for further developments into a fused ring system as well as a biphenyl system, whereas evybactin was identified as an antimycobacterial compound before the target was elucidated. [a] A different assay was used to test the IC_{50} of **9** and **10** compared to all other compounds noted here [44], which utilised fluorescence spectroscopy in comparison to an agarose-gel-based assay used to test compounds **7**, **8**, **11–13**. As a result, they appear to be comparatively better, but are noted as being less potent than thiophene **8**, which was found to have an IC_{50} value of 0.04 μM when tested in the same assay conditions as **9** and **10**.

The series was altered to incorporate fused ring systems in place of the thiophene-amide motif, with the authors hypothesising that constraining the core would improve activity in biochemical assays. Moderate effects were observed in some analogues in the series [13]. Of note, benzisoxazole **9** (Figure 5) showed weak activity against *P. aeruginosa* and *K. pneumoniae*, whilst the primary amine analogue **10** (Figure 5) showed modest improvements against hERG and $Na_V1.5$ channels when compared to **8**. However, the reduction in cytotoxic properties was not significant, with **10** still showing hERG and $Na_V1.5$ IC_{50} values of 23 μM and 22 μM, respectively. Two crystal structures were generated of compound **9** in the allosteric site of topoisomerase with slight variations in the binding modes (PDB IDs 6QX1 and 6QX2). Of note, the chiral phenyl ring was found to adopt a rotated conformation within the pocket compared to parent structure **8**.

A further chemical series derived from thiophene **8** by Orritt et al. replaced the 5-arylthiophene with a biphenyl moiety, and explored changes to the chiral phenyl ring [45]. Although these compounds were not as potent as compound **8** on DNA gyrase (activity for lead compound **11** was 17 μM, cf. 0.3 μM for **8**), compound **11** (Figure 5) was found to have dual activity, targeting both DNA gyrase and topo IV (64 μM, cf. >540 μM for **8**). Current allosteric inhibitors of topoisomerases are designed around knowledge of the DNA gyrase crystal structures (PDB IDs 5NPK, 5NPP), but less is understood about the structure of the equivalent allosteric pocket in topo IV, other than a high degree of amino acid identity. Therefore, SAR data around allosteric inhibition of topo IV could prove beneficial in the development of dual-targeting topoisomerase inhibitors.

It was also observed that whilst the *R*-enantiomers of this series were more potent than the *S*- or racemic counterparts, enantiomeric *S*- and racemic compounds tested showed reasonable activity, with some activities of *S*-enantiomers in the same order of magnitude as their *R*-counterparts. Moreover, functionalising the chiral phenyl ring with a phenol moiety did not give any benefit in either the *meta*- or *para*-position (**12** and **13**, Figure 5). It was predicted that a phenol group would create an additional interaction with Arg^{630} through an intermediary water, but activity was found to drop, though **12** still retained micromolar activity [45].

Imai et al. utilised a different method to identify a novel natural product capable of inhibiting *Mycobacterium tuberculosis* (Mtb) DNA gyrase [43]. By screening extracts taken from bacterial strains symbiotic with a nematode host, the researchers were able to identify and characterise the depsipeptide evybactin (**14**, Figure 5), a cyclic peptide of 12 amino acids, which was shown to have good activity against Mtb (IC_{50} = 1 μM Mtb DNA gyrase, MIC = 0.0625 μg mL^{-1}), but not eukaryotic cell lines. Subsequent analysis and X-ray crystallography showed evybactin to inhibit DNA synthesis by binding to Mtb DNA gyrase at a site overlapping with the allosteric pocket identified by Chan et al. (PDB ID 7UGW) and showed no cross-resistance with known FQ sites. Interestingly, evybactin was found to target Mtb specifically due to its transport into the cell. The transport protein BacA, capable of moving highly hydrophilic molecules across the outer membrane [46], was able to move evybactin into the cytoplasm. In cells lacking BacA or its homolog SbmA, antibacterial and antimycobacterial activities were weak. Even in other organisms bearing SbmA, such as *E. coli*, the presence of other efflux mechanisms such as TolC nullifies the activity of evybactin, making it specific to mycobacteria (MIC *E. coli* ATCC 25922 = 8 μg mL^{-1}, MIC *E. coli* Δ*tolC* = 0.25 μg mL^{-1}). Given the complex structure of evybactin, small molecules which can mimic its binding interactions would be highly desirable to combat Mtb.

3. Inhibitors of the ATPase Site of GyrB

Inhibition of the ATPase activity of the GyrB subunit was identified as a potential route to new antibiotics as early as the 1950s. The discovery of the aminocoumarin group of natural product-derived antibiotics led to the approval of novobiocin (**15**, Figure 6), which reversibly inhibits DNA gyrase and, to a lesser extent, topo IV, through blocking of the ATPase site. This prevents the binding of ATP, metabolism and release of ADP, and the actions associated with this in the catalytic cycle. To date, novobiocin is the only aminocoumarin approved for clinical use, although it has now been withdrawn due to safer, more potent alternatives being developed, particularly later-generation penicillins. Interestingly, novobiocin has recently been studied for its anti-tumour effects as a result of its strong binding to the ATPase domain in DNA polymerase θ, with specific activity over other eukaryotic proteins bearing homologous ATPase domains including HSP90, suggesting a potential repurposing route for the drug [47].

Novobiocin's validation as a topoisomerase inhibitor has, however, enabled the development of numerous subsequent GyrB/ParE inhibitors, many of which were based upon conserved motifs. The historic development of coumarins and other GyrB/ParE inhibitors has been reviewed by Bisacchi et al., which highlights the level of work which has gone into inhibiting this target [48].

Heterocycles bearing a *N*-ethylurea moiety were identified by Vertex in 2002 as a potent motif for binding to Asp^{73} in the adenine-binding region of DNA gyrase [49]. Gradual optimisation of the Vertex scaffold produced compound **16**, referred to as SPR719 (Figure 6), which displayed strong antibacterial properties (MIC_{90} = 0.032 μg mL^{-1} of *S. aureus*) [50], particularly against mycobacteria including Mtb (MIC_{90} = 0.125 μg mL^{-1} of extensively drug-resistant Mtb strain XDR 5) [51], though poor solubility limited its application. Development of prodrug **17** (SPR720, Figure 6), however, overcame these solubility issues, and a promising phase I trial showed minimal side-effects and good tolerability in healthy subjects, with the results indicating that **17** shows potential as an oral treatment of Mtb subject to further trials [52]. Compound **17** progressed to Phase II trials, and is currently being tested for treatment in mycobacterium avium complex pulmonary disease (NCT number NCT05496374).

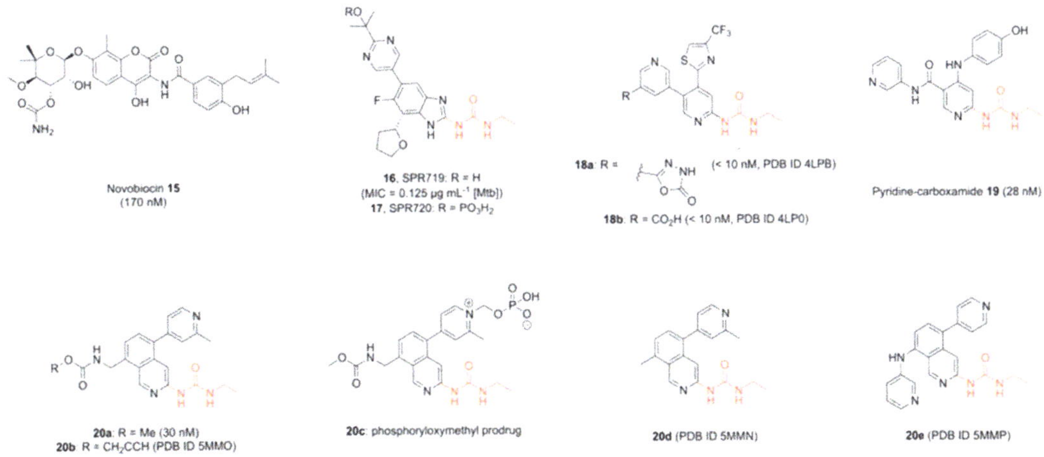

Figure 6. Examples of topoisomerase ATPase domain inhibitors with IC_{50} values against *E. coli* DNA gyrase unless stated otherwise. An ethyl urea moiety (red) was identified as an important motif capable of forming two key interactions with a conserved aspartate residue, and was subsequently used by multiple researchers in the development of novel topoisomerase inhibitors.

Since this first application of the *N*-ethylurea motif, it has been used frequently in a range of gyrase ATPase inhibitors. A fragment-to-lead program by Basarab et al. incorporated this motif into a thiazole-bearing compound **18a** (Figure 6), which showed nanomolar activity against both Gram-positive and Gram-negative bacteria in a protein-based assay (IC_{50} <10 nM *S. aureus* and *E. coli* GyrB), though was only able to kill *E. coli* in an efflux pump mutant (MIC = 0.15 μg mL^{-1} of *E. coli* Δ*tolC*, MIC >64 μg mL^{-1} of *E. coli* wildtype) [53]. Significantly though, **18a** showed high potency against a range of *S. aureus* strains, including one methicillin- and quinolone-resistant (MIC = 0.06 μg mL^{-1}).

Yule et al. further utilised the *N*-ethylurea motif to design and optimise a series of potent GyrB inhibitors. Using an in silico de novo design, the authors produced a pyridine-carboxamide (**19**, Figure 6), which showed nanomolar activity against GyrB (IC_{50} = 24 nM *S. aureus* GyrB and 28 nM *E. coli* GyrB in a protein-based assay) and ParE (IC_{50} = 86 nM *S. aureus* ParE and 940 nM *E. coli* ParE) [54]. In all cases listed above, bioengineered mutants confirmed GyrB/ParE activity, with GyrB mutation T173N (*S. aureus* numbering) frequently found to reduce potency across the described series. GyrA/ParC mutations did not compromise activity, confirming the target to be the GyrB/ParE subunit. Moreover, an assay monitoring the frequency of spontaneous mutation of **18a** in *S. aureus* found mutation T173N to spontaneously occur in two of the three experimental conditions [53].

In another fragment-based approach, Panchaud et al. varied the scaffold by replacing the key pyridine group of compounds **18a**, **18b** and **19** with an isoquinoline, whilst maintaining the *N*-ethylurea moiety (**20a**, Figure 6) [55]. Although the project direction saw optimisation geared towards gyrase inhibition, topo IV was further inhibited significantly (**20a** IC_{50} *E. coli* DNA gyrase = 0.03 μM, IC_{50} *E. coli* topo IV = 8 μM). In both Basarab and Panchaud's work, X-ray crystal structures were obtained of key compounds in the target active site. In the case of Basarab, compounds **18a** and **18b** were crystallised in the ParE subunit (PDB IDs 4LPB, 4LP0, Figure 6), whilst in that of Panchaud, compounds **20b**, **20d** and **20e** were visualised in the GyrB subunit, allowing for key structural data to be observed in topo IV and DNA gyrase, respectively (PDB IDs 5MMO, 5MMN, 5MMP,

Figure 6). In both cases, key interactions were observed between the pyridyl nitrogen and a water molecule, whilst the *N*-ethylurea moiety formed two interactions with a conserved GyrB aspartate residue.

The solubility of many of these ATPase inhibitors was found to be a hurdle in the development of the series. Hit compound **18a** showed an aqueous solubility of approximately 9 μg mL^{-1}, and whilst replacement of the oxadiazolone head with a carboxylic acid (**18b**) improved the solubility 40-fold, the solubility was still below 1 mg mL^{-1} [53]. Panchaud et al. similarly found that their hit compound **20a** had a solubility of 20 μg mL^{-1} in a phosphate buffer, but development of the compound into a prodrug through the introduction of a phosphoryl group improved the solubility to 12.7 mg mL^{-1} (**20c**, Figure 6) [55]. Although similar to the method of optimising SPR719, these two prodrugs were developed independently, suggesting that this could be a promising approach to improving the physicochemical properties of GyrB ATPase inhibitors.

Further developments moved away from the previously prominent *N*-ethylurea moiety, with other researchers looking towards tricyclic cores. A fragment-based drug discovery program by Trius Therapeutics produced two potent pyrimido-indoles bearing a methylamine functionality, capable of mimicking the interaction between the *N*-ethylurea motif previously utilised (**21a** and **21b**, Figure 7) [56]. These were found to be highly potent against Gram-positive bacteria, as well as additionally retaining good activity against a range of Gram-negative bacteria. This included strong potency against *S. aureus* (MIC_{90} **21a** = 0.06 μg mL^{-1}, MIC_{90} **21b** = 0.008 μg mL^{-1}) and *E. coli* (MIC_{90} **21a** = 1.0 μg mL^{-1}, MIC_{90} **21b** = 0.5 μg mL^{-1}). Compound **21a** was crystallised in the GyrB subunit of *E. coli* and *Enterococcus faecalis* (PDB IDs 4KFG, 4K4O) and the ParE subunit of *Francisella tularensis* (PDB ID 4KQV), whilst **21b** was crystallised in the GyrB subunit of *E. faecalis* (PDB ID 4KSG).

Hu et al. developed a *N*-ethylurea-containing scaffold into a pyrido-indole scaffold, which displayed dual activity against both GyrB and ParE [57]. This led to the generation of **22** (Figure 7), which had excellent potency against Gram-negative strains (MIC = 0.31 μg mL^{-1} of *E. coli*), and showed promise in a neutropenic mouse thigh infection model. A further effort to dual-target GyrB and ParE by Durcik et al. stemmed from a benzothiazole core identified previously [58,59]. The lead compound **23a** (Figure 7) showed excellent activity against both DNA gyrase and topo IV across a range of Gram-positive and Gram-negative bacteria (IC_{50} <10 nM *E. coli* and *S. aureus* DNA gyrase, IC_{50} = 54 nM *E. coli* topo IV and 84 nM *S. aureus* topo IV), whilst also showing an excellent cytotoxicity profile. Further development of this series yielded compound **23b** (Figure 7), which showed excellent activity against Gram-negative bacteria [60]. The lead compound was highly active against both DNA gyrase and topo IV in a range of Gram-negative bacteria and performed excellently against wildtype Gram-negatives in addition to efflux-deficient strains in killing assays (MIC = 1.0 μg mL^{-1} of *E. coli*, *K. pneumoniae*, and *P. aeruginosa* wildtypes, MIC = 0.5 μg mL^{-1} of *A. baumannii* wildtype). The authors note, however, that attempts to optimise physicochemical and ADME properties proved difficult, meaning properties such as solubility and plasma-free fraction were not suitable for in vivo testing. Crystal structures were obtained, though, for compound **23b** in the GyrB subunit of *A. baumannii* (PDB IDs 7PQL [racemate] and 7PQM [(*S*)-enantiomer]), as well as *P. aeruginosa* (PDB ID 7PTG [(*S*)-enantiomer]).

McGarry et al. developed a tricyclic GyrB inhibitor, which instead featured a pyrimido[4,5-b]indol-8-amine core [61], derived from previous work into GyrB inhibitors (PDB ID 4K4O) [62]. Mimicking the interactions between the urea moiety and GyrB-Asp^{73} of compounds **16–20**, the authors utilised the 8-methylamino group and *N*-indole functionalities of **24a** and **24b** (Figure 7) to create two hydrogen bonds, as well as creating the hydrogen bond between the previously identified water molecule and pyrimidyl moiety. In addition to identifying a novel GyrB inhibitor scaffold, the authors showed strong activity against Mtb (MIC **24b** = 0.06 μg mL^{-1}), and improved activity within the series by extending into a previously unexplored pocket in the direction of Asp^{106}. The SAR series showed that a carboxamide-oxazole side-chain was preferable, and although the addition of a methyl group at either free site on the oxazole improved efficacy, a molecular dynamic simulation suggested that **24a** displays fewer steric clashes than **24b**, and, hence, a 2-methyl moiety is preferential.

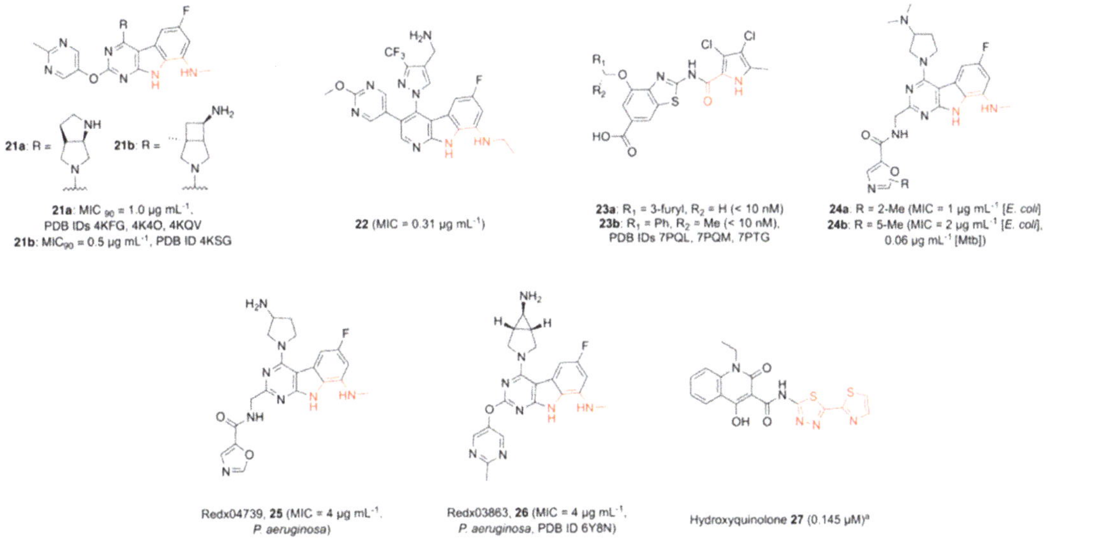

Figure 7. Development of bacterial topoisomerase ATPase inhibitors featuring novel core scaffolds, including tricyclic cores. Motifs binding to a conserved aspartate (red) branched away from the previously identified *N*-ethyl urea moiety. Values in brackets are IC_{50} values measured against *E. coli* DNA gyrase unless otherwise stated. Although no crystal structures were obtained for many of these compounds, conserved motifs were inferred due to comparisons with successfully crystallised analogues: **22** with PDB IDs 6M1S, 6M1J; **23** with PDB ID 6TTG, **24–26** with PDB ID 6Y8N. [a]No crystal structure of **25** or similar analogues have been obtained, so the key binding atoms were not fully elucidated. As such, the entire thiadiazole-thiazole motif is highlighted to indicate that this region binds to the key aspartate residue, though the exact atoms remain undetermined.

However, subsequent work by Henderson et al. determined compound **25** (Redx04739, Figure 7) of the same series to be slightly less potent than a previously described analogue **26** (Redx03863, Figure 7) across a range of Gram-positive and Gram-negative bacterial strains. **26** showed notable activity against *P. aeruginosa*, with an MIC value of 4 μg mL^{-1} [63]. Although this value is higher than activities against other tested strains, *P. aeruginosa* employs a range of defence mechanisms, particularly multiple efflux pumps, which significantly reduces the impact of tested drugs. As such, reasonable activity against this bacterium is a marked improvement over several other topoisomerase inhibitors, including novobiocin and several tested oxazole-bearing tricycles [61]. Moreover, the authors successfully crystallised **26** and novobiocin in the *Mycobacterium thermoresistibile* GyrB ATPase sub-domain, as well as novobiocin in the *Mycobacterium smegmatis* GyrB ATPase sub-domain, allowing them to determine that the two compounds occupy a similar binding site with some overlap, but in general, the tricycle binds deeper into the pocket (PDB IDs 6Y8N, 6Y8L, 6Y8O, respectively). It also makes distinct interactions with GyrB compared to novobiocin, leading to differences in resistance patterns. For example, novobiocin forms a key bonding interaction with Arg^{141}, but site-directed mutagenesis proved that interaction with this residue is not a requirement for **26** to maximise its potency.

Fragment screens have been successfully deployed multiple times in the search for GyrB/ParE inhibitors, in part due to the presence of numerous hydrophilic and charged residues in the active site, as well as the availability of crystal structures of the GyrB/ParE subunits. Huang et al. employed a fragment screening approach to identify novel motifs capable of binding to the GyrB subunit, and novel druggable sites [64]. Through a combination of a thermal-shift assay and an ATPase-activity screen, 49 fragments were identified from a library of 486, of which 10 were successfully co-crystallised into the *E. coli* GyrB ATPase sub-domain, highlighting the high hit-rate a fragment-based approach can yield (Figure 8).

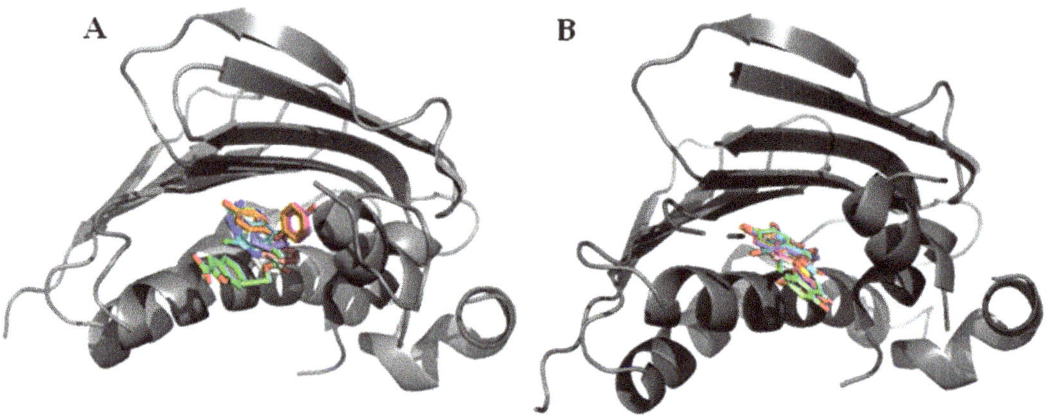

Figure 8. Superimposed co-crystal structures of fragments bound to the ATPase sub-domain of DNA gyrase identified by (**A**) Huang et al. and (**B**) Yu et al. [64,65]. A wider range of positions and functional groups were found in crystals produced by Huang et al., including one fragment (green) which filled a similar space to novobiocin. In comparison, the work by Yu et al. found one dominant binding mode, but multiple examples of phenols in this position suggest it may be a favourable binding mode for fragments of this class (PDB IDs (**A**): 5Z9N, 5Z9N, 5Z4H, 5Z9L, 5Z9P, 5Z9Q, 5Z9F, 5Z4O, 5Z9E, 5Z9B; PDB IDs (**B**): 7DOR, 7DPR, 7DPS, 7DQF, 7DQH, 7DQI, 7DQJ, 7DQL, 7DQM, 7DQS, 7DQU, 7DQW).

Multiple binding modes were identified from the 10 crystallised fragments, with some showing micromolar activity and others showing no activity up to 1 mM. Some active compounds were found to overlap with novobiocin's binding mode, whilst others bound deeper into the pocket. This research could provide a basis for a fragment merging approach in the future to explore the effect of combining these alternate binding sites to explore both the active and inactive fragments as potential sites within topoisomerases to boost activity.

Yu et al. adopted a similar approach to identify and co-crystallise 12 ATPase inhibiting fragments, of which the two most active compounds were found to have high micromolar activity (Figure 8) [65]. Phenols were a common motif in this screen, with all 12 successfully crystallised ATPase binders containing at least one. Whilst this does indicate that a phenol is a suitable fragment core for the binding site, the fragment screen led to limited modes of binding being identified, with eight of the fragments adopting a similar binding pose deep in the ATPase binding pocket.

Interestingly, Xue et al. utilised a different technique to identify a suitable fragment to build from. Instead of screening a fragment library, the group progressed from one single active molecule, identified through their previous work, and after predicting its interactions in the GyrB catalytic site, fragmented that to a 4-hydroxy-2-quinolone core [66,67]. Working from this lone fragment, the authors then performed a virtual screen on compounds bearing this motif and ultimately identified several compounds displaying sub-micromolar activity, including **27** (IC_{50} DNA gyrase = 0.145 µM, Figure 7). The hit compounds tested showed reasonable antibacterial activity against a range of *S. aureus* strains, though optimisation of solubility alongside other physicochemical properties is desirable. The redox potential of the scaffold may further warrant exploration into potential off-target effects.

4. Miscellaneous Topoisomerase Inhibitors

Several natural products have been previously identified, which branch away from the modes of inhibition outlined above. The first, a polythioamide produced by *Clostridium cellulolyticum* called closthioamide (**28**, Figure 9), was found by Chiriac et al. to be a strong DNA gyrase inhibitor (IC_{50} *E. coli* = 1.4 µM), with moderate potency against topo IV (IC_{50} *E. coli* = 113 µM) [68,69]. The authors' findings show that closthioamide likely has a separate mode of inhibition to both FQs and novobiocin. However, their suggestion is that it is inhibiting the ATPase activity of gyrase, though more likely at a site distal to the novobiocin binding site. Comparisons were drawn with a further natural product kibdelomycin (**29**, Figure 9), derived from *Kibdelosporangium* sp. (MA7385), which also shows ATPase inhibitory activity, whilst acting with a different mode of inhibition to novobiocin [70]. Kibdelomycin occupies a similar region to novobiocin in the ATPase domain, but shows no cross-resistance. This is because of a unique binding mode, which does not utilise GyrB-Asp^{89} (*S. aureus* numbering) to form a bonding interaction, which is key to many other DNA gyrase ATPase inhibitors. The natural product displays a U-shaped conformation, which shares no common features to the way in which novobiocin binds to the same region, and hence its ATPase inhibition differs from that of novobiocin [71]. Kibdelomycin features an interesting dichloropyrrole motif, which was identified by researchers at AstraZeneca as a potent binder to the adenine pocket of GyrB/ParE, and subsequently benzothiazole **23a** and **23b**, highlighting its versatility across scaffolds [59,72]. Kibdelomycin displays strong inhibition against DNA gyrase in proteinogenic assays (IC_{50} = 60 nM *E. coli*), though this does not translate to potency in Gram-negative bacteria (MIC >64 µg mL^{-1} of *E. coli*) [70].

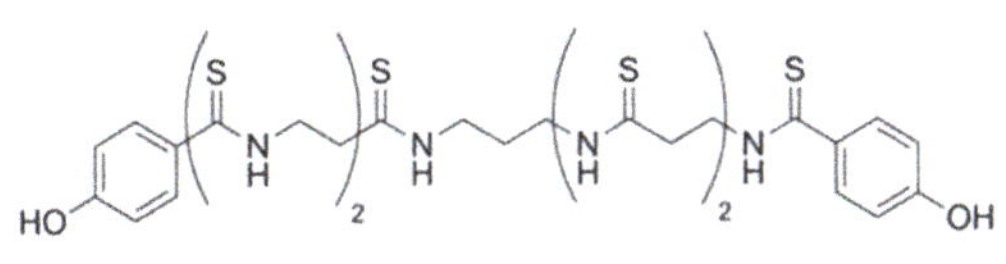

Closthioamide, **28** (1.4 µM)

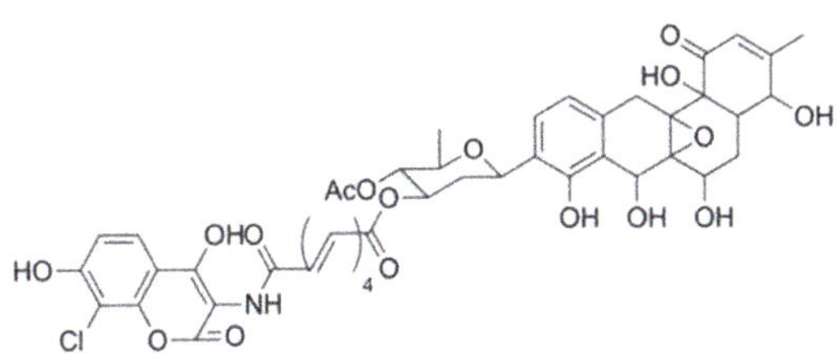

Simocyclinone D8, **30** (0.41 µM)

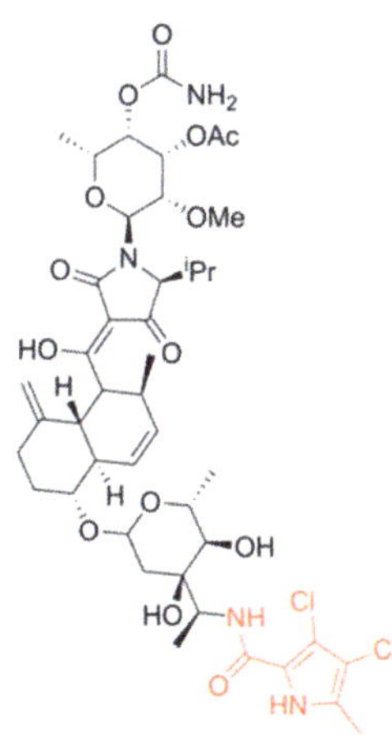

Kibdelomycin, **29** (60 nM)

Figure 9. Bacterial topoisomerase inhibitors displaying unique or unusual modes of inhibition, with IC_{50} values against *E. coli* DNA gyrase in brackets. Closthioamide and kibdelomycin are ATPase inhibitors, though display different modes of inhibition to known ATPase inhibitor novobiocin. Simocyclinone D8 is further found to bind to a region of GyrB; hence, it displays a novel mechanism of action where it is capable of interacting with two regions of DNA gyrase simultaneously.

The natural product simocyclinone D8 (**30**, Figure 9), derived from *Streptomyces antibioticus* Tü 6040, has further been the subject of much research. This bifunctional molecule features both an aminocoumarin head and polyketide, and studies have found that it binds in two sites on DNA gyrase [73]. The primary site on the N-terminal domain of GyrA is occupied by the polyketide head, whilst the aminocoumarin head sits on the C-terminal domain of GyrB [74]. However, simocyclinone D8 is unable to competitively inhibit ATPase activity in GyrB, nor does it promote cleavage complex formation. Therefore, it has a unique mechanism of action with respect to both binding sites.

X-ray crystallography shows that the polyketide head does indeed occupy a space adjacent to ciprofloxacin, and hence likely intercalates DNA, but the site is nonetheless distinct. In fact, the dominant mechanism of action of simocyclinone D8 is to prevent cleavage complex formation, in stark contrast to FQs which allow formation and then stabilise the cleavage complex. Although simocyclinone D8 does also bind to a site on GyrB, this is the minor binding site, and whilst it cooperatively functions with binding on GyrA, the GyrB binding is calculated to be approximately 1000-fold weaker [75,76]. It further shows good activity against *E. coli* in in vitro testing (IC_{50} = 0.41 μM *E. coli* DNA gyrase) [77].

The natural products reported in this section have all been characterised as bacterial topoisomerase inhibitors via macromolecular biosynthesis and supercoiling DNA gel-based assays with ciprofloxacin and novobiocin as controls. To our knowledge, their potential off-target effects have not been characterised with respect to structural features such as Michael acceptors and redox-active aromatics. Therefore, follow-up characterisation is warranted to rule out chemical promiscuity or pan-assay interference before pursuing further as bacterial topoisomerase inhibitors [78].

5. Conclusions

Although improvement upon and development of new FQs will prove fruitful in the generation of new antibiotics, other approaches are required to prevent the antibiotic pipeline from drying up. Topoisomerases are a known target, which have been validated in both Gram-positive and Gram-negative bacteria, making them an excellent drug target. This review has summarised key research into topoisomerase inhibitors with different modes of action to FQs, particularly those which target a recently discovered allosteric site in the GyrA/ParC subunit, and GyrB/ParE inhibitors which bind to the ATPase site. Moreover, a discussion of compounds with novel modes of action highlighted that other DNA-intercalating compounds exist which do not proceed via the FQ mechanism of action, nor that of NBTIs. The research highlighted above shows that the scope for novel topoisomerase inhibitors is broad, and future research may produce new life-saving antibiotics, with some research mentioned already in clinical trials.

Author Contributions: Writing-original draft preparation, S.G.; writing- review and editing, S.G. and M.J.M.; funding and acquisition, C.W.G.F. and M.J.M. All authors have read and agreed to the published version of the manuscript.

Funding: The authors are funded by the BBSRC, grant number BB/V007041/1.

Institutional Review Board Statement: Not applicable.

Informed Consent Statement: Not applicable.

Data Availability Statement: Data sharing not applicable.

Acknowledgments: The authors wish to thank the BBSRC and University of Leeds.

Conflicts of Interest: The authors declare no conflict of interest.

References

1. GBD 2019 Diseases and Injuries Collaborators. Global burden of 369 diseases and injuries in 204 countries and territories, 1990–2019: A systematic analysis for the Global Burden of Disease Study 2019. *Lancet* **2020**, *396*, 1204–1222. [CrossRef] [PubMed]
2. O'Neill, J. *Tackling Drug–Resistant Infections Globally: Final Report and Recommendations*; Government of the UK: London, UK, 2016.
3. Antimicrobial Resistance Collaborators. Global burden of bacterial antimicrobial resistance in 2019: A systematic analysis. *Lancet* **2022**, *399*, 629–655. [CrossRef] [PubMed]
4. Manyi-Loh, C.; Mamphweli, S.; Meyer, E.; Okoh, A. Antibiotic Use in Agriculture and Its Consequential Resistance in Environmental Sources: Potential Public Health Implications. *Molecules* **2018**, *23*, 795. [CrossRef] [PubMed]
5. McCullough, A.R.; Parekh, S.; Rathbone, J.; Del Mar, C.B.; Hoffmann, T.C. A systematic review of the public's knowledge and beliefs about antibiotic resistance. *J. Antimicrob. Chemother.* **2016**, *71*, 27–33. [CrossRef]
6. Fleming-Dutra, K.E.; Hersh, A.L.; Shapiro, D.J.; Bartoces, M.; Enns, E.A.; File, T.M., Jr.; Finkelstein, J.A.; Gerber, J.S.; Hyun, D.Y.; Linder, J.A.; et al. Prevalence of Inappropriate Antibiotic Prescriptions Among US Ambulatory Care Visits, 2010–2011. *JAMA* **2016**, *315*, 1864–1873. [CrossRef]
7. Forterre, P.; Gribaldo, S.; Gadelle, D.; Serre, M.-C. Origin and evolution of DNA topoisomerases. *Biochimie* **2007**, *89*, 427–446. [CrossRef]
8. Hirsch, J.; Klostermeier, D. What makes a type IIA topoisomerase a gyrase or a Topo IV? *Nucleic Acids Res.* **2021**, *49*, 6027–6042. [CrossRef]
9. Vos, S.M.; Tretter, E.M.; Schmidt, B.H.; Berger, J.M. All tangled up: How cells direct, manage and exploit topoisomerase function. *Nat. Rev. Mol. Cell Biol.* **2011**, *12*, 827–841. [CrossRef]
10. Hooper, D.C. Bacterial Topoisomerases, Anti-Topoisomerases, and Anti-Topoisomerase Resistance. *Clin. Infect. Dis.* **1998**, *27*, S54–S63. [CrossRef]
11. Khan, T.; Sankhe, K.; Suvarna, V.; Sherje, A.; Patel, K.; Dravyakar, B. DNA gyrase inhibitors: Progress and synthesis of potent compounds as antibacterial agents. *Biomed. Pharmacother.* **2018**, *103*, 923–938. [CrossRef]
12. Champoux, J.J. DNA Topoisomerases: Structure, Function, and Mechanism. *Annu. Rev. Biochem.* **2001**, *70*, 369–413. [CrossRef]
13. Thalji, R.K.; Raha, K.; Andreotti, D.; Checchia, A.; Cui, H.; Meneghelli, G.; Profeta, R.; Tonelli, F.; Tommasi, S.; Bakshi, T.; et al. Structure-guided design of antibacterials that allosterically inhibit DNA gyrase. *Bioorganic. Med. Chem. Lett.* **2019**, *29*, 1407–1412. [CrossRef] [PubMed]
14. Bates, A.D.; Maxwell, A. Energy Coupling in Type II Topoisomerases: Why Do They Hydrolyze ATP? *Biochemistry* **2007**, *46*, 7929–7941. [CrossRef]
15. Hooper, D.C.; Jacoby, G.A. Topoisomerase Inhibitors: Fluoroquinolone Mechanisms of Action and Resistance. *Cold Spring Harb. Perspect. Med.* **2016**, *6*, a025320. [CrossRef]
16. Bratsman, A.; Mathias, K.; Laubscher, R.; Grigoryan, L.; Rose, S. Outpatient fluoroquinolone prescribing patterns before and after US FDA boxed warning. *Pharmacoepidemiol. Drug Saf.* **2020**, *29*, 701–707. [CrossRef]
17. Bisacchi, G.S. Origins of the Quinolone Class of Antibacterials: An Expanded "Discovery Story". *J. Med. Chem.* **2015**, *58*, 4874–4882. [CrossRef]
18. Concia, E.; Bragantini, D.; Mazzaferri, F. Clinical evaluation of guidelines and therapeutic approaches in multi drug-resistant urinary tract infections. *J. Chemother.* **2017**, *29*, 19–28. [CrossRef]
19. Alves, C.; Mendes, D.; Marques, F.B. Fluoroquinolones and the risk of tendon injury: A systematic review and meta-analysis. *Eur. J. Clin. Pharmacol.* **2019**, *75*, 1431–1443. [CrossRef]
20. Sankar, A.; Swanson, K.M.; Zhou, J.; Jena, A.B.; Ross, J.S.; Shah, N.D.; Karaca-Mandic, P. Association of Fluoroquinolone Prescribing Rates with Black Box Warnings from the US Food and Drug Administration. *JAMA Netw. Open* **2021**, *4*, e2136662. [CrossRef]
21. Iacobino, A.; Piccaro, G.; Pardini, M.; Fattorini, L.; Giannoni, F. Moxifloxacin Activates the SOS Response in *Mycobacterium tuberculosis* in a Dose- and Time-Dependent Manner. *Microorganisms* **2021**, *9*, 255. [CrossRef]
22. Barrett, T.C.; Mok, W.W.K.; Murawski, A.M.; Brynildsen, M.P. Enhanced antibiotic resistance development from fluoroquinolone persisters after a single exposure to antibiotic. *Nat. Commun.* **2019**, *10*, 1177. [CrossRef]
23. Browne, A.J.; Chipeta, M.G.; Haines-Woodhouse, G.; Kumaran, E.P.A.; Hamadani, B.H.K.; Zaraa, S.; Henry, N.J.; Deshpande, A.; Reiner, R.C.; Day, N.P.J.; et al. Global antibiotic consumption and usage in humans, 2000–2018: A spatial modelling study. *Lancet Planet. Health* **2021**, *5*, e893–e904. [CrossRef] [PubMed]
24. Hamed, S.M.; Elkhatib, W.F.; El-Mahallawy, H.A.; Helmy, M.M.; Ashour, M.S.; Aboshanab, K.M.A. Multiple mechanisms contributing to ciprofloxacin resistance among Gram negative bacteria causing infections to cancer patients. *Sci. Rep.* **2018**, *8*, 12268. [CrossRef] [PubMed]
25. Flamm, R.K.; Vojtko, C.; Chu, D.T.; Li, Q.; Beyer, J.; Hensey, D.; Ramer, N.; Clement, J.J.; Tanaka, S.K. In vitro evaluation of ABT-719, a novel DNA gyrase inhibitor. *Antimicrob. Agents Chemother.* **1995**, *39*, 964–970. [CrossRef]
26. Pan, X.-S.; Gould Katherine, A.; Fisher, L.M. Probing the Differential Interactions of Quinazolinedione PD 0305970 and Quinolones with Gyrase and Topoisomerase IV. *Antimicrob. Agents Chemother.* **2009**, *53*, 3822–3831. [CrossRef]
27. Pucci Michael, J.; Podos Steven, D.; Thanassi Jane, A.; Leggio Melissa, J.; Bradbury Barton, J.; Deshpande, M. In Vitro and In Vivo Profiles of ACH-702, an Isothiazoloquinolone, against Bacterial Pathogens. *Antimicrob. Agents Chemother.* **2011**, *55*, 2860–2871. [CrossRef]

28. Savage, V.J.; Charrier, C.; Salisbury, A.-M.; Moyo, E.; Forward, H.; Chaffer-Malam, N.; Metzger, R.; Huxley, A.; Kirk, R.; Uosis-Martin, M.; et al. Biological profiling of novel tricyclic inhibitors of bacterial DNA gyrase and topoisomerase IV. *J. Antimicrob. Chemother.* **2016**, *71*, 1905–1913. [CrossRef]
29. Desai, J.; Sachchidanand, S.; Kumar, S.; Sharma, R. Novel Bacterial Topoisomerase inhibitors (NBTIs)—A comprehensive review. *Eur. J. Med. Chem. Rep.* **2021**, *3*, 100017. [CrossRef]
30. Kokot, M.; Anderluh, M.; Hrast, M.; Minovski, N. The Structural Features of Novel Bacterial Topoisomerase Inhibitors That Define Their Activity on Topoisomerase IV. *J. Med. Chem.* **2022**, *65*, 6431–6440. [CrossRef]
31. Kolarič, A.; Anderluh, M.; Minovski, N. Two Decades of Successful SAR-Grounded Stories of the Novel Bacterial Topoisomerase Inhibitors (NBTIs). *J. Med. Chem.* **2020**, *63*, 5664–5674. [CrossRef]
32. Basarab, G.S.; Kern, G.H.; McNulty, J.; Mueller, J.P.; Lawrence, K.; Vishwanathan, K.; Alm, R.A.; Barvian, K.; Doig, P.; Galullo, V.; et al. Responding to the challenge of untreatable gonorrhea: ETX0914, a first-in-class agent with a distinct mechanism-of-action against bacterial Type II topoisomerases. *Sci. Rep.* **2015**, *5*, 11827. [CrossRef] [PubMed]
33. Bradford, P.A.; Miller, A.A.; O'Donnell, J.; Mueller, J.P. Zoliflodacin: An Oral Spiropyrimidinetrione Antibiotic for the Treatment of *Neisseria gonorrheae*, Including Multi-Drug-Resistant Isolates. *ACS Infect. Dis.* **2020**, *6*, 1332–1345. [CrossRef] [PubMed]
34. Damião Gouveia, A.C.; Unemo, M.; Jensen, J.S. In vitro activity of zoliflodacin (ETX0914) against macrolide-resistant, fluoroquinolone-resistant and antimicrobial-susceptible *Mycoplasma genitalium* strains. *J. Antimicrob. Chemother.* **2018**, *73*, 1291–1294. [CrossRef] [PubMed]
35. Giacobbe, R.A.; Huband, M.D.; de Jonge, B.L.M.; Bradford, P.A. Effect of Susceptibility Testing Conditions on the In Vitro Antibacterial Activity of ETX0914. *Diagn. Microbiol. Infect. Dis.* **2017**, *87*, 139–142. [CrossRef] [PubMed]
36. Basarab, G.S.; Doig, P.; Galullo, V.; Kern, G.; Kimzey, A.; Kutschke, A.; Newman, J.P.; Morningstar, M.; Mueller, J.; Otterson, L.; et al. Discovery of Novel DNA Gyrase Inhibiting Spiropyrimidinetriones: Benzisoxazole Fusion with N-Linked Oxazolidinone Substituents Leading to a Clinical Candidate (ETX0914). *J. Med. Chem.* **2015**, *58*, 6264–6282. [CrossRef]
37. Chan, P.F.; Srikannathasan, V.; Huang, J.; Cui, H.; Fosberry, A.P.; Gu, M.; Hann, M.M.; Hibbs, M.; Homes, P.; Ingraham, K.; et al. Structural basis of DNA gyrase inhibition by antibacterial QPT-1, anticancer drug etoposide and moxifloxacin. *Nat. Commun.* **2015**, *6*, 10048. [CrossRef]
38. Morgan, H.; Lipka-Lloyd, M.; Warren, A.; Hughes, N.; Holmes, J.; Burton, N.; Mahenthiralingam, E.; Bax, B. A 2.8 Å structure of zoliflodacin in a DNA-cleavage complex with *Staphylococcus aureus* DNA gyrase. *bioRxiv*, 2022; *preprint*. [CrossRef]
39. Jones, J.A.; Virga, K.G.; Gumina, G.; Hevener, K.E. Recent advances in the rational design and optimization of antibacterial agents. *MedChemComm* **2016**, *7*, 1694–1715. [CrossRef]
40. Badshah, S.L.; Ullah, A. New developments in non-quinolone-based antibiotics for the inhibiton of bacterial gyrase and topoisomerase IV. *Eur. J. Med. Chem.* **2018**, *152*, 393–400. [CrossRef]
41. Chan, P.F.; Germe, T.; Bax, B.D.; Huang, J.; Thalji, R.K.; Bacqué, E.; Checchia, A.; Chen, D.; Cui, H.; Ding, X.; et al. Thiophene antibacterials that allosterically stabilize DNA-cleavage complexes with DNA gyrase. *Proc. Natl. Acad. Sci.* **2017**, *114*, E4492–E4500. [CrossRef]
42. Bax, B.D.; Chan, P.F.; Eggleston, D.S.; Fosberry, A.; Gentry, D.R.; Gorrec, F.; Giordano, I.; Hann, M.M.; Hennessy, A.; Hibbs, M.; et al. Type IIA topoisomerase inhibition by a new class of antibacterial agents. *Nature* **2010**, *466*, 935–940. [CrossRef]
43. Imai, Y.; Hauk, G.; Quigley, J.; Liang, L.; Son, S.; Ghiglieri, M.; Gates, M.F.; Morrissette, M.; Shahsavari, N.; Niles, S.; et al. Evybactin is a DNA gyrase inhibitor that selectively kills *Mycobacterium tuberculosis*. *Nat. Chem. Biol.* **2022**, *18*, 1236–1244. [CrossRef]
44. Asha, M.K.; Debraj, D.; Prashanth, D.s.; Edwin, J.R.; Srikanth, H.S.; Muruganantham, N.; Dethe, S.M.; Anirban, B.; Jaya, B.; Deepak, M.; et al. In vitro anti-*Helicobacter pylori* activity of a flavonoid rich extract of *Glycyrrhiza glabra* and its probable mechanisms of action. *J. Ethnopharmacol.* **2013**, *145*, 581–586. [CrossRef]
45. Orritt, K.M.; Feng, L.; Newell, J.F.; Sutton, J.N.; Grossman, S.; Germe, T.; Abbott, L.R.; Jackson, H.L.; Bury, B.K.L.; Maxwell, A.; et al. De novo design of type II topoisomerase inhibitors as potential antimicrobial agents targeting a novel binding region. *RSC Med. Chem.* **2022**, *13*, 831–839. [CrossRef]
46. Domenech, P.; Kobayashi, H.; LeVier, K.; Walker Graham, C.; Barry Clifton, E. BacA, an ABC Transporter Involved in Maintenance of Chronic Murine Infections with *Mycobacterium tuberculosis*. *J. Bacteriol.* **2009**, *191*, 477–485. [CrossRef]
47. Zhou, J.; Gelot, C.; Pantelidou, C.; Li, A.; Yücel, H.; Davis, R.E.; Färkkilä, A.; Kochupurakkal, B.; Syed, A.; Shapiro, G.I.; et al. A first-in-class polymerase theta inhibitor selectively targets homologous-recombination-deficient tumors. *Nat. Cancer* **2021**, *2*, 598–610. [CrossRef]
48. Bisacchi, G.S.; Manchester, J.I. A New-Class Antibacterial—Almost. Lessons in Drug Discovery and Development: A Critical Analysis of More than 50 Years of Effort toward ATPase Inhibitors of DNA Gyrase and Topoisomerase IV. *ACS Infect. Dis.* **2015**, *1*, 4–41. [CrossRef]
49. Grillot, A.-L.; Charifson, P.; Stamos, D.; Liao, Y.; Badia, M.; Trudeau, M. Bacterial Gyrase Inhibitors and Uses Thereof. WIPO Patent WO2002060879A3, 27 March 2003.
50. Grillot, A.-L.; Tiran, A.L.; Shannon, D.; Krueger, E.; Liao, Y.; O'Dowd, H.; Tang, Q.; Ronkin, S.; Wang, T.; Waal, N.; et al. Second-Generation Antibacterial Benzimidazole Ureas: Discovery of a Preclinical Candidate with Reduced Metabolic Liability. *J. Med. Chem.* **2014**, *57*, 8792–8816. [CrossRef]

51. Locher Christopher, P.; Jones Steven, M.; Hanzelka Brian, L.; Perola, E.; Shoen Carolyn, M.; Cynamon Michael, H.; Ngwane Andile, H.; Wiid Ian, J.; van Helden Paul, D.; Betoudji, F.; et al. A Novel Inhibitor of Gyrase B Is a Potent Drug Candidate for Treatment of Tuberculosis and Nontuberculosis Mycobacterial Infections. *Antimicrob. Agents Chemother.* **2015**, *59*, 1455–1465. [CrossRef]
52. Talley Angela, K.; Thurston, A.; Moore, G.; Gupta Vipul, K.; Satterfield, M.; Manyak, E.; Stokes, S.; Dane, A.; Melnick, D. First-in-Human Evaluation of the Safety, Tolerability, and Pharmacokinetics of SPR720, a Novel Oral Bacterial DNA Gyrase (GyrB) Inhibitor for Mycobacterial Infections. *Antimicrob. Agents Chemother.* **2021**, *65*, e01208–e01221. [CrossRef]
53. Basarab, G.S.; Manchester, J.I.; Bist, S.; Boriack-Sjodin, P.A.; Dangel, B.; Illingworth, R.; Sherer, B.A.; Sriram, S.; Uria-Nickelsen, M.; Eakin, A.E. Fragment-to-Hit-to-Lead Discovery of a Novel Pyridylurea Scaffold of ATP Competitive Dual Targeting Type II Topoisomerase Inhibiting Antibacterial Agents. *J. Med. Chem.* **2013**, *56*, 8712–8735. [CrossRef]
54. Yule, I.A.; Czaplewski, L.G.; Pommier, S.; Davies, D.T.; Narramore, S.K.; Fishwick, C.W.G. Pyridine-3-carboxamide-6-yl-ureas as novel inhibitors of bacterial DNA gyrase: Structure based design, synthesis, SAR and antimicrobial activity. *Eur. J. Med. Chem.* **2014**, *86*, 31–38. [CrossRef] [PubMed]
55. Panchaud, P.; Bruyère, T.; Blumstein, A.-C.; Bur, D.; Chambovey, A.; Ertel, E.A.; Gude, M.; Hubschwerlen, C.; Jacob, L.; Kimmerlin, T.; et al. Discovery and Optimization of Isoquinoline Ethyl Ureas as Antibacterial Agents. *J. Med. Chem.* **2017**, *60*, 3755–3775. [CrossRef] [PubMed]
56. Tari, L.W.; Li, X.; Trzoss, M.; Bensen, D.C.; Chen, Z.; Lam, T.; Zhang, J.; Lee, S.J.; Hough, G.; Phillipson, D.; et al. Tricyclic GyrB/ParE (TriBE) Inhibitors: A New Class of Broad-Spectrum Dual-Targeting Antibacterial Agents. *PLoS ONE* **2013**, *8*, e84409. [CrossRef] [PubMed]
57. Hu, Y.; Shi, H.; Zhou, M.; Ren, Q.; Zhu, W.; Zhang, W.; Zhang, Z.; Zhou, C.; Liu, Y.; Ding, X.; et al. Discovery of Pyrido [2,3-b]indole Derivatives with Gram-Negative Activity Targeting Both DNA Gyrase and Topoisomerase IV. *J. Med. Chem.* **2020**, *63*, 9623–9649. [CrossRef] [PubMed]
58. Nyerges, A.; Tomašič, T.; Durcik, M.; Revesz, T.; Szili, P.; Draskovits, G.; Bogar, F.; Skok, Ž.; Zidar, N.; Ilaš, J.; et al. Rational design of balanced dual-targeting antibiotics with limited resistance. *PLoS Biol.* **2020**, *18*, e3000819. [CrossRef]
59. Durcik, M.; Nyerges, Á.; Skok, Ž.; Skledar, D.G.; Trontelj, J.; Zidar, N.; Ilaš, J.; Zega, A.; Cruz, C.D.; Tammela, P.; et al. New dual ATP-competitive inhibitors of bacterial DNA gyrase and topoisomerase IV active against ESKAPE pathogens. *Eur. J. Med. Chem.* **2021**, *213*, 113200. [CrossRef]
60. Cotman, A.E.; Durcik, M.; Benedetto Tiz, D.; Fulgheri, F.; Secci, D.; Sterle, M.; Možina, Š.; Skok, Ž.; Zidar, N.; Zega, A.; et al. Discovery and Hit-to-Lead Optimization of Benzothiazole Scaffold-Based DNA Gyrase Inhibitors with Potent Activity against *Acinetobacter baumannii* and *Pseudomonas aeruginosa*. *J. Med. Chem.* **2023**, *66*, 1380–1425. [CrossRef]
61. McGarry, D.H.; Cooper, I.R.; Walker, R.; Warrilow, C.E.; Pichowicz, M.; Ratcliffe, A.J.; Salisbury, A.-M.; Savage, V.J.; Moyo, E.; Maclean, J.; et al. Design, synthesis and antibacterial properties of pyrimido [4,5-b]indol-8-amine inhibitors of DNA gyrase. *Bioorganic Med. Chem. Lett.* **2018**, *28*, 2998–3003. [CrossRef]
62. Cooper, I.; Pichowicz, M.; Stokes, N. Compounds with Activity against Bacteria and Mycobacteria. WIPO Patent WO2016067009A1, 6 May 2016.
63. Henderson, S.R.; Stevenson, C.E.M.; Malone, B.; Zholnerovych, Y.; Mitchenall, L.A.; Pichowicz, M.; McGarry, D.H.; Cooper, I.R.; Charrier, C.; Salisbury, A.-M.; et al. Structural and mechanistic analysis of ATPase inhibitors targeting mycobacterial DNA gyrase. *J. Antimicrob. Chemother.* **2020**, *75*, 2835–2842. [CrossRef]
64. Huang, X.; Guo, J.; Liu, Q.; Gu, Q.; Xu, J.; Zhou, H. Identification of an auxiliary druggable pocket in the DNA gyrase ATPase domain using fragment probes. *MedChemComm* **2018**, *9*, 1619–1629. [CrossRef]
65. Yu, Y.; Guo, J.; Cai, Z.; Ju, Y.; Xu, J.; Gu, Q.; Zhou, H. Identification of new building blocks by fragment screening for discovering GyrB inhibitors. *Bioorganic Chem.* **2021**, *114*, 105040. [CrossRef]
66. Xue, W.; Li, X.; Ma, G.; Zhang, H.; Chen, Y.; Kirchmair, J.; Xia, J.; Wu, S. N-thiadiazole-4-hydroxy-2-quinolone-3-carboxamides bearing heteroaromatic rings as novel antibacterial agents: Design, synthesis, biological evaluation and target identification. *Eur. J. Med. Chem.* **2020**, *188*, 112022. [CrossRef]
67. Xue, W.; Wang, Y.; Lian, X.; Li, X.; Pang, J.; Kirchmair, J.; Wu, K.; Han, Z.; You, X.; Zhang, H.; et al. Discovery of N-quinazolinone-4-hydroxy-2-quinolone-3-carboxamides as DNA gyrase B-targeted antibacterial agents. *J. Enzym. Inhib. Med. Chem.* **2022**, *37*, 1620–1631. [CrossRef]
68. Chiriac, A.I.; Kloss, F.; Krämer, J.; Vuong, C.; Hertweck, C.; Sahl, H.-G. Mode of action of closthioamide: The first member of the polythioamide class of bacterial DNA gyrase inhibitors. *J. Antimicrob. Chemother.* **2015**, *70*, 2576–2588. [CrossRef]
69. Lincke, T.; Behnken, S.; Ishida, K.; Roth, M.; Hertweck, C. Closthioamide: An Unprecedented Polythioamide Antibiotic from the Strictly Anaerobic Bacterium *Clostridium cellulolyticum*. *Angew. Chem.* **2010**, *122*, 2055–2057. [CrossRef]
70. Phillips, J.W.; Goetz, M.A.; Smith, S.K.; Zink, D.L.; Polishook, J.; Onishi, R.; Salowe, S.; Wiltsie, J.; Allocco, J.; Sigmund, J.; et al. Discovery of Kibdelomycin, A Potent New Class of Bacterial Type II Topoisomerase Inhibitor by Chemical-Genetic Profiling in *Staphylococcus aureus*. *Chem. Biol.* **2011**, *18*, 955–965. [CrossRef]
71. Lu, J.; Patel, S.; Sharma, N.; Soisson, S.M.; Kishii, R.; Takei, M.; Fukuda, Y.; Lumb, K.J.; Singh, S.B. Structures of Kibdelomycin Bound to *Staphylococcus aureus* GyrB and ParE Showed a Novel U-Shaped Binding Mode. *ACS Chem. Biol.* **2014**, *9*, 2023–2031. [CrossRef]

72. Eakin Ann, E.; Green, O.; Hales, N.; Walkup Grant, K.; Bist, S.; Singh, A.; Mullen, G.; Bryant, J.; Embrey, K.; Gao, N.; et al. Pyrrolamide DNA Gyrase Inhibitors: Fragment-Based Nuclear Magnetic Resonance Screening to Identify Antibacterial Agents. *Antimicrob. Agents Chemother.* **2012**, *56*, 1240–1246. [CrossRef]
73. Flatman Ruth, H.; Howells Alison, J.; Heide, L.; Fiedler, H.-P.; Maxwell, A. Simocyclinone D8, an Inhibitor of DNA Gyrase with a Novel Mode of Action. *Antimicrob. Agents Chemother.* **2005**, *49*, 1093–1100. [CrossRef]
74. Sissi, C.; Vazquez, E.; Chemello, A.; Mitchenall, L.A.; Maxwell, A.; Palumbo, M. Mapping Simocyclinone D8 Interaction with DNA Gyrase: Evidence for a New Binding Site on GyrB. *Antimicrob. Agents Chemother.* **2010**, *54*, 213–220. [CrossRef]
75. Edwards, M.J.; Williams, M.A.; Maxwell, A.; McKay, A.R. Mass Spectrometry Reveals That the Antibiotic Simocyclinone D8 Binds to DNA Gyrase in a "Bent-Over" Conformation: Evidence of Positive Cooperativity in Binding. *Biochemistry* **2011**, *50*, 3432–3440. [CrossRef] [PubMed]
76. Hearnshaw, S.J.; Edwards, M.J.; Stevenson, C.E.; Lawson, D.M.; Maxwell, A. A New Crystal Structure of the Bifunctional Antibiotic Simocyclinone D8 Bound to DNA Gyrase Gives Fresh Insight into the Mechanism of Inhibition. *J. Mol. Biol.* **2014**, *426*, 2023–2033. [CrossRef] [PubMed]
77. Oppegard, L.M.; Hamann, B.L.; Streck, K.R.; Ellis, K.C.; Fiedler, H.-P.; Khodursky, A.B.; Hiasa, H. In vivo and in vitro patterns of the activity of simocyclinone D8, an angucyclinone antibiotic from *Streptomyces antibioticus*. *Antimicrob. Agents Chemother.* **2009**, *53*, 2110–2119. [CrossRef] [PubMed]
78. Baell, J.B.; Nissink, J.W.M. Seven Year Itch: Pan-Assay Interference Compounds (PAINS) in 2017—Utility and Limitations. *ACS Chem. Biol.* **2018**, *13*, 36–44. [CrossRef]

pharmaceuticals

MDPI

Review

Managing Irinotecan-Induced Diarrhea: A Comprehensive Review of Therapeutic Interventions in Cancer Treatment

Xiaoqin Yang [1], Jiamei Chen [1], Yitao Wang [2], Yihan Wu [1,*] and Jinming Zhang [1,*]

[1] State Key Laboratory of Southwestern Chinese Medicine Resources, School of Pharmacy, Chengdu University of Traditional Chinese Medicine, Chengdu 611137, China; yangxiaoqin202033@126.com (X.Y.); cobby0945@163.com (J.C.)

[2] State Key Laboratory of Quality Research in Traditional Chinese Medicine, Institute of Chinese Medical Sciences, University of Macau, Macau SAR 999078, China; ytwang@umac.mo

* Correspondence: yihanwuone@126.com (Y.W.); cdutcmzjm@126.com (J.Z.)

Abstract: Irinotecan (CPT-11), an inhibitor of DNA topoisomerase I, stands as a pivotal therapeutic agent in oncology. However, its use is primarily constrained by side effects such as neutropenia and the onset of delayed diarrhea. Despite the effective management of neutropenia, CPT-11-induced diarrhea (CID) is often severe, leading to hospitalization, dosage adjustments, and in some cases, treatment discontinuation, which can significantly impact therapeutic outcomes. A multitude of pharmacological agents have been investigated in preclinical and clinical studies with the aim of reducing or preventing the onset of delayed diarrhea associated with CPT-11. This comprehensive review examines the underlying mechanisms of CPT-11-triggered delayed diarrhea and discusses the experimental medications and strategies that have been utilized to combat this adverse effect. This review encompasses an exploration of chemical formulations, the application of traditional Chinese medicine, and the advent of innovative drug delivery systems. It is anticipated that this article will serve as a valuable resource for both novice researchers in the realm of irinotecan chemotherapy and for those who are well-versed in the field, including experts and practicing clinicians.

Keywords: CPT-11(irinotecan); SN-38; CPT-11-induced diarrhea (CID); diarrhea prevention; clinical treatment

1. Introduction

Irinotecan (CPT-11), the inaugural representative of topoisomerase I inhibitors extracted from the camptotheca acuminata plant, has garnered acclaim for its broad-spectrum efficacy against various cancers, including colorectal, lung, breast, and malignant lymphoma. Its clinical prominence as an antineoplastic agent, however, is tempered by the significant adverse effects, most notably diarrhea, which have impeded its widespread adoption in oncology [1,2]. Diarrhea, a prominent and dose-limiting toxicity associated with CPT-11, manifests in two forms: acute and delayed. Acute diarrhea, often observed early in the course of CPT-11 administration, is linked to the inhibitory effect of drugs on cholinergic activity. This inhibition leads to an accumulation of acetylcholine, which in turn increases intestinal motility, disrupts absorption, and provokes contractility—effects that can be mitigated with anticholinergic agents such as atropine [3]. In contrast, delayed diarrhea, which presents over 24 h post-administration, is more severe and persistent, and is often challenging to manage. Characterized by its non-cumulative nature, it affects up to 87% of patients, with 30–40% experiencing severe grade 3 or 4 diarrhea [4,5]. Prolonged diarrhea can lead to renal insufficiency and electrolyte disturbances due to fluid and electrolyte losses and may provoke cardiovascular complications as a result of intravascular volume fluctuations. These complications can necessitate the cessation of CPT-11 treatment and pose life-threatening risks. Furthermore, severe diarrhea significantly diminishes patients' quality of life, disrupting their social roles, interpersonal relationships, and leading to a sense of social isolation.

Indeed, the pathophysiological mechanisms underlying delayed diarrhea induced by CPT-11 have garnered significant interest and have been the focus of extensive experimental research. The severity of this iatrogenic condition is predominantly attributed to the accumulation of the active metabolite, SN-38, derived from CPT-11. The metabolic conversion of CPT-11 to SN-38 is facilitated by carboxylesterases, specifically CES1 and CES2 [6,7]. Thereafter, SN-38 undergoes further metabolism to SN-38G by the enzyme uridine diphosphate glucuronosyltransferase 1A1 (UGT1A1) [8,9]. SN-38G is excreted into the bile, where it can be acted upon by β-glucuronidase (β-GD) present in the intestinal lumen, leading to the release and accumulation of SN-38, which is a primary culprit in the etiology of CID [10]. Despite this, the precise mechanisms of SN-38-induced diarrhea remain a subject of ongoing debate. Additional research has highlighted that CPT-11 can alter the intestinal luminal environment, potentially promoting the proliferation of diverse bacterial species. The escalating activity of bacterial β-glucuronidase can deconjugate SN-38G, reinstating the active SN-38 form, which can precipitate severe intestinal damage and consequent diarrhea. Furthermore, CPT-11 has been shown to cause significant colonic damage, characterized by increased apoptosis, crypt hypoplasia, and dilation, as well as excessive mucus secretion and damage to the small intestine. These histopathological changes, coupled with alterations in the number of goblet cells, contribute to the development of diarrhea.

To date, a myriad of therapeutic strategies and pharmacological agents have been deployed to address the challenge of delayed diarrhea induced by CPT-11. These encompass the inhibition of SN-38 production, enhancement of SN-38 adsorption, and suppression of β-glucuronidase (β-GD) activity. Modern medicinal approaches include the use of loperamide, a synthetic opioid that moderates intestinal motility, extends transit time, and promotes fluid reabsorption, as well as its role in inhibiting the expression of TX-A2, thereby exerting an antisecretory effect [11]. Atropine, a competitive antagonist of muscarinic acetylcholine receptors, and targeted antidiarrheal therapies such as the somatostatin analog octreotide and nonsteroidal anti-inflammatory drugs (NSAIDs) are also utilized clinically to mitigate CID [12,13]. Furthermore, the rich tapestry of traditional Chinese medicine has been harnessed, with herbal formulas, extracts, and phytochemical substances being applied to alleviate diarrheal symptoms. Notable among these are Huangqin Decoction, Hange-Shashin-To, Sairei-To, Shengjiang Xiexin Decoction, and Gegen Qinlian Decoction, which have demonstrated efficacy in reducing diarrhea associated with CPT-11. The quest to enhance the management of CID continues to grow, with a keen interest in developing novel and effective treatment modalities to bolster therapeutic success.

This comprehensive review examines the underlying mechanisms of CPT-11-triggered delayed diarrhea and discusses the experimental medications and strategies that have been utilized to combat this adverse effect. The review encompasses an exploration of chemical formulations, the application of traditional Chinese medicine, and the advent of innovative drug delivery systems. We aim to provide a balanced overview of all major therapeutic interventions, highlighting their strengths and limitations. We acknowledge that the mechanisms and treatments for diarrhea discussed in this review represent just the beginning of a much larger body of knowledge. As biotechnology and genetic research continue to make groundbreaking strides, it is plausible that future developments will yield even more potent methods for managing recalcitrant chemotherapy-induced diarrhea. These advancements will not only enhance our understanding of the condition but also contribute significantly to the improvement of cancer treatment with chemotherapeutic agents.

2. Mechanisms of CPT-11-Induced Diarrhea

2.1. Mechanism of Pharmacokinetic

The metabolic pathway of CPT-11 is intricate, involving a multitude of enzymes responsible for drug metabolism. CPT-11, functioning as a prodrug, undergoes biotransformation into its active cellular metabolite, SN-38 (7-ethyl-10-hydroxycamptothecin), via the ester bond at the C-10 position. SN-38 demonstrates cytotoxic activity that is 100 to 1000 times more potent than that of CPT-11 itself [6,7]. Over recent years, the etiology of intestinal toxicity has been closely linked to the metabolic processes of CPT-11 and its active metabolite, SN-38, within the body. Figure 1 provides an overview of the metabolism of irinotecan, highlighting the key metabolites and enzymes involved in the onset of delayed diarrhea.

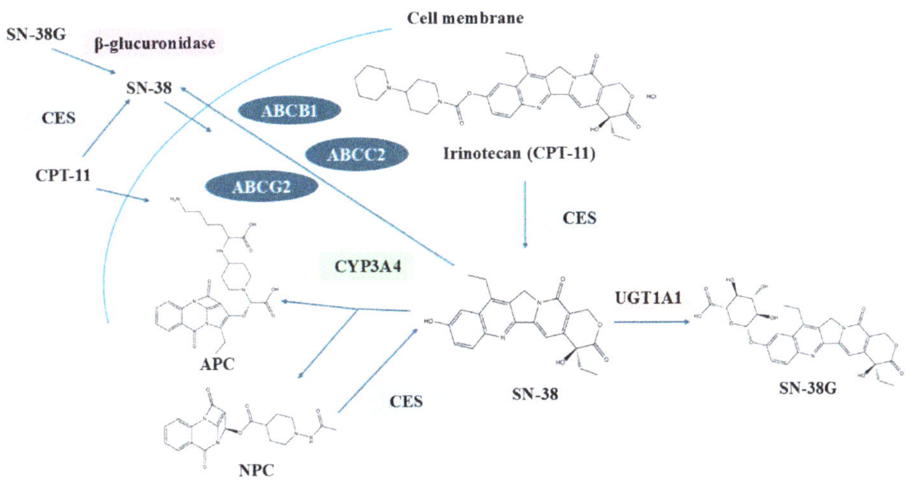

Figure 1. Structure and conformation of irinotecan. The main enzymes implicated in the conversion of irinotecan into its active metabolite SN-38 and the inactive product SN-38G are indicated.

Upon intravenous administration, CPT-11 is rapidly bioactivated to SN-38 in the plasma and liver. This initial conversion is predominantly facilitated by carboxylesterases (CEs), which are also found in the small intestine, liver, and colon. The expression levels and activity of these CEs can significantly influence the balance between CPT-11 and SN-38, subsequently impacting the circulating concentrations and antineoplastic efficacy of SN-38 [14,15]. Additionally, CPT-11 can be metabolized into 7-ethyl-10-[4-N-(5-aminopentanoic acid)-1-piperidino]-carbonyloxy-camptothecine (APC) and 7-ethyl-10-(4-amino-1-piperidino)-carbonyloxy-camptothecine (NPC) by the cytochrome P450 enzyme system. Emerging research indicates that NPC serves as a substrate for carboxylesterase, being nearly entirely converted to SN-38, albeit with a minimal impact on the overall antitumor efficacy due to the scarce quantity of NPC produced [16,17].

Prior to its excretion into the bile, SN-38 is inactivated by conversion to SN-38-glucuronide (SN-38G) through the action of glucuronyl transferases (UGTs). Irinotecan and its metabolites, including SN-38, SN-38G, and APC, are then secreted via the bile into the small intestine, where they may undergo enterohepatic recirculation back to the liver. This process is mediated by transport proteins such as multidrug resistance-associated protein 1 (MDR1/ABCB1), multidrug resistance-associated protein 2 (MRP2/ABCC2), and breast cancer resistance protein (BCRP/ABCG2). Moreover, SN-38G can be reconverted to active SN-38 by β-glucuronidase, an enzyme produced by the intestinal bacterial flora. Concurrently, CPT-11 is re-hydrolyzed to SN-38 by CEs, allowing for the reabsorption of both CPT-11 and SN-38 through the intestinal lining. These compounds subsequently enter the bloodstream and are transported to the liver under the auspices of organic anion transporter polypeptides (OATPs), thereby completing the hepatic–intestinal cycle [18]. Figure 2 provides a detailed overview of the metabolic pathway of irinotecan, highlighting the key compounds and enzymes involved in its metabolism and the formation of active metabolites such as SN-38.

2.2. CPT-11 Factors Leading to Diarrhea

2.2.1. The Effect of UGT1A1 Polymorphisms

Under the catalytic influence of carboxylesterase, CPT-11 is converted into its active metabolite, SN-38 [6]. Subsequently, SN-38 is metabolized into an inactive form, glucuronidated SN-38 (SN-38G), by the liver enzyme UGT1A1 [19]. These metabolites are ultimately excreted via the bile and eliminated from the body through feces. Currently, UGT1A1 is recognized as a critical determinant of SN-38 concentrations, which in turn influence the severity of intestinal toxicity [20]. There is a well-documented correlation between UGT1A1 gene polymorphisms and the toxicity associated with CPT-11 use both domestically and internationally.

UGT1A1 exhibits a high degree of polymorphism, with over 100 genetic variants identified to date [21]. The most pertinent UGT1A1 polymorphisms in relation to the pharmacokinetics and pharmacodynamics of CPT-11 are UGT1A1*6 and UGT1A1*28 [22]. The promoter region of the UGT1A1 gene features an atypical TATA domain comprising five to eight thymine–adenine (TA) repeats, with the six-repeat genotype being the most prevalent. An increase in the number of TA repeats is associated with reduced UGT1A1 expression. Notably, the UGT1A1*28 variant incorporates an additional TA repeat, which significantly diminishes UGT1A1 transcription and expression by approximately 70%, leading to decreased levels of SN-38G [19]. This genetic variant is more prevalent among Caucasians and Africans/African Americans, yet it is less common in the Asian population [23]. Individuals with the UGT1A1*28 non-wild type genotype exhibit a significantly higher incidence of diarrhea compared to those with the wild type, emphasizing the utility of gene polymorphism screening prior to CPT-11 chemotherapy to identify high-risk groups and anticipate potential adverse effects, thereby informing clinical decision-making [24].

Given the lower prevalence of UGT1A1*28 in Asians, its impact on toxicity outcomes is less pronounced in this demographic. Several meta-analyses have indicated that the UGT1A16 polymorphism may serve as a potential biomarker for predicting CPT-11-related toxicity in the Asian population [20,25]. Both Caucasian and Asian patients who are homozygous or heterozygous for the UGT1A1*28 variant are at an elevated risk for severe diarrhea following CPT-11 administration compared to wild-type patients, with a dose-dependent effect observed in a meta-analysis of Caucasian carriers [26]. In Asian patients, the UGT1A1*6 polymorphism is closely linked to the risk of CPT-11-induced neutropenia and is also significantly associated with severe diarrhea [20,27]. However, the dose dependency of this association remains unclear, as dose-subgroup analyses have not been conducted [27].

Beyond UGT1A1*6 and UGT1A1*28, other UGT1A1 polymorphisms may theoretically impact CPT-11-related toxicity [28]. For instance, UGT1A1*60, which is in linkage with UGT1A1*28, is associated with reduced transcriptional activity [29]. Nevertheless, UGT1A1*60 status has not been significantly correlated with CPT-11-related toxicities or pharmacokinetics in clinical studies [30]. Similarly, UGT1A1*93, also in linkage disequilibrium with UGT1A1*28, has been associated with increased SN-38 area under the curve (AUC), reduced neutrophil counts, hematological toxicity, diarrhea, and grade 3 vomiting [31].

Genetic polymorphisms in UGT1A9 and UGT1A7 are likewise intimately connected with the risk of diarrhea. Individuals with the UGT1A9*22 genotype demonstrate higher enzyme expression and increased SN-38 glucuronidation compared to those with UGT1A9*1/*1, thus facing a higher risk of diarrhea [32,33]. Conversely, the UGT1A7*3 and UGT1A7*4 polymorphisms are characterized by diminished enzyme activity and SN-38 binding, with UGT1A7*3/*3 carriers being at a greater risk for adverse events during CPT-11 chemotherapy [34].

2.2.2. The Effect of Drug Transporter Polymorphisms

The ATP-binding cassette (ABC) membrane transporters play a pivotal role in the multidrug resistance observed in tumor cells. Members of the ABC family, including ABCB1, ABCC1, ABCC2, and ABCG2, are integral to the transport pathway of drugs within the body [35]. Given that both CPT-11 and its active metabolite, SN-38, are substrates for ABC transporters, polymorphisms in these transporter genes may significantly influence the pharmacokinetics and toxicity profiles of CPT-11. Clinical analyses have established a clear link between ABCB1 polymorphisms (specifically SNPs 1128503, rs2032582, and rs1045642) and the toxicities associated with CPT-11 treatment [36]. Individuals harboring these SNPs have been shown to exhibit a poorer response to CPT-11-based therapies and a shorter survival rate in advanced colorectal cancer [37].

Multivariate analyses have revealed associations between ABCC1 single nucleotide polymorphisms (SNPs rs6498588 and rs1750133) and increased plasma concentrations of SN-38, as well as decreased absolute neutrophil counts [38]. Conversely, the ABCB1 variant rs12720066 has been associated with reduced SN-38 exposure and elevated neutrophil levels. Beyond ABCB1 and ABCC1, polymorphisms in ABCC2 (rs3740066) and ABCG2 (rs2231137) have been identified as independent predictors of toxicity [34]. However, the impact of the ABCG2 (421C > A) polymorphism on CPT-11 exposure appears to be more limited [39]. The ABCC2 gene, which encodes an extrahepatic transporter, may confer a protective effect against diarrhea, potentially through reduced hepatobiliary transport of CPT-11, thereby lessening its intestinal exposure [40]. Although their specific role in CPT-11 efflux remains to be determined, ABCC5 and ABCG1 may also participate in this process, as several SNPs associated with these transporters have been linked to severe diarrhea [41].

Within the OATP family of genes, OATP1B1, encoded by the SLCO1B1 gene, is noted for its high uptake of SN-38 in the liver. Mutations in the SLCO1B1 gene are hypothesized to impact the transport activity of OATP1B1, consequently affecting the hepatic clearance of SN-38. Studies have highlighted that two prevalent SLCO1B1 gene mutations can influence the transport and metabolism of CPT-11, albeit through distinct mechanisms [42]. The SLCO1B1 (T521C) mutation predominantly reduces the affinity of OATP1B1 for its substrate, thereby impairing its transport efficiency. In contrast, the SLCO1B1 (A388G) mutation predominantly leads to diminished expression levels of OATP1B1, compromising its transport capacity and impacting the hepatic metabolism of SN-38, which can result in more severe adverse reactions [43]. Clinical observations have indicated that patients with the SLCO1B1**15 allele are at a significantly higher risk of experiencing diarrhea and neutropenia in the week following chemotherapy. The increased toxicity of CPT-11 in these patients may be attributable to enhanced distribution and bioavailability conferred by the SLCO1B1**15 variant [44,45].

2.2.3. The Effect of CYP3A Polymorphisms

The CYP3A enzyme, a member of the cytochrome P450 superfamily, is encoded by the CYP3A gene located on human chromosome 7. It plays a crucial role in the metabolism of numerous drugs, including CPT-11. Theoretical considerations suggest that diminished CYP3A activity or expression could reduce the synthesis of APC and NPC, shunting the metabolic pathway towards increased conversion of CPT-11 to SN-38 by carboxylesterase (CES). This metabolic shift could potentially elevate the risk of adverse reactions due to heightened SN-38 production. Consequently, polymorphisms in the CYP3A gene are hypothesized to be associated with the variability in CPT-11's adverse reaction profiles [46].

A number of single nucleotide polymorphisms (SNPs) in the CYP3A4 gene have been documented, with their prevalence varying significantly across different ethnic groups. Notably, the CYP3A15 variant (485G > A [Arg162Gln]) is observed in 2–4% of African Americans, while other variants such as CYP3A2 (664T > C [Ser222Pro]), CYP3A10 (520G > C [Asp174His]), and CYP3A17 (566T > C [Phe189Ser]) are more prevalent among Caucasians and Mexicans, affecting 2–5% of individuals. In East Asian populations, the CYP3A16 (554C > G [Thr185Ser]) and CYP3A18 (878T > C [Leu293Pro]) variants are more commonly encountered, affecting 1–10% of individuals [47,48].

Research has indicated that variations in the CYP3A4 genotype can influence the clearance rate of CPT-11. However, the clinical relevance of CYP3A4 SNPs is often overshadowed by the low frequency of these variants and the substantial influence of exogenous and endogenous factors on enzyme activity. Inter-individual differences in enzyme activity are more likely to be attributable to environmental and physiological factors, including drug interactions, nutritional status, alterations in liver function, and the overall health condition of the patient, rather than genetic polymorphisms alone. As a result, the inclusion of CYP3A4 gene detection in routine clinical practice remains limited [49].

2.2.4. Pathophysiology of CPT-11-Induced Diarrhea

The alteration of the intestinal milieu post-CPT-11 administration is a recognized mechanism that precipitates diarrhea [50]. Treatment with CPT-11 can lead to significant damage to both the colonic and small intestinal tissues. This damage is characterized by heightened apoptosis, a diminished villi-to-crypt ratio, dilated crypts, increased lymphatic infiltration within the mucosa, and excessive mucus secretion accompanied by villous atrophy [51]. Moreover, the active metabolite SN-38 induces direct mucosal injury, which manifests as malabsorption of water and electrolytes and heightened mucosal secretion [52].

Following CPT-11 administration, there are notable shifts in fecal sodium and potassium levels, prompting an osmotic flow of water into the intestinal lumen to restore electrolyte balance, a process that culminates in diarrhea. Concurrently, there are significant alterations in serum sodium, chloride, and osmolality levels. In vivo studies have shown that CPT-11 can result in a thinner intestinal wall and ileal epithelial vacuolization associated with apoptosis, further exacerbating malabsorption. Additionally, CPT-11 induces goblet cell hyperplasia and an excess of sulfomucin in the cecum, pointing to increased mucin secretion [53].

The epithelial barrier is crucial for intestinal function, and CPT-11 has been shown to compromise the integrity of tight junctions, including the vital protein components claudin-1 and occludin [50]. This disruption can lead to bacterial translocation and exacerbate diarrhea. Furthermore, CPT-11 induces alterations in the gut microbiota, which can contribute to chemotherapy-induced mucositis. The changed luminal environment can also promote the proliferation of bacteria capable of producing β-glucuronidase, an enzyme that can convert the inactive SN-38 glucuronide (SN-38G) back to the active SN-38, thereby augmenting intestinal damage and diarrhea [51].

The direct role of the intestinal microbiota in CID has been highlighted by studies such as those by Giovanni Brandi and colleagues [54]. Using germ-free and holoxenic mice, they demonstrated that the presence of a normal gut microbiota can significantly influence the severity of CID and intestinal epithelial damage. Notably, holoxenic mice treated with a lethal dose of CPT-11 exhibited severe intestinal mucosal damage, whereas germ-free mice showed no such damage, underscoring the role of bacterial factors in the pathogenesis of CID.

2.2.5. The Dose-Dependent Relationship Between CPT-11 Dosage and Diarrhea Incidence

The incidence and severity of diarrhea associated with irinotecan (CPT-11) treatment have been well-documented in numerous clinical studies, highlighting a clear dose-dependent relationship. Higher doses of CPT-11 are consistently linked to increased frequency and severity of both acute and delayed diarrhea, posing significant challenges in clinical management.

In a phase II study by Shimada et al. [4], the administration of escalating doses of CPT-11 in patients with metastatic colorectal cancer revealed a direct correlation between dose intensity and the incidence of diarrhea. Specifically, patients receiving higher doses exhibited a significantly higher rate of grade 3 and 4 diarrheas compared to those on lower dosing regimens. This observation underscores the need for meticulous dose titration to balance therapeutic efficacy and adverse effects.

Further supporting this dose-dependent relationship, Abigerges et al. [11] demonstrated that high-dose CPT-11 regimens necessitated intensive management with high-dose loperamide to control severe diarrhea. Their findings indicated that the severity of diarrhea increased proportionally with higher doses of CPT-11, necessitating careful monitoring and intervention to prevent life-threatening complications.

Moreover, the pharmacokinetic and pharmacodynamic properties of CPT-11 have been extensively studied, revealing that its active metabolite, SN-38, is primarily responsible for the observed gastrointestinal toxicity [47]. The conversion of CPT-11 to SN-38 is facilitated by carboxylesterases, and the subsequent metabolism of SN-38 to its inactive form, SN-38G, is mediated by UGT1A1 [31]. Genetic polymorphisms in UGT1A1, such as UGT1A1*28, have been identified as significant predictors of increased SN-38 levels and, consequently, higher incidence of severe diarrhea [55]. This genetic variability further complicates the dose-dependent relationship, as individuals with certain UGT1A1 genotypes may be at higher risk for adverse events at standard doses.

In addition to genetic factors, clinical studies have shown that the timing and schedule of CPT-11 administration also influence the incidence of diarrhea. For instance, a study by JP Armand [5] demonstrated that the frequency of severe diarrhea was significantly higher in patients receiving CPT-11 every 3 weeks compared to those on a more frequent dosing schedule. This suggests that the dosing interval may modulate the accumulation and systemic exposure of SN-38, thereby affecting the severity of gastrointestinal toxicity.

In summary, the dose-dependent relationship between CPT-11 and diarrhea incidence is well-established through clinical and pharmacogenetic studies. Higher doses of CPT-11 are associated with increased frequency and severity of diarrhea, necessitating careful dose adjustment and patient monitoring. Future research should focus on personalized dosing strategies that account for genetic variability and pharmacokinetic profiles to optimize therapeutic outcomes while minimizing adverse effects.

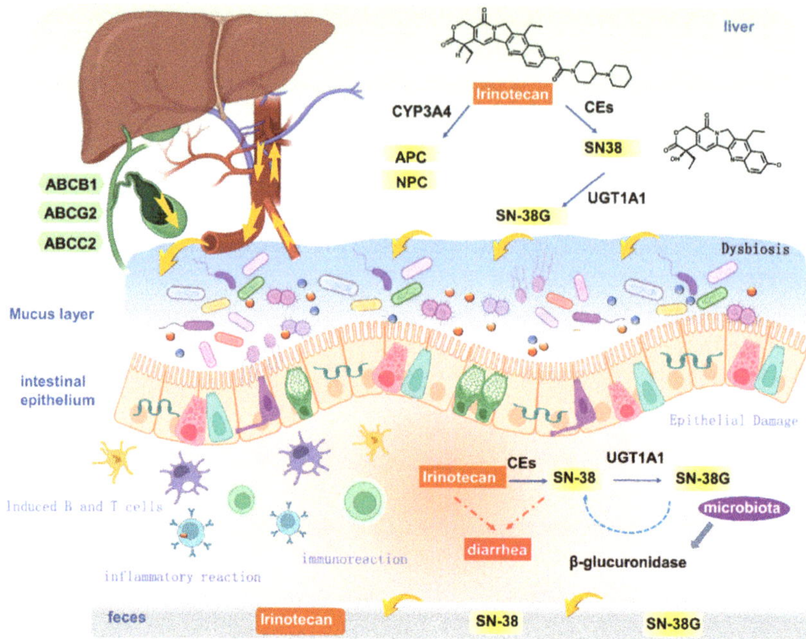

Figure 2. Overview of irinotecan metabolism. CPT-11 is a prodrug that is converted to active metabolite ethyl-10-hydroxy-camptothecin (SN-38) by liver carboxylesterase-converting enzymes (CES1/2) and is then transported to the liver by 1B1 polypeptide (OATP1B1) and inactivated by microsomal uridine 5′-diphospho-glucuronosyltransferase enzymes (UGTs): UGT1A1. Irinotecan is transported to the bile by a group of ATP-binding cassette transporters (ABC transporters): ABCB1, ABCC2, and ABCG2. Irinotecan is efficiently metabolized by cytochrome P450 enzymes: CYP3A4 and CYP3A5. This results in the generation of less active metabolites APC (7-ethyl-10-[4-N-(5-aminopentanoic acid)-1-piperidino] carbonyloxycamptothecin) and NPC (7-ethyl-10-[4-amino-1-piperidino] carbonyloxycamptothecin). NPC (but not APC) can be further converted to SN-38 by CES1, and CES2 gut microbiota may also participate in irinotecan metabolism by the production of β-glucuronidase, which catalyzes the breakdown of SN-38G into SN-38.

3. Strategies to Block or Treat Delayed Diarrhea

A spectrum of preventive and therapeutic strategies has been theorized and evaluated in both animal models and clinical studies, aimed at mitigating or counteracting CID [56]. The proactive approach of diarrhea prevention is widely regarded as the optimal strategy to avert this grave drug-related complication. Implementing such a strategy not only enhances the safety profile of the drug but also has the potential to decrease healthcare costs associated with hospitalization, elevate the quality of life for patients, and possibly facilitate escalated dosing regimens. This could, in turn, augment the therapeutic efficacy of the treatment through an improved tumor response [57]. Beyond the initial preventive steps endorsed by clinical guidelines, which encompass adjustments to the treatment schedule and dosage, as well as genetic screening, this review delves into the utilization of traditional Chinese medicine and modern pharmaceuticals. It specifically highlights their application in either forestalling or treating diarrhea that arises as a consequence of CPT-11 administration.

3.1. Chemical Drug Treatment of Delayed Diarrhea

3.1.1. Antidiarrheal Therapy

In accordance with the current clinical guidelines, loperamide is widely acknowledged as the first-line therapeutic agent for the management of CID. Loperamide functions as an inhibitor of intestinal peristalsis by activating opioid receptors within the intestinal plexus and blocking the release of acetylcholine, thereby reducing intestinal motility and secretion [58]. Research has demonstrated that while diarrhea poses a dose-limiting toxicity at a dosage of 350 mg/m^2 for CPT-11 administered every three weeks, the concurrent use of high-dose loperamide can potentially escalate the CPT-11 dosage to 750 mg/m^2. Nonetheless, loperamide is not without side effects and has been noted to have a relatively high failure rate in treating CID [59].

Octreotide, a somatostatin analog and an inhibitor of intestinal secretion, is proposed as a second-line treatment for CID following the inadequacy of loperamide. It mitigates diarrhea by curbing gastrointestinal hormone secretion, diminishing intestinal peristalsis, extending gastrointestinal transit time, and enhancing the reabsorption of water and electrolytes while reducing secretion [60]. Although octreotide has shown promise in the secondary prevention of refractory CID in a limited case series, there is an acknowledged scarcity of robust research on its efficacy in the context of CID [61,62].

Acetorphan, an oral active enkephalinase inhibitor marketed as Racecadotril or Tiorfan®, has been investigated as an alternative to loperamide due to its antidiarrheal and antisecretory properties [63]. A low-dose escalation study indicated that prophylactic acetylmorphine could significantly reduce the incidence of diarrhea without inducing constipation. However, a subsequent randomized, open-label, multicenter phase II trial revealed that prophylactic administration of acetorphan at 300 mg/day did not exert a beneficial effect on CID [64].

Budesonide, typically utilized in the treatment of inflammatory bowel diseases, has been explored for its potential role in CID management. In a phase I trial, fourteen patients experiencing stage 4 diarrhea due to CPT-11 were administered budesonide (9 mg oral dose each morning). Budesonide was observed to reduce the severity of diarrhea by at least two levels in 86% of the CPT-11-treated patients [65]. However, in a phase III randomized, double-blind, placebo-controlled trial involving patients with advanced colorectal cancer (CRC), oral budesonide did not manifest significant advantages in the prevention of CID, although the findings were somewhat encouraging [66].

3.1.2. Intestinal Alkalization

CPT-11, its active metabolite SN-38, and the glucuronide conjugate SN-38G all possess an unstable α-hydroxy-δ-lactone ring, the integrity of which is susceptible to pH-dependent hydrolysis. The detrimental impact of SN-38 on intestinal epithelial cells is postulated to be the primary cause of delayed diarrhea associated with CPT-11 use. At physiological pH levels or higher, the less harmful carboxylic acid form of the compound predominates, whereas in an acidic environment, the more toxic lactone form is favored [67]. Consequently, modulating the equilibrium between these carboxylate isomers can attenuate the toxicity of CPT-11. Given that the intestinal absorption rates of CPT-11 and SN-38 are pH-sensitive, with absorption decreasing by over 65% when the pH exceeds 6.8, an alkaline intestinal milieu can effectively diminish the absorption of CPT-11 and SN-38 by the intestinal cells.

A case-control study involving lung cancer patients treated with CPT-11 utilized an alkalinization regimen consisting of sodium bicarbonate, magnesium oxide, water, and ursodeoxycholic acid administered for four days following CPT-11 treatment. This approach, which included intestinal alkalinization and controlled defecation, was found to markedly enhance the management of delayed diarrhea, as well as other side effects such as nausea, vomiting, and neutropenia [68]. Subsequent research involving patients with advanced gastrointestinal cancer who were on a CPT-11 treatment regimen revealed that the daily intake of 2 g of primary sodium bicarbonate powder, diluted in 250 mL of water for four days post-CPT-11 injection, resulted in a reduced incidence of grade 3 to 4 diarrhea to 16%, as opposed to the anticipated rate of 24% in large clinical phase III trials [69]. Nonetheless, a case report highlighted that intestinal alkalinization could significantly lower plasma levels of SN-38 and CPT-11 [70]. Moreover, the preventive regimen, while potentially effective, is notably cumbersome, necessitating the consumption of over 2–3 L of highly alkalized water daily throughout the treatment period [68,71].

3.1.3. Transporter Inhibition

As previously discussed, the pharmacokinetics of CPT-11 are influenced by a multitude of transporters, including ABCB1, ABCC1, ABCC2, and ABCG2 [18]. The modulation of these transporters through inhibition by various compounds can significantly impact the pharmacokinetic profile of CPT-11, the biliary concentration of SN-38, and consequently, the risk of intestinal toxicity.

The diarrhea induced by CPT-11 is predominantly attributed to the direct cytotoxic effects of SN-38 on the intestinal tract. SN-38 and its glucuronide conjugate, SN-38G, are transported from systemic circulation into the biliary tract via ABCB1 and ABCC2, subsequently entering the intrahepatic circulation [18]. Probenecid, an established inhibitor of ABCB1 and ABCC2, has demonstrated potential in reducing biliary excretion of CPT-11 metabolites in preclinical studies [72,73]. In a phase I clinical study involving 37 CPT-11-treated patients who received intravenous cyclosporine (5 mg/kg over 3 days), only one case of grade 3 diarrhea was reported. The subsequent phase of the study proposed a dosing regimen of 100 mg/m^2 of CPT-11 every two weeks [49]. When 34 patients were treated at this recommended dose, the incidence of grade 4 diarrhea was observed to be 3%. However, the study also noted that some patients had to discontinue participation due to cyclosporine-induced toxicity [74].

The transporter MRP2/ABCC2 plays a pivotal role in the biliary excretion of SN-38 and SN-38G, whereas its involvement in CPT-11 transport is considerably less [47]. Probenecid, an MRP2 inhibitor, when co-administered with CPT-11, has been shown to decrease the biliary excretion of CPT-11, SN-38, and SN-38G, thereby increasing their systemic concentrations. A reduction in the dosage of CPT-11 resulted in decreased intestinal SN-38 levels and a reduced incidence of CID, without significantly altering plasma SN-38 levels or the occurrence of myelosuppression compared to the control group [75].

3.1.4. Enzyme Induction and Inhibition

β-Glucuronidase Inhibition

The active metabolite SN-38, derived from CPT-11, can be metabolized into its inactive glucuronide form, SN-38G, by the action of hepatic uridine diphosphate glucosyltransferase [34,76]. A body of research has shed light on the pivotal role of bacterial β-glucuronidase in the pathogenesis of delayed diarrhea. The intestinal microbiota has the enzymatic capability to convert SN-38G back into the active and potentially damaging SN-38, a process that can result in significant intestinal mucosal injury [77].

D-saccharic acid 1,4-lactone (SAL), recognized for its potential as a therapeutic agent for cerebral ischemia–reperfusion injury in rats, has also been investigated for its capacity to inhibit β-glucuronidase. Studies have shown that the concurrent use of SAL can markedly decrease the intestinal mucosal damage induced by CPT-11 in preclinical rat models [78].

There is an ongoing effort to develop targeted methods that selectively inhibit bacterial β-glucuronidase without adversely affecting the intestinal symbiotes or the activity of mammalian β-glucuronidase. Rasmussen et al. [79] synthesized a novel compound, uronic acid-noan glucoside, which has demonstrated a competitive inhibitory effect against Escherichia coli β-glucuronidase. However, the compound exhibited minimal inhibitory activity against mammalian β-glucuronidase derived from bovine liver. More recently, nicotinamide, isoniazid, and amoxapine have been identified as inhibitors of bacterial β-glucuronidase, with negligible effects on the mammalian counterpart. Despite these promising findings, the clinical efficacy and safety of these compounds await further evaluation in human trials [80].

UGT1A1 Induction

As previously mentioned, UGT1A1 plays a crucial role in the metabolic pathway of CPT-11, facilitating the conversion of the cytotoxic metabolite SN-38 into a less harmful glucuronide form, SN-38G. Chrysin, despite its low oral bioavailability [81], has demonstrated the ability to upregulate UGT1A1 activity [82]. Consequently, chrysin can selectively enhance the glucuronidation of SN-38 to SN-38G within the gastrointestinal tract through the induction of UGT1A1. This process may mitigate intestinal mucosal damage and the onset of delayed diarrhea without impacting the systemic levels of SN-38 or its concentration within tumors. Other agents known to induce UGT1A1, including phenobarbital—as previously discussed—and glucocorticoids, have been the subject of clinical investigations. Notably, dexamethasone has not been found to significantly alter the area under the curve (AUC) for SN-38 and CPT-11, suggesting minimal influence on the enzymes CYP3A4 and UGT1A1 through which these substances are metabolized [83,84].

Carboxylesterases Inhibition and Activation

Intestinal carboxylesterase (hiCE) is instrumental in the conversion of CPT-11 to its active metabolite, SN-38, within the intestinal lumen. Evidence from human intestinal biopsies has confirmed the presence of hiCE, and in vitro assays have substantiated the direct conversion of CPT-11 to SN-38 by this enzyme. It is hypothesized that the inhibition of hiCE, thereby diminishing the levels of the active metabolite SN-38, could potentially lower the incidence of delayed diarrhea [85].

In this regard, Wadkins et al. [86] have synthesized a suite of seven hiCE inhibitors, all sulfonamide-based, which exhibit over 200-fold selectivity for hiCE relative to hepatic carboxylesterase. These inhibitors do not adversely affect human acetylcholinesterase or butyrylcholinesterase activities. Among the lead compounds developed, four nitrophenol derivatives have shown particular promise, with compound 3 demonstrating superior inhibitory efficacy against hiCE compared to rabbit liver carboxylesterase, leaving 14% of CES activity intact, as reported by Yoon et al. [87]. More recently, a new class of fluorene analogues has been introduced as hiCE inhibitors, showcasing enhanced potency and efficiency [85]. The efficacy and safety profiles of these novel inhibitors in animal models and, ultimately, in human studies remain to be established.

CYP3A4 Inducers

Anti-epileptic medications, such as phenytoin, carbamazepine, and phenobarbital, which are known to induce the activity of the cytochrome P450 3A4 (CYP3A4) enzyme, have been utilized in combination with or without dexamethasone. This treatment approach has been observed to elevate the recommended dosage of CPT-11 for patients who have not previously received these anti-epileptic drugs. Specifically, the dose of CPT-11 can be escalated from 350 mg/m^2 to 750 mg/m^2 when administered every three weeks, due to the increased metabolic clearance induced by these CYP3A4 inducers [83]. This induction effect results in an accelerated elimination of CPT-11 and a subsequent reduction in the area under the curve (AUC) for its active metabolite, SN-38.

In a separate study, a parallel increase in the clearance rate of CPT-11 and the dosage of temozolomide was noted. The escalated dosage regimen of temozolomide was 500 mg/m^2, administered every fifteen days over a 28-day cycle, in patients undergoing treatment with enzyme-inducing anti-epileptic drugs [88]. These findings underscore the importance of considering the impact of concomitant medications on the pharmacokinetics and optimal dosing strategies of chemotherapeutic agents.

COX-2 Inhibition

Cyclooxygenase-2 (COX-2) is notably overexpressed in metastatic colorectal cancer, where it contributes to tumorigenesis through multiple pathways, notably the synthesis of prostaglandins E2 (PGE2) and thromboxane A2 (TXA2) [89]. The administration of CPT-11 has been linked to upregulation of COX-2 in intestinal epithelial cells, which leads to elevated PGE2 levels. This increase in PGE2 induces heightened chloride secretion and diminished sodium absorption in the intestinal cells, a sequence of events that culminates in diarrhea. Research by Trifan et al. [90] has shown that celecoxib, a COX-2 inhibitor, can curtail the production of prostaglandins within the intestinal mucosa. It also lessens the colon's inflammatory response and markedly reduces the frequency of diarrhea in mice, with the severity of diarrhea correlated to the administered dose of celecoxib.

Further investigation in Ward CRC rat models demonstrated that oral celecoxib, administered at a dosage of 30 mg/kg daily in two divided doses, could lessen the toxicity associated with CPT-11 and bolster antitumor efficacy, alongside improved survival rates at doses that would otherwise be lethal [91]. However, subsequent clinical studies did not replicate these beneficial effects, failing to demonstrate any significant amelioration in the management of CID or in the treatment of colorectal cancer [92]. Diarrhea persisted as the primary non-hematological side effect. The lack of therapeutic improvement may stem from the possibility that COX-2 inhibition alone is insufficient to mitigate CID, or potentially due to inadequate delivery of celecoxib to achieve the desired protective effect on the target tissue [92].

3.1.5. Alteration of Intestinal Microflora

Prebiotics and Antibiotics

A study has demonstrated the potential benefits of Lactobacillus casei strain Shirota (LcS) in a rat model. The rats were administered LcS intragastrically at a dosage of 1.64×10^{11} colony-forming units (CFU) per 0.5 g in 3 mL saline for a period of 28 days. Commencing from the 14th day, the rats were also given CPT-11 at a dosage of 100 mg/kg for four consecutive days. A control group was simultaneously treated with CPT-11 and an equivalent volume of saline. The findings indicated that LcS treatment led to a significant reduction in weight loss, and the rats exhibited a markedly higher food intake compared to the control group. Moreover, the LcS group showed an improvement in the symptoms of delayed diarrhea associated with CPT-11, a beneficial effect that may be attributed to the inhibition of β-glucuronidase activity by LcS, an enzyme produced by the intestinal flora [93].

In another investigation, VSL#3, a commercially available probiotic formulation comprising Lactobacillus, Bifidobacterium, and Streptococcus species, was assessed for its effects on CPT-11-treated rats. The study found that VSL#3 could enhance glandular proliferation, mitigate weight loss, and reduce the severity of diarrhea, intestinal cell apoptosis, mucin secretion, and the increase of goblet cells in the jejunal crypts induced by CPT-11. However, the protective effects of VSL#3 were most pronounced when the probiotic was administered both prior to and following the chemotherapy regimen [94].

Antibiotics

The clinical use of antibiotics serves a dual purpose in the management of CID by targeting microorganisms that produce β-glucuronidase, an enzyme that can convert the inactive metabolite SN-38G back to its active form, SN-38, thereby exacerbating diarrhea. By reducing the intestinal bacterial load and β-glucuronidase activity, the concentration of SN-38 in the gut can be diminished, which may alleviate diarrheal symptoms [95]. In a clinical study involving patients with colorectal and small cell lung cancer treated with CPT-11, the administration of neomycin (500 mg, twice daily) in subsequent treatment cycles effectively prevented the recurrence of severe diarrhea [96].

In vivo research has shown that amoxicillin can mitigate the toxicity of CPT-11 and enhance its antitumor efficacy in tumor-bearing mice to a certain degree. This effect is associated with the inhibition of β-glucuronidase activity in various bacterial strains, including Escherichia coli, enterococci, streptococci, and staphylococci [97]. However, the alleviation of CID by antibiotics is not solely attributable to the inhibition of microbial β-glucuronidase. An animal study utilizing three different diarrhea models demonstrated that while streptomycin could ameliorate intestinal toxicity, it did not inhibit β-glucuronidase activity [98]. This discrepancy may stem from the significant variability in the catalytic efficiency, substrate binding, and reaction rates of β-glucuronidase among different bacterial species [99]. Other antibiotics, such as levofloxacin [100] and cefaclor [101], have also been utilized clinically to treat CID.

3.1.6. Prevention of Direct Intestinal Exposure

Activated charcoal (AC) is a widely utilized oral adsorbent with the capacity to adsorb the active metabolite SN-38 within the intestine, thereby mitigating intestinal mucosal injury and reducing the incidence of CID. AC also augments the clearance of SN-38 by engaging with intestinal capillaries and impeding the enterohepatic recirculation of the drug [102]. In a pediatric study involving patients undergoing CPT-11 chemotherapy, those in the intervention group were administered 250 mg of AC three times daily concurrently with CPT-11. This regimen resulted in a significant reduction of grade 3 and 4 diarrhea cases to 4.4% within the AC group, as opposed to the control group. Additionally, the intervention group experienced fewer chemotherapy discontinuations (6.6% vs. 52.3%). Consequently, AC has been shown to improve CPT-11 compliance and diminish both the frequency and severity of CID. Nonetheless, AC's efficacy is not absolute, it can also adsorb other concurrently administered oral medications, and the requirement for thrice-daily administration can be burdensome [103].

AST-120 (clemizine) is an alternative oral carbonaceous adsorbent that has been investigated for its potential to prevent delayed diarrhea. Clemizine has demonstrated a significant adsorption capacity for CPT-11 both in vitro and within the rat gastrointestinal tract. In rats administered with clemizine, there was approximately a 50% reduction in the frequency of diarrhea compared to untreated controls, with a corresponding decrease in severity [104]. In a clinical trial, the administration of two grams of clemizine, given in three divided doses daily, was found to decrease the occurrence of CID during and post-CPT-11 treatment, with minimal impact on the pharmacokinetics of CPT-11 and its metabolites [105].

3.1.7. Cytokine and Growth Factors Induction and Inhibition

Thalidomide, a synthetic derivative of glutamic acid, has been recognized for its potential to mitigate the intestinal pathological changes induced by CPT-11. It exerts its effects by inhibiting the production of inflammatory cytokines within the intestine, reducing apoptosis of intestinal epithelial cells, and modulating immune and angiogenic responses [106]. Preliminary clinical studies have affirmed thalidomide's utility beyond its antitumor properties, highlighting its anti-angiogenic and immunomodulatory capabilities, and its capacity to alleviate CID [107]. While one study has reported a significant reduction in the metabolic conversion of CPT-11 to SN-38 with thalidomide treatment, others have not observed a substantial impact on the pharmacokinetics of CPT-11 [108,109].

In an additional study, Velafermin, a fibroblast growth factor-20, was administered at a dosage of 16 mg/kg prior to CPT-11 treatment. This approach was found to ameliorate gastrointestinal mucositis, diarrhea, and mortality associated with CPT-11 in tumor-bearing DA rats. Although rats treated with verapamil experienced severe or moderate diarrhea, the onset was delayed, and the condition was less severe and shorter in duration. Notably, verapamil did not influence tumor proliferation. Various doses of verapamil were assessed, with some showing diminished effectiveness in reducing the severity and mortality of CID. Interestingly, certain dosages were associated with increased diarrhea and mortality [110].

Interleukin-15 (IL-15) is a cytokine that has demonstrated a pronounced protective effect against CPT-11-induced intestinal toxicity and has the potential to moderately boost the antitumor efficacy in advanced colorectal cancer models [111]. JBT3002, a novel synthetic bacterial lipopeptide, has been shown to stimulate the production of IL-15. This compound can protect the integrity of the intestinal epithelium, prevent damage to the intestinal epithelium and mucosal lamina propria, and is tolerable at higher intravenous doses [112].

3.1.8. Other Chemical Drug Treatment Options

In the realm of preclinical research, the supplementation of diets with low concentrations of fish oil—specifically at levels of 3% or 6%—has been observed to enhance the regression of MCF7 human breast cancer xenografts in nude mice, both prior to and during the administration of CPT-11. This augmentation of CPT-11's therapeutic efficacy is accompanied by a reduction in the pathological damage to the intestinal tissue [113].

L-Glutamine is categorized as a conditionally essential amino acid, particularly in conditions of stress where the body's synthesis may not meet the increased demands [114]. It is an indispensable nutrient that supports the growth, differentiation, and the maintenance of integrity and barrier function of the intestinal mucosal epithelium. Studies have demonstrated its role in facilitating electrolyte absorption in animals subjected to experimentally induced diarrhea [94].

Phloroglucinol, a myophilic smooth muscle antispasmodic agent, has been identified in a study to ameliorate diarrhea and normalize electrolyte levels [115]. This drug does not exert anticholinergic effects and is less likely to cause adverse cardiovascular effects such as tachycardia, hypotension, or arrhythmia. Given these attributes, phloroglucinol is deemed more appropriate for use in elderly patients with pre-existing cardiovascular or cerebrovascular conditions [116]. Figure 3 illustrates the major chemotherapeutic agents for the treatment of CPT-11-induced diarrhea. (A summary of the mechanisms of action and limitations of major chemotherapeutic agents for the treatment of CPT-11-induced diarrhea can be found in Supplementary Material Table S1).

3.2. *Traditional Chinese Medicine Treatment of Delayed Diarrhea*

Despite the availability of numerous chemical drugs for the prevention and treatment of CID, there remains no universally recognized and consistently effective standard treatment. The efficacy of individual drugs in isolation is challenging to ascertain. However, modern pharmacological studies have established that traditional Chinese medicine (TCM) possesses a unique multi-target and multi-pathway approach. TCM is tailored to individual patient profiles, which often results in favorable outcomes in the treatment of diarrhea caused by CPT-11. Figure 4 outlines the major traditional Chinese medicine treatment of delayed diarrhea.

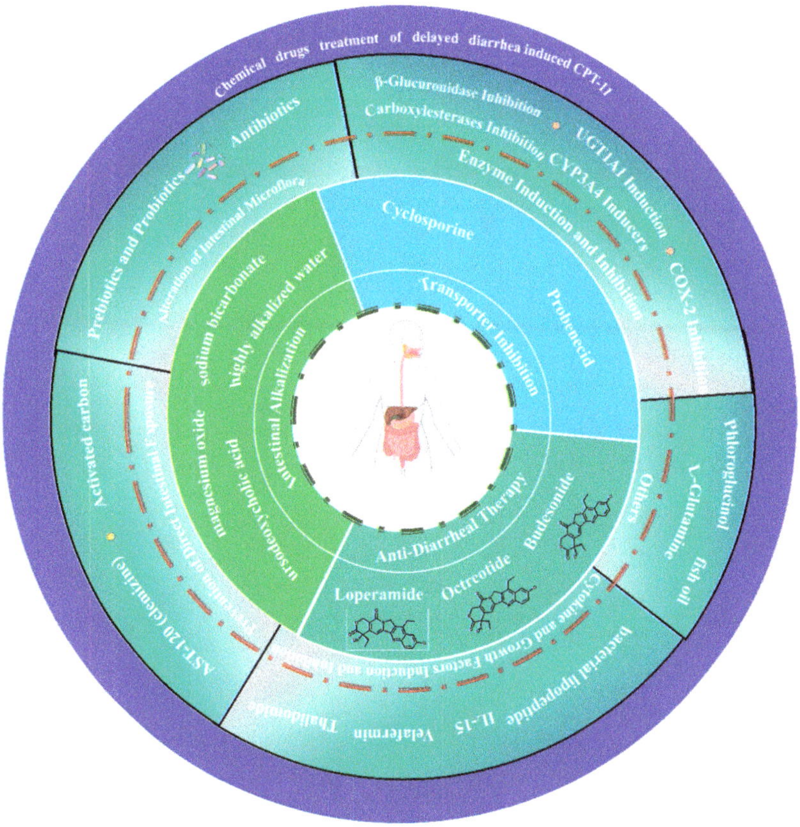

Figure 3. The major chemotherapeutic agents for the treatment of CPT-11-induced diarrhea.

3.2.1. Single Component and Active Components of Traditional Chinese Medicine

Hypericum perforatum L. (St. John's Wort) Extract

Hypericum perforatum L., commonly referred to as St. John's wort, is a member of the Guttiferae family and has been traditionally used for its diarrhea-treating properties [117]. Experimental rat studies have substantiated the efficacy of St. John's wort in this context. In these experiments, a control group was treated solely with CPT-11, whereas a treatment group additionally received oral St. John's wort. The results indicated a significant reduction in the expression of the inflammatory marker TNF-α mRNA in rats administered with St. John's wort. Furthermore, it was observed to partially inhibit apoptosis in intestinal epithelial cells, thereby mitigating the intestinal damage caused by CPT-11 [118].

Berberine

Berberine, a principal active constituent of the traditional Chinese medicinal herb *Coptis chinensis* (huanglian), has been ascribed with a spectrum of pharmacological activities, including anti-inflammatory, antioxidant, anticancer properties, lipid metabolism regulation, and energy balance maintenance [119]. In an animal model, berberine was found to mitigate the mucosal structural damage, ulceration, and neutrophil infiltration induced by CPT-11. It also enhances mucosal barrier function by increasing the number of goblet cells, preserving transepithelial electrical resistance (TEER), reducing permeability, and upregulating tight junction proteins. Berberine was noted to reduce fecal SN-38 levels, which may be linked to a decrease in the activity of β-glucuronidase and the corresponding bacterial population. Notably, berberine preserves the anticancer efficacy of CPT-11 while simultaneously reducing its intestinal toxicity in xenograft tumor models [120].

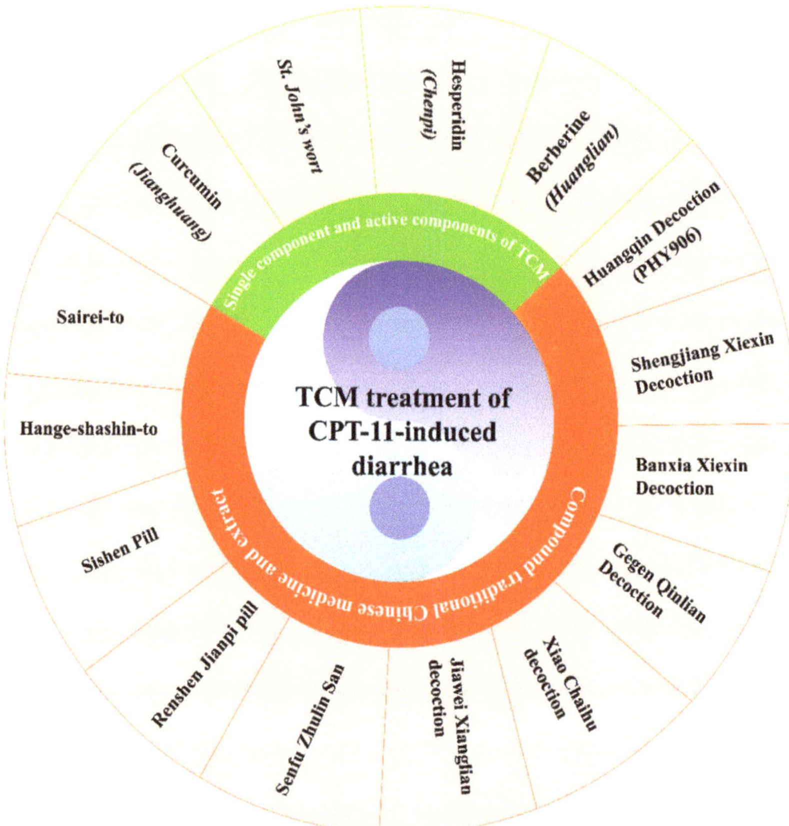

Figure 4. The major traditional Chinese medicine treatment of delayed diarrhea.

Curcumin

Curcumin, a polyphenolic compound derived from the rhizome of *Curcuma longa* L. (turmeric), is renowned for its diverse pharmacological effects, including anti-inflammatory, antioxidant, and antitumor activities. In the establishment of a mouse model for delayed diarrhea induced by CPT-11, a group treated with curcumin prophylactically showed protective effects against the symptoms and pathophysiology of the condition [121]. Further animal studies demonstrated that curcumin could effectively alleviate CID symptoms and intestinal mucosal structural aberrations in nude mice. Curcumin was found to upregulate the expression of P4HB and PRDX4 in the small intestine, enhance cell morphology, inhibit apoptosis, maintain mitochondrial membrane potential, and reduce the rise of reactive oxygen species (ROS) levels provoked by CPT-11 (20 μg/mL) in vitro. Additionally, curcumin increases the expression of molecular chaperone proteins such as GRP78, P4HB, and PrDX4, and suppresses the expression of apoptosis-related proteins like CHOP and cleaved caspase-3, thereby interrupting the NF-κB signaling pathway and safeguarding cells against CPT-11-induced apoptosis [122].

Hesperidins

Hesperidin, the predominant flavonoid in *Citrus reticulata Blanco* (chenpi), is recognized for its vascular protective properties, enhanced lymphatic circulation, and demonstrated anti-inflammatory and antiviral activities. It is postulated that CPT-11's dose-restricted diarrhea is associated with the exposure of the active metabolite SN-38 to the intestinal tract. Hesperidin has been shown to modulate the biliary excretion transporters of CPT-11 and its metabolites, thereby influencing the pharmacokinetics of both CPT-11 and SN-38 [123,124].

3.2.2. Compound Traditional Chinese Medicine and Extract

Huangqin Decoction

Huangqin Decoction (HQD) is a complex traditional Chinese medicinal formula consisting of *Scutellaria baicalensis* Georgi, *Glycyrrhiza uralensis* Fisch, *Paeonia lactiflora* Pall, and *Ziziphus jujube* Mill in a dry weight ratio of 3:2:2:2. With a history spanning over 1800 years, HQD has been widely utilized in China for the treatment of gastrointestinal disorders characterized by diarrhea, nausea, abdominal cramps, and vomiting [125]. In a study involving ICR rats with CID, HQD was administered one day prior to the experiment and continued for eight days. The findings indicated that HQD-treated rats exhibited a reduced rate of body weight loss and intestinal mucosal injury. There was a significant increase in the expression of nitric oxide (NO) in the colon, a decrease in proliferating cell nuclear antigen (PCNA) expression, and an elevated count of blood neutrophils [126]. In another animal experiment, the concurrent administration of HQD (10 g/kg, twice daily) at the onset of CPT-11 significantly mitigated delayed diarrhea in rats, although it was ineffective in preventing acute diarrhea during the initial two days. Researchers employed GC/MS and LC/MS metabolomics to analyze serum metabolite changes in male SD rats before and after HQD treatment, suggesting that HQD could mediate metabolic alterations by normalizing amino acid, lipid, and bile acid metabolic pathways [127]. PHY906, a derivative of HQD, has demonstrated its utility as a modulator of chemotherapeutic agents, particularly in alleviating cancer therapy-induced nausea, vomiting, and diarrhea [128,129]. In a mouse model of allogeneic colon transplantation, oral PHY906 (dosed at 50, 500, or 1000 mg/kg, twice daily) was shown to attenuate CPT-11-induced gastrointestinal toxicity through multiple mechanisms, including the inhibition of NF-κB, COX-2, and iNOS inflammatory pathways, and the promotion of intestinal progenitor cell regeneration via the upregulation of Wnt signaling components, with a particular emphasis on Wnt3a [130]. A phase I, multicenter, double-blind, randomized, placebo-controlled crossover study involving 17 patients with advanced colorectal cancer treated with PHY906 in combination with CPT-11 and 5-FU/IFL regimens showed a reduction in the overall incidence of grade 3 or 4 diarrhea and decreased reliance on antidiarrheal medications such as loperamide and lomotil. Notably, PHY906 did not alter the pharmacokinetics of CPT-11 or its metabolite SN-38. However, given the small patient sample size in this trial, further large-scale randomized trials are necessary to fully assess the benefits of PHY906 in the context of CID [131].

Shengjiang Xiexin Decoction

Shengjiang Xiexin Decoction (SXD), a classic compound from the traditional text Shang Han Lun, is composed of eight Chinese herbal medicines and is extensively used in contemporary clinical practice for the treatment of gastroenteritis, ulcerative colitis, and diarrhea [132]. In a rat model of diarrhea, SXD, administered at dosages of 5, 10, or 15 g/kg per day, was found to promote intestinal cell proliferation while inhibiting intestinal cell apoptosis and β-glucuronidase activity, thereby preventing delayed diarrhea induced by CPT-11 in a dose-dependent manner [133]. In an experiment utilizing a CT26 colon cancer mouse model, mice in the experimental group were given SXD (10 g/kg, twice daily) three days prior to CPT-11 administration for a total of eight days. In comparison to the model and control groups, which received equivalent volumes of normal saline, CPT-11 injection led to significant diarrhea in the model group. The SXD group exhibited significantly lower diarrhea scores, less severe intestinal mucosal damage under light microscopy, decreased levels of TNF-α, and increased levels of IL-10, effectively alleviating neutropenia [134]. In a randomized controlled trial involving 115 patients treated with CPT-11 combined with 5-fluorouracil and calcium 1-folinate, SXD (100 mL, twice daily) was shown to significantly reduce the incidence of delayed diarrhea in patients with UGT1A1*28 or UGT1A1*6 mutations, without compromising the clinical response to chemotherapy [135].

Banxia Xiexin Decoction

Banxia Xiexin Decoction (BXD), a formulation comprising seven medicinal ingredients, is a potent compound prescription in traditional Chinese medicine used to address conditions such as gastroenteritis, ulcerative colitis, vomiting, and diarrhea [136]. In experiments involving mice with small-cell lung cancer (SCLC) induced to experience diarrhea by CPT-11, BXD was observed to ameliorate the condition. This improvement is hypothesized to be due to the inhibition of COX-2 expression in the colonic tissue and a reduction in the local concentration of SN-38 [137]. A clinical study involving 27 patients with recurrent SCLC undergoing CPT-11 chemotherapy reported that BXD, administered prior to the second chemotherapy cycle, effectively prevented and treated delayed diarrhea caused by CPT-11. Out of six patients who developed delayed diarrhea, four were relieved after BXD treatment. However, the small sample size in this study necessitates further evaluation in larger, high-quality, randomized controlled trials to confirm the efficacy of BXD in managing CPT-11-induced delayed diarrhea [138].

Gegen Qinlian Decoction

Gegen Qinlian Decoction (GQD) is a traditional Chinese medicinal formula that *includes Pueraria* lobata (Gegen), *Scutellaria baicalensis* (Huangqin), *Coptis chinensis* (Huanglian), and *Glycyrrhiza uralensis* (Gancao). Originating from the Shan Han Lun, a text dating back to the Han Dynasty (202–220 BC), GQD is widely utilized for gastrointestinal disorders, particularly diarrhea [139]. In a mouse model of CID, GQD extract administration for five days led to a significant decrease in the levels of inflammatory cytokines such as IL-1β, COX-2, ICAM-1, and tumor necrosis factor-α within the intestinal tissue. Additionally, GQD demonstrated antioxidant properties, activated the Keap1/Nrf2 pathway, and enhanced the intestinal barrier function by upregulating the expression of tight junction proteins like ZO-1, HO-1, and occludin. The GQD extract also exhibited a potent inhibitory effect on hCE2 in vitro, with an IC50 value of 0.187 mg/mL, suggesting its potential in mitigating hCE2-mediated severe diarrhea [140]. Moreover, the extract of GQD has been shown to synergistically inhibit the growth of colon cancer when used in conjunction with CPT-11. In vitro studies have also highlighted GQD's significant inhibitory effect on CES2 [141].

Xiao Chaihu Decoction

Xiao Chaihu Decoction (XCD), a blend of seven Chinese medicinal herbs, is the principal prescription for Shaoyang disease as described in the Shang Han Lun of the Han Dynasty [142]. Extensive clinical and experimental studies have validated XCD's efficacy in treating liver and digestive system diseases. In an animal study, the experimental group received XCD at a dosage of 1500 mg/kg (based on crude drug) once daily for 17 days, while other groups, except for the normal control, were injected with CPT-11 to induce delayed diarrhea from the 4th to the 10th day. The results indicated that XCD significantly reduced the rate of hematochezia and improved intestinal mucosal injury in the treated mice [143].

Other TCM Preparations for Treatment

Sishen Pill, a classic formula in traditional Chinese medicine, is recognized for its effectiveness in treating diarrhea and has shown promise in managing irritable bowel syndrome and ulcerative colitis. In an animal model using ICR mice, Sishen Pill demonstrated preventive and therapeutic effects on CPT-11-induced delayed diarrhea, potentially through reducing intestinal β-glucuronidase activity and the levels of IL-1β and TNF-α [144].

Jiawei Xianglian Decoction (JWXLD), a combination of six medicinal ingredients, is a clinically used drug in China effective against diarrhea. Studies have found that JWXLD at dosages of 0.12, 0.23, and 0.46 g significantly mitigated the severity of CID and altered the levels of Lactobacillus and Bifidobacterium in mice, effects that were reversible with JWXLD. Furthermore, JWXLD was shown to reduce β-glucuronidase activity. Histopathological assessments revealed that JWXLD could significantly lessen the severity of intestinal mucosal injury caused by CPT-11 in rats [145]. Senfu Zhulin San [146] and Renshen Jianpi Pill [147] are also utilized in Chinese clinical practice to treat CID, likely through mechanisms involving the regulation of intestinal flora and the inhibition of inflammation.

Hange-shashin-to, a formula consisting of seven Chinese herbal medicines, is used in Japan for treating diarrhea and acute gastroenteritis. In animal studies, Hange-shashin-to (1 g/kg, twice daily) showed protective effects against CPT-11-induced intestinal toxicity by inhibiting β-glucuronidase activity, leading to reduced weight loss, improved anorexia, and delayed onset of diarrhea symptoms [148]. Sairei-to, a preparation of 12 traditional Chinese herbs, is used in Japan for severe diarrhea and various inflammatory conditions, including rheumatoid arthritis, systemic lupus erythematosus, and nephrotic syndrome [149]. A preclinical study in male Wistar rats indicated that Sairei-to could alleviate CPT-11-induced delayed diarrhea, possibly through the inhibition of bacterial β-glucuronidase [150]. (A summary of the proved effects of single Chinese herbs and traditional Chinese medicine compound prescriptions on CPT-11-induced diarrhea can be found in Supplementary Material Tables S2 and S3).

3.3. Structural/Chemical Modification and Novel Drug Delivery Methods

To address the challenges of low bioavailability and enterotoxicity associated with the anticancer drug CPT-11 and its active metabolite SN-38, various novel dosage forms and drug delivery systems have been explored. These innovations include the development of liposomal formulations, polymer conjugates, nanoparticles, dendrimers, and the utilization of peptides and carbohydrates. Figure 5 summarizes the structural modifications and novel drug delivery methods explored to enhance the efficacy and reduce the toxicity of irinotecan and its metabolites.

3.3.1. Novel Drug Delivery of CPT-11

Liposomal Formulations of CPT-11

Irinotecan liposomes, such as Nal-IRI (ONIVYDE), were approved in 2015 [151] and have since been employed as a second-line treatment for metastatic pancreatic cancer [152,153]. New liposomal preparations of CPT-11, like CPX-1, have demonstrated antitumor activity in phase I studies by combining CPT-11 with fluorouracil, improving outcomes in patients with advanced solid tumors [154]. Additionally, highly stable nanoparticles/liposomes integrated with convection-enhanced delivery (CED) have been recognized for their potential to prolong drug retention in tissues and reduce toxicity [155]. These liposomal formulations are designed to enhance target specificity, thereby minimizing systemic toxicity. Despite the potential shown in vitro, many of these carriers have yet to fulfill their promise in vivo.

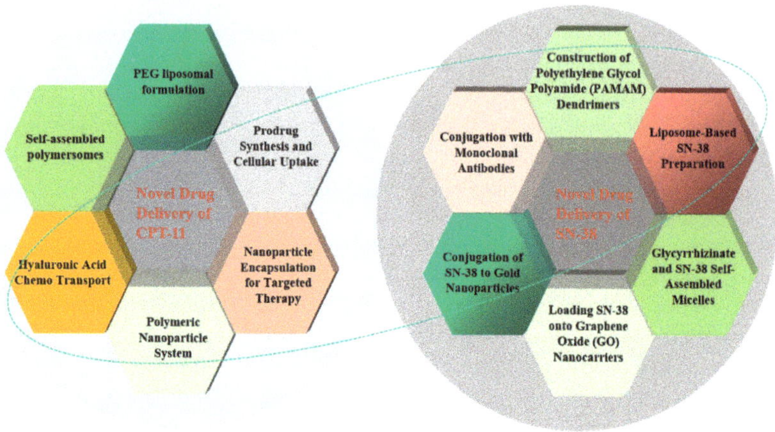

Figure 5. Structural/Chemical modification and novel drug delivery methods.

Nanoparticles of CPT-11

Wang et al. have developed novel nanoparticles using hyaluronic acid, poly (lactic acid-glycolic acid), chitosan, and Pluronic F-127 as carriers for CPT-11 and doxorubicin. Hyaluronic acid chemical transport (HyACT®) has emerged as a carrier system for CPT-11, enhancing the reactivity with CD44-positive tumor cells and improving progression-free survival rates in metastatic colorectal cancer when used in combination therapies [156,157]. In a randomized phase II trial, hyaluronic acid-modified CPT-11 was found to enhance progression-free survival in patients with metastatic colorectal cancer resistant to 5-fluorouracil [158].

Polymer Conjugates of CPT-11

The covalent attachment of polyethylene glycol (PEG) molecules has been utilized to enhance the systemic circulation and reduce clearance of CPT-11. PEGylated liposomal CPT-11 preparations, such as MM-398, have shown improved cytotoxicity in mouse brain metastasis models compared to CPT-11 monotherapy [159,160]. In a phase I clinical and pharmacokinetic study, IHL-305, a novel PEGylated liposome formulation of CPT-11, demonstrated a safe repeat dosing regimen over 4 or 2 weeks [161]. Zhang et al. have explored the combination of CPT-11 with fatty acids to increase lipophilicity and facilitate self-assembly in an aqueous environment, protecting CPT-11 from carboxylesterase-mediated hydrolysis and enhancing intracellular accumulation and cytotoxicity [162]. Zashikhina et al. have developed self-assembled poly (l-lysine)-b-poly (l-leucine) (plys-b-pleu) polymers that exhibit no cytotoxicity in tested cell lines and maintain in vitro antitumor activity similar to that of free CPT-11 [163].

Strategies to enhance CPT-11 activity, such as adenovirus-mediated β-glucuronidase expression in tumors [164] and gene-directed enzyme/prodrug therapy (CES/CPT-11) [165], may indirectly improve antitumor efficacy and ameliorate CID. The hypothesis is that targeted administration and the use of liposomes could reduce the concentration of CPT-11 and SN-38 in bile, thereby minimizing intestinal damage.

3.3.2. Novel Drug Delivery of SN-38

Liposomal Formulations of SN-38

Machmudi et al. have constructed polyethylene glycol polyamide (PAMAM) dendrimers conjugated with SN-38 and targeting moieties BR2 and CyLoP1. These dendrimers have shown enhanced cytotoxicity and cellular uptake in mouse colon cancer (CT26) cell lines compared to native SN-38, with in vivo studies demonstrating improved drug accumulation at the tumor site and increased antitumor efficacy [143]. Zhang et al. have developed a liposome-based SN-38 preparation (LE-SN-38) that has exhibited superior efficacy in enhancing cytotoxicity against various tumor cell lines and in treating xenogeneic mice models [166]. Sun et al. have created amorphous solid disodium glycyrrhizinate and SN-38 self-assembled micelles (Na2GA/SN-38-BM) with favorable pharmacokinetics and distribution properties, demonstrating enhanced cytotoxicity against tumor cells and significant tumor growth inhibition [167].

Nanoparticles of SN-38

Karki et al. have investigated the loading of SN-38 onto graphene oxide (GOS) modified with polyvinylpyrrolidone (PVP) or β-cyclodextrin (β-CD). Their studies revealed that SN-38 loaded onto GO-PVP nanocarriers exhibited higher cytotoxic activity against human breast cancer cells (MCF-7) than GO-β-CD nanocarriers, suggesting the potential of GO-PVP as an effective drug delivery system [168]. Naumann et al. have conjugated SN-38 to gold nanoparticles using oligonucleotides specific to Ewing's sarcoma cells, allowing for the targeted release of SN-38 and its subsequent inhibition of topoisomerase, with effective and selective drug release observed in both in vitro and in vivo conditions [169]. Furthermore, coupling was performed with monoclonal antibodies (Labetuzumab-SN-38 immunoconjugates [170]), along with a variety of other preparations and methods [171] (A summary of the mechanisms of action and effects on diarrhea relief of new drug delivery systems for CPT-11 and SN-38 can be found in Supplementary Material Tables S4 and S5).

3.3.3. Impact on Diarrhea Risk

Despite the promising strategies offered by advanced drug delivery systems for enhancing the therapeutic efficacy of CPT-11 and SN-38 and reducing systemic toxicity, their impact on diarrhea risk remains a significant consideration. Systems such as liposomal formulations and nanoparticles, which reduce systemic exposure to the active metabolite SN-38, may potentially lower the incidence of CID. However, clinical study results have been inconsistent, with some trials reporting reduced severe diarrhea and others showing no significant difference compared to conventional formulations [151]. Moreover, the long-term clinical outcomes and specific impact on diarrhea incidence require further elucidation.

In summary, while novel drug delivery systems show promise in enhancing the therapeutic efficacy of CPT-11 and SN-38 and reducing systemic toxicity, their impact on diarrhea risk remains a critical factor. Further research is needed to fully understand the long-term effects of these delivery systems on CID risk and to optimize their clinical application [154,158,172].

4. Discussion

4.1. Clinical and Preclinical Strategies for Blocking or Treating

The management of irinotecan-induced diarrhea (CID) encompasses a diverse array of therapeutic approaches, each with varying levels of clinical validation and developmental maturity. Clinically validated treatments, such as loperamide and octreotide, have demonstrated robust efficacy in managing acute symptoms and are widely adopted in clinical practice. These treatments have undergone rigorous clinical trials, ensuring their safety and effectiveness are well-established [11,12]. For instance, loperamide, a synthetic opioid, effectively reduces intestinal motility and promotes fluid reabsorption, thereby alleviating acute diarrhea episodes. Similarly, octreotide, a somatostatin analog, mitigates gastrointestinal secretion and motility, offering significant relief in cases of severe diarrhea [13].

In contrast, several innovative strategies remain in the preclinical stage, showcasing promising potential but requiring further validation. Enzyme inhibitors targeting β-glucuronidase and UGT1A1 inducers have demonstrated substantial efficacy in reducing the severity and frequency of CID in preclinical models [31,128]. These approaches aim to modulate the metabolism of irinotecan's active metabolite, SN-38, thereby mitigating its toxic effects on the intestinal mucosa. Additionally, advanced drug delivery systems, such as liposomal formulations and nanoparticles, are being explored to enhance the targeted delivery of irinotecan while minimizing systemic toxicity [47,156]. These systems hold the potential to reduce the incidence of CID by limiting the exposure of the gastrointestinal tract to SN-38.

Traditional Chinese medicine (TCM) also presents a unique avenue for CID management, with formulations like Huangqin Decoction and Shengjiang Xiexin Decoction demonstrating efficacy in both preclinical and clinical settings [129,173]. These herbal preparations leverage a multi-target approach, modulating inflammation, enhancing intestinal barrier function, and reducing mucosal injury. However, their long-term safety and efficacy profiles are still under evaluation, and standardization of these formulations remains an ongoing challenge [50].

Distinguishing between clinically validated and preclinical strategies is essential for guiding clinical practice and future research. While clinically validated treatments provide immediate relief and are supported by substantial evidence, preclinical strategies offer innovative solutions that could significantly improve patient outcomes once validated. Future research should focus on bridging the gap between these categories, ensuring that emerging therapies are rigorously tested and standardized before clinical application.

4.2. Long-Term Efficacy and Safety Considerations

The long-term efficacy and safety of therapeutic interventions for irinotecan-induced diarrhea (CID) are critical factors in optimizing patient outcomes and ensuring sustained therapeutic benefits. While many experimental treatments show promise in short-term studies, the availability of long-term data remains limited for several emerging strategies. For instance, conventional treatments such as loperamide and octreotide have demonstrated consistent efficacy in managing acute symptoms but may face challenges in maintaining long-term effectiveness due to potential tolerance development or side effects [11,12]. In contrast, novel approaches like enzyme inhibitors and advanced drug delivery systems, although highly effective in preclinical models, are still in the early stages of clinical evaluation, and their long-term impacts on patient health are yet to be fully elucidated [31,47].

Traditional Chinese medicine (TCM) formulations, which have garnered attention for their holistic and multi-target effects, also present unique considerations for long-term use. While TCM has shown potential in reducing the severity and frequency of CID in both preclinical and clinical settings [128,129], the long-term safety profile remains an area of active investigation. The complexity of herbal compositions and the potential for interactions with other medications necessitate careful monitoring and further research to establish their long-term efficacy and safety [173].

Moreover, the potential for delayed side effects cannot be overlooked. For example, prolonged use of certain medications may lead to gastrointestinal dysbiosis or other systemic complications, which could exacerbate the overall burden of treatment [50]. Therefore, long-term monitoring and follow-up studies are essential to identify and mitigate any adverse effects that may emerge over time.

In summary, while significant progress has been made in developing treatments for CID, the long-term efficacy and safety of these interventions remain areas of ongoing research. Future studies should prioritize the collection of long-term data to inform clinical practice and guide the development of sustainable, patient-centered treatment strategies. This approach will be crucial in addressing the multifaceted challenges of managing CID and improving the quality of life for patients undergoing irinotecan therapy.

4.3. Comparative Analysis of Therapeutic Approaches for Irinotecan-Induced Diarrhea

The management of irinotecan-induced diarrhea (CID) necessitates a thorough evaluation of the diverse therapeutic strategies available, each with distinct advantages and disadvantages. Conventional treatments, such as loperamide and octreotide, remain pivotal for their rapid alleviation of acute symptoms, supported by extensive clinical validation. However, their efficacy may wane over time due to potential drug tolerance and side effects, and they predominantly address symptoms rather than underlying mechanisms.

Traditional Chinese medicine (TCM) provides a holistic, multi-target approach, modulating inflammation and enhancing intestinal barrier function. While TCM formulations like Huangqin Decoction and Shengjiang Xiexin Decoction have shown efficacy in both preclinical and clinical settings, challenges remain in standardization and long-term safety assessment.

Novel drug delivery systems, including liposomal formulations and nanoparticles, offer enhanced targeted delivery and reduced systemic toxicity, potentially lowering the incidence and severity of CID. Despite promising preclinical results, clinical outcomes have been inconsistent, highlighting the need for further validation. (The advantages and disadvantages of different treatment methods can be found in Table 1).

Table 1. Comparative analysis of the major therapeutic approaches.

Approach	Strengths	Limitations
Conventional Treatments (Loperamide, Octreotide)	- Well-established in clinical practice - Immediate symptom relief	- High failure rate in severe cases - Potential side effects (e.g., constipation, hyperglycemia)
Traditional Chinese Medicine (TCM)	- Holistic approach - Potential for long - Term benefits - Modulation of gut microbiota	- Standardization challenges - Limited long-term clinical data - Potential drug interactions
Novel Drug Delivery Systems	- Enhanced efficacy - Reduced toxicity - Targeted delivery	- Limited clinical validation - Manufacturing complexities - High costs

Future research should focus on integrating these approaches to develop personalized treatment strategies that optimize patient outcomes.

5. Conclusions

It is undeniable that nearly two decades after its introduction, CPT-11 remains a cornerstone in the arsenal of cytotoxic anticancer drugs, particularly for the treatment of advanced colon cancer and certain solid tumors [174]. The metabolic pathway of CPT-11 is intricate, involving numerous factors, leading to considerable inter-individual variability in pharmacokinetics and posing challenges for personalized therapeutic regimens [173]. Despite this, recent advancements have been made in tailoring treatment combinations to better suit diverse patient populations. The oral formulation of CPT-11 is also gaining traction. Yet, the drug is not without serious side effects, most notably neutropenia and diarrhea. CID is often severe, necessitating dose reductions, treatment omissions, and hospitalizations. These complications diminish the therapeutic efficacy of CPT-11, escalate healthcare costs, and degrade the quality of life for patients. Our understanding of the pathophysiology of CID has advanced, highlighting the interplay of inflammation and dysbiosis. Although numerous products are in development to treat or prevent CID, delayed diarrhea continues to be a significant challenge. Grade 3–4 diarrhea is frequently observed even with premedication, underscoring the limitations of current treatments. The complex metabolism of CPT-11, involving various enzymes, metabolites, transporters, and hepatointestinal circulation, coupled with patient-specific genetic profiles and clinical risk factors, complicates the issue. The journey toward personalized CPT-11 chemotherapy is ongoing, yet the drug's role in treating advanced solid tumors is indispensable.

In the contemporary era of targeted cancer therapy and the surge of immunobiological treatments, it is somewhat paradoxical that cytotoxic agents like CPT-11 continue to be indispensable in oncology. The imperative for clinicians is to optimize these treatments, leveraging medical informatics and cutting-edge science for the ultimate benefit of patients. As previously discussed, preclinical and clinical studies suggest that traditional Chinese medicine (TCM) formulations, herbal prescriptions, and their active constituents may hold potential in preventing or mitigating the symptoms of chronic diarrhea associated with CPT-11-based chemotherapy. The mechanisms by which TCM exerts its effects are multifaceted, primarily targeting the metabolic pathways of CPT-11 through interactions with specific enzymes, metabolites, and transporters. While TCM offers certain advantages in managing CID, the complexity and heterogeneity of its components necessitate further research to elucidate the active ingredients and underlying mechanisms [173]. The integration of novel drug delivery systems with TCM-derived active ingredients may represent a promising frontier in the treatment of CID.

Regardless of the therapeutic approach, the overarching goal is to bolster the tumor's response to CPT-11, effectively mitigate CID, elevate patient quality of life, and curtail medical expenditures. There is an imperative for sustained research into the etiology of CID and the refinement of targeted treatment strategies. Additionally, vigilant polypharmacy management and the judicious application of CPT-11 in specific clinical contexts are pivotal in reducing the prevalence of CID.

In summary, while CPT-11 remains a cornerstone in cancer treatment, managing its associated diarrhea is a multifaceted challenge. Both conventional and traditional approaches have shown promise, but each comes with its own set of limitations. Future research should focus on developing personalized treatment strategies that integrate the best of both worlds, ensuring optimal patient outcomes.

Supplementary Materials: The following supporting information can be downloaded at: https://www.mdpi.com/article/10.3390/ph18030359/s1, Table S1. Major chemotherapeutic agents for the treatment of CPT-11-induced diarrhea (Guide recommendation). Table S2. Summary of the proved effects of Single TCM and active ingredients in CPT-11-induced diarrhea. Table S3. Summary of the proved effects of in TCM compound prescription CPT-11-induced diarrhea. Table S4. New irinotecan formulations with effects of modifications and references. Table S5. New SN-38 formulations with effects of modifications and references.

Author Contributions: X.Y. and J.C. made significant contributions to the literature collection and organization. The primary manuscript writing was undertaken by X.Y.; Y.W. (Yitao Wang) revised the manuscript. J.Z. and Y.W. (Yihan Wu) made substantial contributions to the structure and argumentation of the review. All authors have read and approved the final manuscript.

Funding: This work was supported by Innovation Team and Talents Cultivation Program of National Administration of Traditional Chinese Medicine (No: ZYYCXTD-D-202209), National Natural Science Foundation of China (NSFC) (No: 82405145), China Postdoctoral Science Foundation (No: GZC20230335, No: 2024m750289).

Conflicts of Interest: The author declares no conflicts of interest associated with this publication and there has been no significant financial support for this work that could have influenced its outcome.

References

1. Wagner, L.M. Fifteen years of irinotecan therapy for pediatric sarcoma: Where to next? *Clin. Sarcoma Res.* **2015**, *5*, 20. [CrossRef]
2. Makimoto, A.; Mugishima, H.; Taga, T.; Ishida, Y.; Nagatoshi, Y.; Ida, K.; Kumagai, M.; Kimura, T.; Ohashi, Y.; Kaneko, M. Registration-directed phase 1/2 trial of irinotecan for pediatric solid tumors. *Pediatr. Int.* **2019**, *61*, 453–458. [CrossRef] [PubMed]
3. Cheng, C.; Lau, J.E.; Earl, M.A. Use of atropine-diphenoxylate compared with hyoscyamine to decrease rates of irinotecan-related cholinergic syndrome. *J. Community Support. Oncol.* **2015**, *13*, 3–7. [CrossRef]
4. Shimada, Y.; Yoshino, M.; Wakui, A.; Nakao, I.; Futatsuki, K.; Sakata, Y.; Kambe, M.; Taguchi, T.; Ogawa, N. Phase II Study of CPT-11, a New Camptothecin Derivative, in Metastatic Colorectal Cancer. *J. Clin. Oncol.* **1993**, *11*, 909–913. [CrossRef]
5. Armand, J. CPT-11: Clinical experience in phase I studies. *Semin. Oncol.* **1996**, *23*, 27–33.
6. Slatter, J.; Su, P.; Sams, J.; Schaaf, L.; Wienkers, L. Bioactivation of the Anticancer Agent CPT-11 to SN-38 by Human Hepatic Microsomal Carboxylesterases and the in Vitro Assessment of Potential Drug Interactions. *Drug Metab. Dispos.* **1997**, *25*, 1157–1164.
7. Morton, C.; Wadkins, R.; Danks, M.; Potter, P. The anticancer prodrug CPT-11 is a potent inhibitor of acetylcholinesterase but is rapidly catalyzed to SN-38 by butyrylcholinesterase. *Cancer Res.* **1999**, *59*, 1458–1463. [PubMed]
8. Rivory, L.; Robert, J. Identification and kinetics of a 13-glucuronide metabolite of SN-38 in human plasma after administration of the camptothecin derivative irinotecan. *Cancer Chemother. Pharmacol.* **1995**, *36*, 176–179. [CrossRef] [PubMed]
9. Haaz, M.C.; Rivory, L.; Jantet, S.; Ratanasavanh, D.; Robert, J. Glucuronidation of SN-38, the Active Metabolite of Irinotecan, by Human Hepatic Microsomes. *Pharmacol. Toxicol.* **2009**, *80*, 91–96. [CrossRef]
10. Younis, I.R.; Malone, S.; Friedman, H.S.; Schaaf, L.J.; Petros, W.P. Enterohepatic recirculation model of irinotecan (CPT-11) and metabolite pharmacokinetics in patients with glioma. *Cancer Chemother. Pharmacol.* **2008**, *63*, 517–524. [CrossRef]
11. Abigerges, D.; Armand, J.; Chabot, G.; Costa, L.D.; Fadel, E.; Cote, C.; Herait, P.; Gandia, D. Irinotecan (CPT-11) high-dose escalation using intensive high-dose loperamide to control diarrhea. *J. Natl. Cancer Inst.* **1993**, *86*, 446–449. [CrossRef] [PubMed]
12. Bleiberg, H.; Cvitkovic, E. Characterisation and Clinical Management of CPT-11 (Irinotecan)-induced Adverse Events: The European Perspective. *Eur. J. Cancer* **1996**, *32A* (Suppl. S3), S18–S23. [CrossRef]
13. Goumas, P.; Naxakis, S.; Christopoulou, A.; Chrysanthopoulos, C.; Nikolopoulou, V.; Kalofonos, H.P. Octreotide Acetate in the Treatment of Fluorouracil-Induced Diarrhea. *Oncologist* **1998**, *3*, 50–53. [CrossRef]
14. Sanghani, S.P.; Quinney, S.K.; Fredenburg, T.B.; Davis, W.I.; Murry, D.J.; Bosron, W.F. Hydrolysis of irinotecan and its oxidative metabolites, 7-ethyl-10-[4-N-(5-aminopentanoic acid)-1-piperidino] carbonyloxycamptothecin and 7-ethyl-10-[4-(1-piperidino)-1-amino]-carbonyloxycamptothecin, by human carboxylesterases CES1A1, CES2, and a newly expressed carboxylesterase isoenzyme, CES3. *Drug Metab. Dispos.* **2004**, *32*, 505–511. [PubMed]
15. Wu, M.H.; Chen, P.; Remo, B.F.; Cook, E.H., Jr.; Das, S.; Dolan, M.E. Characterization of multiple promoters in the human carboxylesterase 2 gene. *Pharmacogenetics* **2003**, *13*, 425–435. [CrossRef] [PubMed]
16. Dodds, H.M.; Haaz, M.; Riou, J.; Robert, J.; Rivory, L. Identification of a new metabolite of CPT-11 (irinotecan): Pharmacological properties and activation to SN-38. *J. Pharmacol. Exp. Ther.* **1998**, *286*, 578–583. [CrossRef]
17. Rivory, L.; Riou, J.; Haaz, M.; Sable, S.; Vuilhorgne, M.; Commercon, A.; Pond, S.; Robert, J. Identification and Properties of a Major Plasma Metabolite of Irinotecan (CPT-11) Isolated from the Plasma of Patients. *Cancer Res.* **1996**, *56*, 3689–3694.
18. Smith, N.F.; Figg, W.D.; Sparreboom, A. Pharmacogenetics of irinotecan metabolism and transport: An update. *Toxicol. Vitr.* **2006**, *20*, 163–175. [CrossRef]
19. Bosma, P.; Chowdhury, J.; Bakker, C.; Gantla, S.; Boer, A.; Oostra, B.; Lindhout, D.; Tytgat, G.; Jansen, P.; Elferink, R.O.; et al. The genetic basis of the reduced expression of bilirubin UDP-glucuronosyltransferase 1 in Gilbert's syndrome. *N. Engl. J. Med.* **1995**, *333*, 1171–1175. [CrossRef]

20. Chen, X.; Liu, L.; Guo, Z.; Liang, W.; He, J.; Huang, L.; Deng, Q.; Tang, H.; Pan, H.; Guo, M.; et al. UGT1A1 polymorphisms with irinotecan-induced toxicities and treatment outcome in Asians with Lung Cancer: A meta-analysis. *Cancer Chemother. Pharmacol.* **2017**, *79*, 1109–1117. [CrossRef]
21. Barbarino, J.M.; Haidar, C.E.; Klein, T.E.; Altman, R.B. PharmGKB summary: Very important pharmacogene information for UGT1A1. *Pharmacogenet. Genom.* **2014**, *24*, 177–183. [CrossRef] [PubMed]
22. Douillard, J.; Cunningham, D.; Roth, A.; Navarro, M.; James, R.; Karasek, P.; Jandik, P.; Iveson, T.; Carmichael, J.; Alakl, M.; et al. Irinotecan combined with fluorouracil compared with fluorouracil alone as first-line treatment for metastatic colorectal cancer: A multicentre randomised trial. *Lancet* **2000**, *355*, 1041–1047. [CrossRef]
23. Hall, D.; Ybazeta, G.; Destro-Bisol, G.; Petzl-Erler, M.; Rienzo, A.D. Variability at the uridine diphosphate glucuronosyltransferase 1A1 promoter in human populations and primates. *Pharmacogenetics* **1999**, *9*, 591–599. [CrossRef] [PubMed]
24. Beutler, E.; Gelbart, T.; Demina, A. Racial variability in the UDP-glucuronosyltransferase 1 (UGT1A1) promoter: A balanced polymorphism for regulation of bilirubin metabolism? *Proc. Natl. Acad. Sci. USA* **1998**, *95*, 8170–8174. [CrossRef]
25. Chen, Y.; Hu, F.; Li, C.; Fang, J.; Chu, L.; Zhang, X.; Xu, Q. The association of UGT1A1*6 and UGT1A1*28 with irinotecan-induced neutropenia in Asians: A meta-analysis. *Biomarkers* **2014**, *19*, 56–62. [CrossRef]
26. Liu, X.; Cheng, D.; Kuang, Q.; Liu, G.; Xu, W. Association of UGT1A1*28 polymorphisms with irinotecan-induced toxicities in colorectal cancer: A meta-analysis in Caucasians. *Pharmacogenom. J.* **2014**, *14*, 120–129. [CrossRef] [PubMed]
27. Cheng, L.; Li, M.; Hu, J.; Ren, W.; Xie, L.; Sun, Z.; Liu, B.; Xu, G.; Dong, X.; Qian, X. UGT1A1*6 polymorphisms are correlated with irinotecan-induced toxicity: A system review and meta-analysis in Asians. *Cancer Chemother. Pharmacol.* **2014**, *73*, 551–560. [CrossRef] [PubMed]
28. Sugatani, J.; Yamakawa, K.; Yoshinari, K.; Machida, Y.; Takagi, H.; Mori, M.; Kakizaki, S.; Sueyoshi, T.; Negishi, M.; Miwa, M. Identification of a Defect in the UGT1A1 Gene Promoter and Its Association with Hyperbilirubinemia. *Biochem. Biophys. Res. Commun.* **2002**, *292*, 492–497. [CrossRef]
29. Kim, S.; Hong, Y.; Shim, E.; Kong, S.-Y.; Shin, A.; Baek, J.; Jung, K. S-1 plus irinotecan and oxaliplatin for the first-line treatment of patients with metastatic colorectal cancer: A prospective phase II study and pharmacogenetic analysis. *Br. J. Cancer* **2013**, *109*, 1420–1427. [CrossRef]
30. Innocenti, F.; Grimsley, C.; Das, S.; Ramírez, J.; Cheng, C.; Kuttab-Boulos, H.; Ratain, M.J.; Rienzo, A.D. Haplotype structure of the UDP-glucuronosyltransferase 1A1 promoter in different ethnic groups. *Pharmacogenetics* **2003**, *12*, 725–733. [CrossRef]
31. Innocenti, F.; Undevia, S.D.; Iyer, L.; Chen, P.X.; Das, S.; Kocherginsky, M.; Karrison, T.; Janisch, L.; Ramírez, J.; Rudin, C.M.; et al. Genetic Variants in the UDP-glucuronosyltransferase 1A1 Gene Predict the Risk of Severe Neutropenia of Irinotecan. *J. Clin. Oncol.* **2004**, *22*, 1382–1388. [CrossRef] [PubMed]
32. Han, J.-Y.; Lim, H.-S.; Park, Y.H.; Lee, S.Y.; Lee, J.S. Integrated pharmacogenetic prediction of irinotecan pharmacokinetics and toxicity in patients with advanced non-small cell lung cancer. *Lung Cancer* **2009**, *63*, 115–120. [CrossRef]
33. Inoue, K.; Sonobe, M.; Kawamura, Y.; Etoh, T.; Takagi, M.; Matsumura, T.; Kikuyama, M.; Kimura, M.; Minami, S.; Utsuki, H.; et al. Polymorphisms of the UDP-glucuronosyl transferase 1A genes are associated with adverse events in cancer patients receiving irinotecan-based chemotherapy. *Tohoku J. Exp. Med.* **2013**, *229*, 107–114. [CrossRef] [PubMed]
34. Gagne, J.-F.; Montminy, V.; Belanger, P.; Journault, K.; Gaucher, G.; Guillemette, C. Common human UGT1A polymorphisms and the altered metabolism of irinotecan active metabolite 7-ethyl-10-hydroxycamptothecin (SN-38). *Mol. Pharmacol.* **2002**, *62*, 608–617. [CrossRef]
35. Dean, M.; Moitra, K.; Allikmets, R. The human ATP-binding cassette (ABC) transporter superfamily. *Genome Res.* **2001**, *11*, 1156–1166. [CrossRef] [PubMed]
36. Hu, Y.; Shi, A.; Fu, D.; Hu, X.; Zhang, J. Impact of genetic polymorphism on efficacy and toxicity irinotecan. *Chin. J. New Drugs* **2008**, *17*, 1298–1302.
37. Glimelius, B.; Garmo, H.; Berglund, A.; Fredriksson, L.A.; Berglund, M.; Kohnke, H.; Bystrom, P.; Sorbye, H.; Wadelius, M. Prediction of irinotecan and 5-fluorouracil toxicity and response in patients with advanced colorectal cancer. *Pharmacogenom. J.* **2010**, *11*, 61–71. [CrossRef]
38. Li, M.; Seiser, E.; Baldwin, R.; Ramirez, J.; Ratain, M.; Innocenti, F.; Kroetz, D. ABC transporter polymorphisms are associated with irinotecan pharmacokinetics and neutropenia. *Pharmacogenom. J.* **2018**, *18*, 35–42. [CrossRef]
39. Jong, F.A.D.; Marsh, S.; Mathijssen, R.H.J.; King, C.; Verweij, J.; Sparreboom, A.; McLeod, H.L. ABCG2 pharmacogenetics: Ethnic differences in allele frequency and assessment of influence on irinotecan disposition. *Clin. Cancer Res.* **2004**, *10*, 5889–5894. [CrossRef]
40. Jong, F.A.D.; Scott-Horton, T.J.; Kroetz, D.L.; McLeod, H.L.; Friberg, L.E.; Mathijssen, R.H.; Verweij, J.; Marsh, S.; Sparreboom, A. Irinotecan-induced diarrhea: Functional significance of the polymorphic ABCC2 transporter protein. *Clin. Pharmacol. Ther.* **2007**, *81*, 42–49. [CrossRef]

41. Chen, S.; Villeneuve, L.; Jonker, D.; Couture, F.; Laverdiere, I.; Cecchin, E.; Innocenti, F.; Toffoli, G.; Levesque, E.; Guillemette, C. ABCC5 and ABCG1 polymorphisms predict irinotecan-induced severe toxicity in metastatic colorectal cancer patients. *Pharmacogenet. Genom.* **2015**, *12*, 573–583. [CrossRef] [PubMed]
42. Gong, I.Y.; Kim, R.B. Impact of Genetic Variation in OATP Transporters to Drug Disposition and Response. *Drug Metab. Pharmacokinet.* **2013**, *28*, 4–18. [CrossRef] [PubMed]
43. Nies, A.T.; Niemi, M.; Burk, O.; Winter, S.; Zanger, U.M.; Stieger, B.; Schwab, M.; Schaeffeler, E. Genetics is a major determinant of expression of the human hepatic uptake transporter OATP1B1, but not of OATP1B3 and OATP2B1. *Genome Med.* **2013**, *5*, 1. [CrossRef]
44. Takane, H.; Miyata, M.; Burioka, N.; Kurai, J.; Fukuoka, Y.; Suyama, H.; Shigeoka, Y.; Otsubo, K.; Ieiri, I.; Shimizu, E. Severe toxicities after irinotecan-based chemotherapy in a patient with lung cancer: A homozygote for the SLCO1B1*15 allele. *Ther. Drug Monit.* **2007**, *29*, 666–668. [CrossRef]
45. Takane, H.; Kawamoto, K.; Sasaki, T.; Moriki, K.; Moriki, K.; Kitano, H.; Higuchi, S.; Otsubo, K.; Ieiri, I. Life-threatening toxicities in a patient with UGT1A1*6/*28 and SLCO1B1*15/*15 genotypes after irinotecan-based chemotherapy. *Cancer Chemother. Pharmacol.* **2009**, *63*, 1165–1169. [CrossRef]
46. Paulik, A.; Nekvindova, J.; Filip, S. Irinotecan toxicity during treatment of metastatic colorectal cancer: Focus on pharmacogenomics and personalized medicine. *Tumori* **2018**, *106*, 87–94. [CrossRef] [PubMed]
47. Mathijssen, R.; Alphen, R.V.; Verweij, J.; Loos, W.; Nooter, K.; Stoter, G.; Sparreboom, A. Clinical pharmacokinetics and metabolism of irinotecan (CPT-11). *Clin. Cancer Res.* **2001**, *7*, 2182–2194.
48. Marsh, S.; Xiao, M.; Yu, J.; Ahluwalia, R.; Minton, M.; Freimuth, R.R.; Kwok, P.-Y.; McLeod, H.L. Pharmacogenomic assessment of carboxylesterases 1 and 2. *Genomics* **2004**, *84*, 661–668. [CrossRef] [PubMed]
49. Mathijssen, R.H.J.; Jong, F.A.D.; Schaik, R.H.N.V.; Lepper, E.R.; Friberg, L.E.; Rietveld, T.; Bruijn, P.D.; Graveland, W.J.; Figg, W.D.; Verweij, J.; et al. Prediction of Irinotecan Pharmacokinetics by Use of Cytochrome P450 3A4 Phenotyping Probes. *J. Natl. Cancer Inst.* **2004**, *96*, 1585–1592. [CrossRef]
50. Stringer, A.M.; Gibson, R.J.; Bowen, J.M.; Logan, R.M.; Ashton, K.; Yeoh, A.S.J.; Al-Dasooqi, N.; Keefe, D.M.K. Irinotecan-induced mucositis manifesting as diarrhoea corresponds with an amended intestinal flora and mucin profile. *Int. J. Exp. Pathol.* **2009**, *90*, 489–499. [CrossRef]
51. Stringer, A.M. Interaction between host cells and microbes in chemotherapy-induced mucositis. *Nutrients* **2013**, *5*, 1488–1499. [CrossRef] [PubMed]
52. Mahdy, M.S.; Azmy, A.F.; Dishisha, T.; Mohamed, W.R.; Ahmed, K.A.; Hassan, A.; Aidy, S.E.; El-Gendy, A.O. Irinotecan-gut microbiota interactions and the capability of probiotics to mitigate Irinotecan-associated toxicity. *BMC Microbiol.* **2023**, *23*, 53. [CrossRef]
53. Stringer, A.M.; Gibson, R.J.; Logan, R.M.; Bowen, J.M.; Yeoh, A.S.J.; Laurence, J.; Keefe, D.M.K. Irinotecan-induced mucositis is associated with changes in intestinal mucins. *Cancer Chemother. Pharmacol.* **2009**, *64*, 123–132. [CrossRef]
54. Brandi, G.; Dabard, J.; Raibaud, P.; Battista, M.D.; Bridonneau, C.; Pisi, A.M.; Labate, A.M.M.; Pantaleo, M.A.; Vivo, A.D.; Biasco, G. Intestinal microflora and digestive toxicity of irinotecan in mice. *Clin. Cancer Res.* **2006**, *12*, 1299–1307. [CrossRef]
55. Fukuda, M.; Oka, M.; Soda, H.; Kinoshita, A.; Fukuda, M.; Nagashima, S.; Kuba, M.; Takatani, H.; Tsurutani, J.; Nakamura, Y.; et al. Phase II study of irinotecan combined with carboplatin in previously untreated non-small-cell lung cancer. *Clin. Trial* **2004**, *54*, 573–577. [CrossRef] [PubMed]
56. Swami, U.; Goel, S.; Mani, S. Therapeutic targeting of CPT-11 induced diarrhea: A case for prophylaxis. *Curr. Drug Targets* **2013**, *14*, 777–797. [CrossRef] [PubMed]
57. Benson, A.B., III; Ajani, J.A.; Catalano, R.B.; Engelking, C.; Kornblau, S.M.; Martenson, J.A., Jr.; McCallum, R.; Mitchell, E.P.; O'Dorisio, T.M.; Vokes, E.E.; et al. Recommended Guidelines for the Treatment of Cancer Treatment-Induced Diarrhea. *J. Clin. Oncol.* **2004**, *22*, 2918–2926. [CrossRef]
58. Lee, K.J. Pharmacologic Agents for Chronic Diarrhea. *Intest. Res.* **2015**, *13*, 306–312. [CrossRef]
59. Abigerges, D.; Chabot, G.; Armand, J.; Herait, P.; Gouyette, A.; Gandia, D. Phase I and pharmacologic studies of the camptothecin analog irinotecan administered every 3 weeks in cancer patients. *J. Clin. Oncol.* **1995**, *13*, 210. [CrossRef]
60. Andreyev, J.; Ross, P.; Donnellan, C.; Lennan, E.; Leonard, P.; Waters, C.; Wedlake, L.; Bridgewater, J.; Glynne-Jones, R.; Allum, W.; et al. Guidance on the management of diarrhoea during cancer chemotherapy. *Lancet Oncol.* **2014**, *15*, e447–e460. [CrossRef]
61. Barbounis, V.; Koumakis, G.; Vassilomanolakis, M.; Demiri, M.; Efremidis, A. Control of irinotecan-induced diarrhea by octreotide after loperamide failure. *Support. Care Cancer* **2001**, *9*, 258–260. [CrossRef] [PubMed]
62. Rosenoff, S. Octreotide LAR resolves severe chemotherapy-induced diarrhoea (CID) and allows continuation of full-dose therapy. *Eur. J. Cancer Care* **2004**, *13*, 380–383. [CrossRef] [PubMed]
63. Goncalves, E.; Costa, L.D.; Abigerges, D.; Armand, J.P. A new enkephalinase inhibitor as an alternative to loperamide in the prevention of diarrhea induced by CPT-11. *J. Clin. Oncol.* **1995**, *13*, 2144–2146. [CrossRef]

64. Ychou, M.; Douillard, J.; Rougier, P.; Adenis, A.; Mousseau, M.; Dufour, P.; Wendling, J.; Burki, F.; Mignard, D.; Marty, M. Randomized comparison of prophylactic antidiarrheal treatment versus no prophylactic antidiarrheal treatment in patients receiving CPT-11 (irinotecan) for advanced 5-FU-resistant colorectal cancer: An open-label multicenter phase II study. *Am. J. Clin. Oncol.* **2000**, *23*, 143–148. [CrossRef] [PubMed]
65. Lenfers, B.; Loeffler, T.; Droege, C.; Hausamen, T. Substantial activity of budesonide in patients with irinotecan (CPT-11) and 5-fluorouracil induced diarrhea and failure of loperamide treatment. *Ann. Oncol.* **1999**, *10*, 1251–1253. [CrossRef]
66. Karthaus, M.; Ballo, H.; Abenhardt, W.; Steinmetz, T.; Geer, T.; Schimke, J.; Braumann, D.; Behrens, R.; Behringer, D.; Kindler, M.; et al. Prospective, Double-Blind, Placebo-Controlled, Multicenter, Randomized Phase III Study with Orally Administered Budesonide for Prevention of Irinotecan (CPT-11)-Induced Diarrhea in Patients with Advanced Colorectal Cancer. *Oncology* **2005**, *68*, 326–332. [CrossRef]
67. Kobayashi, K.; Bouscarel, B.; Matsuzaki, Y.; Ceryak, S.; Kudoh, S.H. Fromm, pH-dependent uptake of irinotecan and its active metabolite, SN-38, by intestinal cells. *Int. J. Cancer* **1999**, *83*, 491–496. [CrossRef]
68. Takeda, Y.; Kobayashi, K.; Akiyama, Y.; Soma, T.; Handa, S.; Kudoh, S.; Kudo, K. Prevention of irinotecan (CPT-11)-induced diarrhea by oral alkalization combined with control of defecation in cancer patients. *Int. J. Cancer* **2001**, *92*, 269–275. [CrossRef]
69. Moreno, V.V.; Vidal, J.B.; Alemany, H.M.; Salvia, A.S.; Serentill, M.L.; Montero, I.C.; Tormo, S.S.; Bert, E.S.; Padro, J.G. Prevention of irinotecan associated diarrhea by intestinal alkalization. A pilot study in gastrointestinal cancer patients. *Clin. Transl. Oncol.* **2006**, *8*, 208–212. [CrossRef]
70. Hamada, A.; Aoki, H.T.A.; Ito, K.; Yokoo, K.; Sasaki, Y.; Saito, H. Pharmacokinetic changes of irinotecan by intestinal alkalinization in an advanced colorectal cancer patient. *Ther. Drug Monit.* **2005**, *27*, 536–538. [CrossRef]
71. Tamura, T.; Yasutake, K.; Nishisaki, H.; Nakashima, T.; Horita, K.; Hirohata, S.; Ishii, A.; Hamano, K.; Aoyama, N.; Shirasaka, D.; et al. Prevention of Irinotecan-Induced Diarrhea by Oral Sodium Bicarbonate and Influence on Pharmacokinetics. *Oncology* **2005**, *67*, 327–337. [CrossRef]
72. Chester, J.D.; Joel, S.P.; Cheeseman, S.L.; Hall, G.D.; Braun, M.S.; Perry, J.; Davis, T.; Button, C.J.; Seymour, M.T. Phase I and pharmacokinetic study of intravenous irinotecan plus oral ciclosporin in patients with fuorouracil-refractory metastatic colon cancer. *J. Clin. Oncol.* **2003**, *21*, 1125–1132. [CrossRef] [PubMed]
73. Gupta, E.; Safa, A.; Wang, X.; Ratain, M. Pharmacokinetic modulation of irinotecan and metabolites by cyclosporin A. *Cancer Res.* **1996**, *56*, 1309–1314. [PubMed]
74. Vasudev, N.; Jagdev, S.; Anthoney, D.; Seymour, M. Intravenous irinotecan plus oral ciclosporin. *Clin. Oncol.* **2005**, *17*, 646–649. [CrossRef] [PubMed]
75. Horikawa, M.; Kato, Y.; Sugiyama, Y. Reduced gastrointestinal toxicity following inhibition of the biliary excretion of irinotecan and its metabolites by probenecid in rats. *Pharm. Res.* **2002**, *19*, 1345–1353. [CrossRef]
76. Ciotti, M.; Basu, N.; Brangi, M.; Owens, I. Glucuronidation of 7-ethyl-10-hydroxycamptothecin (SN-38) by the human UDP-glucuronosyltransferases encoded at the UGT1 locus. *Biochem. Biophys. Res. Commun.* **1999**, *260*, 199–202. [CrossRef]
77. Kehrer, D.; Sparreboom, A.; Verweij, J.; Bruijn, P.D.; Nierop, C.; Schraaf, J.V.D.; Ruijgrok, E.; Jonge, M.D. Modulation of irinotecan-induced diarrhea by cotreatment with neomycin in cancer patients. *Clin. Cancer Res.* **2001**, *7*, 1136–1141.
78. Fittkau, M.; Voigt, W.; Holzhausen, H.-J.; Schmoll, H.-J. Saccharic acid 1.4-lactone protects against CPT-11-induced mucosa damage in rats. *J. Cancer Res. Clin. Oncol.* **2004**, *130*, 388–394. [CrossRef]
79. Rasmussen, T.S.; Koldso, H.; Nakagawa, S.; Kato, A.; Schiott, B.; Jensen, H.H. Synthesis of uronic-noeurostegine--a potent bacterial β-glucuronidase inhibitor. *Org. Biomol. Chem.* **2011**, *9*, 7807–7813. [CrossRef]
80. Ahmad, S.; Hughes, M.A.; Yeh, L.-A.; Scott, J.E. Potential repurposing of known drugs as potent bacterial β-glucuronidase inhibitors. *J. Biomol. Screen.* **2012**, *17*, 957–965. [CrossRef]
81. Walle, T.; Otake, Y.; Brubaker, J.; Walle, U.; Halushka, P. Disposition and metabolism of the flavonoid chrysin in normal volunteers. *Br. J. Clin. Pharmacol.* **2001**, *51*, 143–146. [CrossRef]
82. Walle, T.; Otake, Y.; Galijatovic, A.; Ritter, J.; Walle, U. Induction of UDP-glucuronosyltransferase UGT1A1 by the flavonoid chrysin in the human hepatoma cell line hep G2. *Drug Metab. Dispos.* **2000**, *28*, 1077–1082. [CrossRef]
83. Prados, M.D.; Yung, W.K.A.; Jaeckle, K.A.; Robins, H.I.; Mehta, M.P.; Fine, H.A.; Wen, P.Y.; Cloughesy, T.F.; Chang, S.M.; Nicholas, M.K.; et al. Phase 1 trial of irinotecan (CPT-11) in patients with recurrent malignant glioma: A North American Brain Tumor Consortium study. *Neuro Oncol.* **2004**, *6*, 44–54. [CrossRef]
84. Innocenti, F.; Undevia, S.D.; Ramírez, J.; Mani, S.; Schilsky, R.L.; Vogelzang, N.J.; Prado, M.; Ratain, M.J. A phase I trial of pharmacologic modulation of irinotecan with cyclosporine and phenobarbital. *Clin. Pharmacol. Ther.* **2004**, *76*, 490–502. [CrossRef]
85. Hicks, L.D.; Hyatt, J.L.; Stoddard, S.; Tsurkan, L.; Edwards, C.C.; Wadkins, R.M.; Potter, P.M. Improved, selective, human intestinal carboxylesterase inhibitors designed to modulate 7-ethyl-10-[4-(1-piperidino)-1-piperidino]carbonyloxycamptothecin (Irinotecan; CPT-11) toxicity. *J. Med. Chem.* **2009**, *52*, 3742–3752. [CrossRef]
86. Khanna, R.; Morton, C.; Danks, M.; Potter, P. Proficient metabolism of irinotecan by a human intestinal carboxylesterase. *Cancer Res.* **2000**, *60*, 4725–4728.

87. Hyatt, J.L.; Tsurkan, L.; Wierdl, M.; Edwards, C.C.; Danks, M.K.; Potter, P.M. Intracellular inhibition of carboxylesterases by benzil: Modulation of CPT-11 cytotoxicity. *Mol. Cancer Ther.* **2006**, *5*, 2281–2288. [CrossRef]
88. Loghin, M.E.; Prados, M.D.; Wen, P.; Junck, L.; Lieberman, F.; Fine, H.; Fink, K.L.; Metha, M.; Kuhn, J.; Lamborn, K.; et al. Phase I study of temozolomide and irinotecan for recurrent malignant gliomas in patients receiving enzyme-inducing antiepileptic drugs: A north american brain tumor consortium study. *Clin. Cancer Res.* **2007**, *13*, 7133–7138. [CrossRef]
89. Voutsadakis, I.A. Pathogenesis of colorectal carcinoma and therapeutic implications: The roles of the ubiquitin-proteasome system and Cox-2. *J. Cell. Mol. Med.* **2007**, *11*, 252–285. [CrossRef]
90. Trifan, O.C.; Durham, W.F.; Salazar, V.S.; Horton, J.; Levine, B.D.; Zweifel, B.S.; Davis, T.W.; Masferrer, J.L. Cyclooxygenase-2 inhibition with celecoxib enhances antitumor efficacy and reduces diarrhea side effect of CPT-11. *Cancer Res.* **2002**, *62*, 5778–5784.
91. Javle, M.M.; Cao, S.; Durrani, F.A.; Pendyala, L.; Lawrence, D.D.; Smith, P.F.; Creaven, P.J.; Noel, D.C.; Iyer, R.V.; Rustum, Y.M. Celecoxib and mucosal protection: Translation from an animal model to a phase I clinical trial of celecoxib, irinotecan, and 5-fluorouracil. *Clin. Cancer Res.* **2007**, *13*, 965–971. [CrossRef]
92. Fakih, M.G.; Rustum, Y.M. Does celecoxib have a role in the treatment of patients with colorectal cancer? *Clin. Colorectal. Cancer* **2009**, *8*, 11–14. [CrossRef]
93. Ooi, K.; Miya, T.; Sasaki, H.; Morimoto, Y. Prevention of irinotecan hydrochloride-induced diarrhea by oral administration of Lactobacillus casei strain Shirota in rats. *Gan Kagaku Ryoho* **2008**, *35*, 951–954.
94. Bowen, J.M.; Stringer, A.M.; Gibson, R.J.; Yeoh, A.S.J.; Hannam, S.; Keefe, D.M.K. VSL#3 probiotic treatment reduces chemotherapy-induced diarrhea and weight loss. *Cancer Biol. Ther.* **2007**, *6*, 1449–1454.
95. Yue, B.; Gao, R.; Wang, Z.; Dou, W. Microbiota-Host-Irinotecan Axis: A New Insight Toward Irinotecan Chemotherapy. *Front. Cell. Infect. Microbiol.* **2021**, *11*, 710945. [CrossRef]
96. Schmittel, A.; Jahnke, K.; Thiel, E.; Keilholz, U. Neomycin as secondary prophylaxis for irinotecan-induced diarrhea. *Ann. Oncol.* **2004**, *15*, 1296. [CrossRef]
97. Kong, R.; Liu, T.; Zhu, X.; Ahmad, S.; Williams, A.L.; Phan, A.T.; Zhao, H.; Scott, J.E.; Yeh, L.-A.; Wong, S.T.C. Old drug new use—Amoxapine and its metabolites as potent bacterial β-glucuronidase inhibitors for alleviating cancer drug toxicity. *Clin. Cancer Res.* **2014**, *20*, 3521–3530. [CrossRef]
98. Kurita, A.; Kado, S.; Matsumoto, T.; Asakawa, N.; Kaneda, N.; Kato, I.; Uchida, K.; Onoue, M.; Yokokura, T. Streptomycin alleviates irinotecan-induced delayed-onset diarrhea in rats by a mechanism other than inhibition of β-glucuronidase activity in intestinal lumen. *Cancer Chemother. Pharmacol.* **2011**, *67*, 201–213. [CrossRef]
99. Wallace, B.D.; Roberts, A.B.; Pollet, R.M.; Ingle, J.D.; Biernat, K.A.; Pellock, S.J.; Venkatesh, M.K.; Guthrie, L.; O'Neal, S.K.; Robinson, S.J.; et al. Structure and Inhibition of Microbiome β-Glucuronidases Essential to the Alleviation of Cancer Drug Toxicity. *Chem. Biol.* **2015**, *22*, 1238–1249. [CrossRef]
100. Flieger, D.; Klassert, C.; Hainke, S.; Keller, R.; Kleinschmidt, R.; Fischbach, W. Phase II clinical trial for prevention of delayed diarrhea with cholestyramine/levofloxacin in the second-line treatment with irinotecan biweekly in patients with metastatic colorectal carcinoma. *Oncology* **2007**, *72*, 10–16. [CrossRef]
101. Furman, W.L.; Crews, K.R.; Billups, C.; Wu, J.; Gajjar, A.J.; Daw, N.C.; Patrick, C.C.; Rodriguez-Galindo, C.; Stewart, C.F.; Dome, J.S.; et al. Cefixime allows greater dose escalation of oral irinotecan: A phase I study in pediatric patients with refractory solid tumors. *J. Clin. Oncol.* **2006**, *24*, 563–570. [CrossRef] [PubMed]
102. Chyka, P. Multiple-dose activated charcoal and enhancement of systemic drug clearance: Summary of studies in animals and human volunteers. *J. Toxicol. Clin. Toxicol.* **1995**, *33*, 399–405. [CrossRef]
103. Balram, C.; Zhou, Q.-Y.; Cheung, Y.B.; Lee, E.J.D. Influence of multiple dose activated charcoal on the disposition kinetics of irinotecan in rats. *Drug Metab. Metabol. Drug Interact.* **2002**, *19*, 137–148. [CrossRef]
104. Sergio, G.-C.; Felix, G.-M.; Luis, J.-V. Activated charcoal to prevent irinotecan-induced diarrhea in children. *Pediatr. Blood Cancer* **2008**, *51*, 49–52. [CrossRef]
105. Maeda, Y.; Ohune, T.; Nakamura, M.; Yamasaki, M.; Kiribayashi, Y.; Murakami, T. Prevention of irinotecan-induced diarrhoea by oral carbonaceous adsorbent (Kremezin) in cancer patients. *Oncol. Rep.* **2004**, *12*, 581–585. [CrossRef]
106. Singhal, S.; Mehta, J.; Desikan, R.; Ayers, D.; Roberson, P.; Eddlemon, P.; Munshi, N.; Anaissie, E.; Wilson, C.; Dhodapkar, M.; et al. Antitumor activity of thalidomide in refractory multiple myeloma. *N. Engl. J. Med.* **1999**, *341*, 1565–1571. [CrossRef] [PubMed]
107. Govindarajan, R.; Heaton, K.; Broadwater, R.; Zeitlin, A.; Lang, N.; Hauer-Jensen, M. Effect of thalidomide on gastrointestinal toxic effects of irinotecan. *Lancet* **2000**, *356*, 566–567. [CrossRef] [PubMed]
108. Ramirez, J.; Wu, K.; Janisch, L.; Karrison, T.; House, L.K.; Innocenti, F.; Cohen, E.E.W.; Ratain, M.J. The effect of thalidomide on the pharmacokinetics of irinotecan and metabolites in advanced solid tumor patients. *Cancer Chemother. Pharmacol.* **2011**, *68*, 1629–1632. [CrossRef] [PubMed]
109. Villalona-Calero, M.; Schaaf, L.; Phillips, G.; Otterson, G.; Panico, K.; Duan, W.; Kleiber, B.; Shah, M.; Young, D.; Wu, W.-H.; et al. Thalidomide and celecoxib as potential modulators of irinotecan's activity in cancer patients. *Cancer Chemother. Pharmacol.* **2007**, *59*, 23–33. [CrossRef]

110. Gibson, R.J.; Stringer, A.M.; Bowen, J.M.; Logan, R.M.; Yeoh, A.S.-J.; Burns, J.; Alvarez, E.; Keefe, D.M.K. Velafermin improves gastrointestinal mucositis following irinotecan treatment in tumor-bearing DA rats. *Cancer Biol. Ther.* **2007**, *6*, 541–547. [CrossRef]
111. Cao, S.; Black, J.; Troutt, A.; Rustum, Y. Interleukin 15 offers selective protection from irinotecan-induced intestinal toxicity in a preclinical animal model. *Cancer Res.* **1998**, *58*, 3270–3274. [PubMed]
112. Shinohara, H.; Bucana, C.; Killion, J.; Fidler, I. Intensified regression of colon cancer liver metastases in mice treated with irinotecan and the immunomodulator JBT 3002. *J. Immunother.* **2000**, *23*, 321–331. [CrossRef] [PubMed]
113. Hardman, W.; Moyer, M.; Cameron, I. Fish oil supplementation enhanced CPT-11 (irinotecan) efficacy against MCF7 breast carcinoma xenografts and ameliorated intestinal side-effects. *Br. J. Cancer* **1999**, *81*, 440–448. [CrossRef]
114. Rao, R.; Samak, G. Role of Glutamine in Protection of Intestinal Epithelial Tight Junctions. *J. Epithel. Biol. Pharmacol.* **2012**, *5* (Suppl. 1-M7), 47–54. [PubMed]
115. He, W.; Zhang, H. Treatment of delayed diarrhea caused by irinotecan. *Guide China Med.* **2013**, *11*, 476–477.
116. Li, X. Clinical observation of Phloroglucinol combined with Octreotide in treatment of severe delayed-onset diarrhea caused by Irinotecan. *China Mod. Dr.* **2012**, *50*, 135–136.
117. Di Carlo, G.; Borrelli, F.; Ernst, E.; Izzo, A.A. St John's wort: Prozac from the plant kingdom. *Trends Pharmacol. Sci.* **2001**, *22*, 292–297. [CrossRef]
118. Fiebich, B.; Hollig, A.; Lieb, K. Inhibition of substance P-induced cytokine synthesis by St. John's wort extracts. *Pharmacopsychiatry* **2001**, *34* (Suppl. S1), S26–S28. [CrossRef]
119. Wang, S.; Xu, Z.; Cai, B.; Chen, Q. Berberine as a Potential Multi-Target Agent for Metabolic Diseases: A Review of Investigations for Berberine. *Endocr. Metab. Immune Disord. Drug Targets* **2021**, *21*, 971–979.
120. Yue, B.; Gao, R.; Lv, C.; Yu, Z.; Wang, H.; Geng, X.; Wang, Z.; Dou, W. Berberine Improves Irinotecan-Induced Intestinal Mucositis Without Impairing the Anti-colorectal Cancer Efficacy of Irinotecan by Inhibiting Bacterial β-glucuronidase. *Front. Pharmacol.* **2021**, *12*, 774560. [CrossRef]
121. Luo, Z.; Chen, X.; Zhu, D.; Wang, G.; Liu, C. The protective effect of curcumin on delayed diarrhea caused by irinotecan chemotherapy. *Guangdong Med. J.* **2016**, *37*, 1462–1466.
122. Ouyang, M.; Luo, Z.; Zhang, W.; Zhu, D.; Lu, Y.; Wu, J.; Yao, X. Protective effect of curcumin against irinotecan-induced intestinal mucosal injury via attenuation of NF-κB activation, oxidative stress and endoplasmic reticulum stress. *Int. J. Oncol.* **2019**, *54*, 1376–1386. [CrossRef] [PubMed]
123. Wang, X.; Rao, Z.; Qin, H.; Zhang, G.; Ma, Y.; Jin, Y.; Han, M.; Shi, A.; Wang, Y.; Wu, X. Effect of hesperidin on the pharmacokinetics of CPT-11 and its active metabolite SN-38 by regulating hepatic Mrp2 in rats. *Biopharm. Drug Dispos.* **2016**, *37*, 421–432. [CrossRef] [PubMed]
124. Yu, Y.; Kong, R.; Cao, H.; Yin, Z.; Liu, J.; Nan, X.; Phan, A.T.; Ding, T.; Zhao, H.; Wong, S.T.C. Two birds, one stone: Hesperetin alleviates chemotherapy-induced diarrhea and potentiates tumor inhibition. *Oncotarget* **2018**, *9*, 27958–27973. [CrossRef]
125. Cui, D.; Wang, X.; Chen, J.; Lv, B.; Zhang, P.; Zhang, W.; Zhang, Z.; Xu, F. Metabolomic study of Chinese medicine Huang Qin decoction as an effective treatment for irinotecan-induced gastrointestinal toxicity. *Rsc Adv.* **2015**, *5*, 26420–26429.
126. Wu, Q.; Ye, H.; Zhu, Y.; Guo, M.; Zheng, X. Experimental Study on Preventive Effect of Huangqin Tang on Irinotecan Induced Delayed Diarrhea. *Chin. J. Exp. Tradit. Med. Formulae* **2013**, *19*, 163–168.
127. Wang, X.; Cui, D.; Dai, X.; Wang, J.; Zhang, W.; Zhang, Z.; Xu, F. HuangQin Decoction Attenuates CPT-11-Induced Gastrointestinal Toxicity by Regulating Bile Acids Metabolism Homeostasis. *Front. Pharmacol.* **2017**, *30*, 156. [CrossRef]
128. Lam, W.; Jiang, Z.; Gu, F.; Huang, X.; Hu, R.; Wang, J.; Bussom, S.; Liu, S.-H.; Zhao, H.; Yen, Y.; et al. PHY906(KD018), an adjuvant based on a 1800-year-old Chinese medicine, enhanced the anti-tumor activity of Sorafenib by changing the tumor microenvironment. *Sci. Rep.* **2015**, *5*, 9384. [CrossRef]
129. Lam, W.; Ren, Y.; Guan, F.; Jiang, Z.; Cheng, W.; Xu, C.; Liu, S.; Cheng, Y. Mechanism Based Quality Control (MBQC) of Herbal Products: A Case Study YIV-906 (PHY906). *Front. Pharmacol.* **2018**, *9*, 1324. [CrossRef]
130. Lam, W.; Bussom, S.; Guan, F.; Jiang, Z.; Zhang, W.; Gullen, E.A.; Liu, S.-H.; Cheng, Y.-C. The four-herb Chinese medicine PHY906 reduces chemotherapy-induced gastrointestinal toxicity. *Sci. Transl. Med.* **2010**, *2*, 45ra59. [CrossRef]
131. Kummar, S.; Copur, M.S.; Rose, M.; Wadler, S.; Stephenson, J.; O'Rourke, M.; Brenckman, W.; Tilton, R.; Liu, S.-H.; Jiang, Z.; et al. A phase I study of the chinese herbal medicine PHY906 as a modulator of irinotecan-based chemotherapy in patients with advanced colorectal cancer. *Clin. Color. Cancer* **2011**, *10*, 85–92. [CrossRef] [PubMed]
132. Wang, J.; Jia, L.; Tan, H.; Pang, L.; Yu, L.; Deng, B. Effect of Shengjiang Xiexin Decoction on the Repair of Damaged Rat intestinal Mucosa after irinotecan Chemotherapy. *Chin. J. Integr. Tradit. West. Med.* **2015**, *35*, 1236–1243.
133. Deng, C.; Deng, B.; Jia, L.; Tan, H.; Zhang, P.; Liu, S.; Zhang, Y.; Song, A.; Pan, L. Preventive Effects of a Chinese Herbal Formula, Shengjiang Xiexin Decoction, on Irinotecan-Induced Delayed-Onset Diarrhea in Rats. *Evid. Based Complement. Altern. Med.* **2017**, *2017*, 7350251. [CrossRef] [PubMed]
134. Wu, M.; Deng, B.; Jia, L. Preventive effects of Shengjiang Xiexin Decoction against irinotecan-induced toxicity in CT26 tumorbearing mice. *J. China-Jpn. Friendsh. Hosp.* **2018**, *32*, 356–359+377+320.

135. Liu, X. Clinical Effect of Shengjiang Xiexin Decoction on Gastrointestinal Reaction after FOLFIRI Chemotherapy. *China J. Pharm. Econ.* **2021**, *16*, 108–110.
136. Wang, X.; Yang, J.; Cao, Q.; Tang, J. Therapeutic efficacy and mechanism of water-soluble extracts of Banxiaxiexin decoction on BALB/c mice with oxazolone-induced colitis. *Exp. Ther. Med.* **2014**, *8*, 1201–1204. [CrossRef]
137. Lu, H.; Ma, S.; Guo, Y.; Cai, J.; Wang, X.; Ye, W. Study on the improvement and mechanism of Banxia Xiexin Decoction against irinotecan-induced diarrhea in nude bearing mice tumor of human small cell lung cancer. *Chin. Arch. Tranditional Chin. Med.* **2009**, *27*, 1082–1084.
138. Lu, H.; Qin, J.; Han, N.; Xie, F.; Gong, L.; Li, C. Banxia Xiexin Decoction Is Effective to Prevent and Control Irinotecan-Induced Delayed Diarrhea in Recurrent Small Cell Lung Cancer. *Integr. Cancer Ther.* **2018**, *17*, 1109–1114. [CrossRef]
139. Wu, Y.; Cheng, Y.; Wang, D.; Yang, X.; Zhong, X.; Lin, J.; Fu, C.; Zhang, J.; Hu, Y. Analysis of Mechanism of Chloroform Extract of Gegen Qinliantang on Alleviating Enterotoxicity Induced by Irinotecan. *Chin. J. Exp. Tradit. Med. Formulae* **2021**, *27*, 16–23.
140. Wu, Y.; Wang, D.; Yang, X.; Fu, C.; Zou, L.; Zhang, J. Traditional Chinese medicine Gegen Qinlian decoction ameliorates irinotecan chemotherapy-induced gut toxicity in mice. *Biomed. Pharmacother.* **2019**, *109*, 2252–2261. [CrossRef]
141. Zhou, M.; Zhang, H.; Zhang, Y.; Qiu, J.; Zhang, J. Determination of chloroform fraction in Gegen Qinlian decoction and its inhibition on carboxylesterase. *Pharm. Clin. Chin. Mater. Medica* **2019**, *10*, 10–13.
142. Fu, Y.; Wang, J. A review of the clinical application of Xiao Chai Hu Tang in the digestive system. *HeiLongJiang Tradit. Chin. Med.* **2016**, *45*, 71–72.
143. Zhou, Y.; Yang, G.; Wan, L. Preventive and Treatment Effects of Xiaochaihu Decoction on Irinotecan- induced Bloodystool in Model Mice with Delayed Diarrhea. *China Pharm.* **2017**, *28*, 1762–1765.
144. Ji, L.; Xie, L.; Zhang, D.; Xian, H.; Yu, L.; Chen, H.; Yan, W.; Hu, W. Effect of Sishen Pill on Delayed Diarrhea-Induced by Irinotecan in Mice. *Liaoning J. Tranditional Chin. Med.* **2017**, *44*, 1294–1297+1777.
145. Lu, J.; Lin, Z.; Huang, S.; Shen, Y.; Jiang, J.; Lin, S. Jiawei Xianglian Decoction (JWXLD), a Traditional Chinese Medicine (TCM), Alleviates CPT-11-Induced Diarrhea in Mice. *Evid. Based Complement. Altern. Med.* **2020**, *2020*, 7901231. [CrossRef]
146. Lei, X.; Yu, S. Modified Shenling Baizhu Powder for treating diarrhea after chemotherapy for malignant tumors. *Chin. Traditonal Med.* **2007**, *29*, 1419–1421.
147. Cui, Q.; Hu, Y.; Cui, Q.; Zhang, X. The effect of Ren Shen Jian Pi Pill on delayed diarrhea caused by irinotecan and its impact on gut microbiota and serum inflammatory factors. *Acta Chin. Med. Pharmacol.* **2021**, *49*, 83–86.
148. Urushiyama, H.; Jo, T.; Yasunaga, H.; Michihata, N.; Yamana, H.; Matsui, H.; Hasegawa, W.; Hiraishi, Y.; Mitani, A.; Fushimi, K.; et al. Effect of Hangeshashin-To (Japanese Herbal Medicine Tj-14) on Tolerability of Irinotecan: Propensity Score and Instrumental Variable Analyses. *J. Clin. Med.* **2018**, *7*, 246. [CrossRef]
149. Kato, S.; Hayashi, S.; Kitahara, Y.; Nagasawa, K.; Aono, H.; Shibata, J.; Utsumi, D.; Amagase, K.; Kadowaki, M. Saireito (TJ-114), a Japanese traditional herbal medicine, reduces 5-fluorouracil-induced intestinal mucositis in mice by inhibiting cytokine-mediated apoptosis in intestinal crypt cells. *PLoS ONE* **2015**, *10*, e0116213. [CrossRef]
150. Takasuna, K.; Kasai, Y.; Kitano, Y.; Mori, K.; Kobayashi, R.; Hagiwara, T.; Kakihata, K.; Hirohashi, M.; Nomura, M.; Nagai, E. Protective effects of kampo medicines and baicalin against intestinal toxicity of a new anticancer camptothecin derivative, irinotecan hydrochloride (CPT-11), in rats. *Cancer Sci.* **2010**, *86*, 978–984. [CrossRef]
151. Bernards, N.; Ventura, M.; Fricke, I.B.; Hendriks, B.S.; Fitzgerald, J.; Lee, H.; Zheng, J. Liposomal Irinotecan Achieves Significant Survival and Tumor Burden Control in a Triple Negative Breast Cancer Model of Spontaneous Metastasis. *Mol. Pharm.* **2018**, *15*, 4132–4138. [CrossRef] [PubMed]
152. Wang-Gillam, A.; Hubner, R.A.; Siveke, J.T.; Hoff, D.D.V.; Belanger, B.; Jong, F.A.D.; Mirakhur, B.; Chen, L.-T. NAPOLI-1 phase 3 study of liposomal irinotecan in metastatic pancreatic cancer: Final overall survival analysis and characteristics of long-term survivors. *Eur. J. Cancer* **2019**, *108*, 78–87. [CrossRef]
153. Woo, W.; Carey, E.T.; Choi, M. Spotlight on liposomal irinotecan for metastatic pancreatic cancer: Patient selection and perspectives. *Onco Targets Ther.* **2019**, *12*, 1455–1463. [CrossRef]
154. Batist, G.; Gelmon, K.A.; Chi, K.N.; Miller, W.H.M., Jr.; Chia, S.K.L.; Mayer, L.D.; Swenson, C.E.; Janoff, A.S.; Louie, A.C. Safety, pharmacokinetics, and efficacy of CPX-1 liposome injection in patients with advanced solid tumors. *Clin. Cancer Res.* **2009**, *15*, 692–700. [CrossRef]
155. Noble, C.O.; Krauze, M.T.; Drummond, D.C.; Yamashita, Y.; Saito, R.; Berger, M.S.; Kirpotin, D.B.; Bankiewicz, K.S.; Park, J.W. Novel Nanoliposomal CPT-11 Infused by Convection-Enhanced Delivery in Intracranial Tumors: Pharmacology and Efficacy. *Cancer Res.* **2006**, *66*, 2801–2806. [CrossRef]
156. Wang, H.; Agarwal, P.; Zhao, S.; Xu, R.X.; Yu, J.; Lu, X.; He, X. Hyaluronic acid-decorated dual responsive nanoparticles of Pluronic F127, PLGA, and chitosan for targeted co-delivery of doxorubicin and irinotecan to eliminate cancer stem-like cells. *Biomaterials* **2015**, *72*, 74–89. [CrossRef]

157. Alamgeer, M.; Watkins, D.N.; Banakh, I.; Kumar, B.; Gough, D.J.; Markman, B.; Ganju, V. A phase IIa study of HA-irinotecan, formulation of hyaluronic acid and irinotecan targeting CD44 in extensive-stage small cell lung cancer. *Investig. New Drugs* **2018**, *36*, 288–298. [CrossRef]
158. Gibbs, P.; Clingan, P.R.; Ganju, V.; Strickland, A.H.; Wong, S.S.; Tebbutt, N.C.; Underhill, C.R.; Fox, R.M.; Clavant, S.P.; Leung, J.; et al. Hyaluronan-Irinotecan improves progression-free survival in 5-fluorouracil refractory patients with metastatic colorectal cancer: A randomized phase II trial. *Cancer Chemother. Pharmacol.* **2011**, *67*, 153–163. [CrossRef]
159. Harris, J.M.; Chess, R.B. Effect of pegylation on pharmaceuticals. *Nat. Rev. Drug Discov.* **2003**, *2*, 214–221. [CrossRef]
160. Mohammad, A.S.; Griffith, J.I.; Adkins, C.E.; Shah, N.; Sechrest, E.; Dolan, E.L.; Terrell-Hall, T.B.; Hendriks, B.S.; Lee, H.; Lockman, P.R. Liposomal Irinotecan Accumulates in Metastatic Lesions, Crosses the Blood-Tumor Barrier (BTB), and Prolongs Survival in an Experimental Model of Brain Metastases of Triple Negative Breast Cancer. *Pharm. Res.* **2018**, *35*, 31. [CrossRef]
161. Infante, J.R.; Keedy, V.L.; Jones, S.F.; Zamboni, W.C.; Chan, E.; Bendell, J.C.; Lee, W.; Wu, H.; Ikeda, S.; Kodaira, H.; et al. Phase I and pharmacokinetic study of IHL-305 (PEGylated liposomal irinotecan) in patients with advanced solid tumors. *Cancer Chemother. Pharmacol.* **2012**, *70*, 699–705. [CrossRef] [PubMed]
162. Zhang, C.; Jin, S.; Xue, X.; Zhang, T.; Jiang, Y.; Wang, P.C.; Liang, X.-J. Tunable self-assembly of Irinotecan-fatty acid prodrugs with increased cytotoxicity to cancer cells. *J. Mater. Chem. B* **2016**, *4*, 3286–3291. [CrossRef] [PubMed]
163. Mahmoudi, A.; Jaafari, M.R.; Ramezanian, N.; Gholami, L.; Malaekeh-Nikouei, B. BR2 and CyLoP1 enhance in-vivo SN38 delivery using pegylated PAMAM dendrimers. *Int. J. Pharm.* **2019**, *564*, 77–89. [CrossRef]
164. Huang, P.; Chen, K.; Prijovich, Z.; Cheng, T.; Leu, Y.; Roffler, S. Enhancement of CPT-11 antitumor activity by adenovirus-mediated expression of beta-glucuronidase in tumors. *Cancer Gene Ther.* **2011**, *18*, 381–389. [CrossRef]
165. Kojima, A.; Hackett, N.R.; Ohwada, A.; Crystal, R.G. In vivo human carboxylesterase cDNA gene transfer to activate the prodrug CPT-11 for local treatment of solid tumors. *J. Clin. Investig.* **1998**, *101*, 1789–1796. [CrossRef]
166. Zhang, J.A.; Xuan, T.; Parmar, M.; Ma, L.; Ugwu, S.; Ali, S.; Ahmad, I. Development and characterization of a novel liposome-based formulation of SN-38. *Int. J. Pharm.* **2004**, *270*, 93–107. [CrossRef]
167. Sun, X.; Zhu, D.; Cai, Y.; Shi, G.; Gao, M.; Zheng, M. One-step mechanochemical preparation and prominent antitumor activity of SN-38 self-micelle solid dispersion. *Int. J. Nanomed.* **2019**, *14*, 2115–2126. [CrossRef]
168. Karki, N.; Tiwari, H.; Pal, M.; Chaurasia, A.; Bal, R.; Joshi, P.; Sahoo, N.G. Functionalized graphene oxides for drug loading, release and delivery of poorly water soluble anticancer drug: A comparative study. *Colloids Surf. B Biointerfaces* **2018**, *169*, 265–272. [CrossRef]
169. Naumann, J.A.; Widen, J.C.; Jonart, L.A.; Ebadi, M.; Tang, J.; Gordon, D.J.; Harki, D.A.; Gordon, P.M. SN-38 Conjugated Gold Nanoparticles Activated by Ewing Sarcoma Specific mRNAs Exhibit In Vitro and In Vivo Efficacy. *Bioconjug. Chem.* **2018**, *29*, 1111–1118. [CrossRef]
170. Govindan, S.V.; Cardillo, T.M.; Moon, S.-J.; Hansen, H.J.; Goldenberg, D.M. CEACAM5-Targeted Therapy of Human Colonic and Pancreatic Cancer Xenografts with Potent Labetuzumab-SN-38 Immunoconjugates. *Clin. Cancer Res.* **2009**, *15*, 6052–6061. [CrossRef]
171. Venditto, V.J.; Simanek, E.E. Cancer Therapies Utilizing the Camptothecins: A Review of theın VivoLiterature. *Mol. Pharm.* **2010**, *7*, 307–349. [CrossRef] [PubMed]
172. Kurzrock, R.; Goel, S.; Wheler, J.; Hong, D.; Fu, S.; Rezai, K.; Morgan-Linnell, S.K.; Urien, S.; Mani, S.; Chaudhary, I.; et al. Safety, pharmacokinetics, and activity of EZN-2208, a novel conjugate of polyethylene glycol and SN38, in patients with advanced malignancies. *Cancer* **2012**, *118*, 6144–6151. [CrossRef] [PubMed]
173. Tang, L.; Li, X.; Wan, L.; Xiao, Y.; Zeng, X.; Ding, H. Herbal Medicines for Irinotecan-Induced Diarrhea. *Front. Pharmacol.* **2019**, *10*, 182. [CrossRef] [PubMed]
174. Bailly, C. Irinotecan: 25 years of cancer treatment. *Pharmacol. Res.* **2019**, *148*, 104398. [CrossRef]

pharmaceuticals

Article

5,8-Dimethyl-9*H*-carbazole Derivatives Blocking hTopo I Activity and Actin Dynamics

Jessica Ceramella [1], Domenico Iacopetta [1,*], Anna Caruso [1], Annaluisa Mariconda [2], Anthi Petrou [3], Athina Geronikaki [3], Camillo Rosano [4], Carmela Saturnino [2], Alessia Catalano [5], Pasquale Longo [6] and Maria Stefania Sinicropi [1]

1 Department of Pharmacy, Health and Nutritional Sciences, University of Calabria, 87036 Arcavacata di Rende, Italy
2 Department of Science, University of Basilicata, 85100 Potenza, Italy
3 Department of Pharmacy, School of Health, Aristotle University of Thessaloniki, 54124 Thessaloniki, Greece
4 U.O. Proteomica e Spettrometria di Massa, IRCCS Ospedale Policlinico San Martino, Largo R. Benzi 10, 1632 Genova, Italy
5 Department of Pharmacy-Drug Sciences, University of Bari "Aldo Moro", 70126 Bari, Italy
6 Department of Chemistry and Biology, University of Salerno, Via Giovanni Paolo II, 132, 84084 Fisciano, Italy
* Correspondence: domenico.iacopetta@unical.it; Tel.: +39-0984-493200

Abstract: Over the years, carbazoles have been largely studied for their numerous biological properties, including antibacterial, antimalarial, antioxidant, antidiabetic, neuroprotective, anticancer, and many more. Some of them have gained great interest for their anticancer activity in breast cancer due to their capability in inhibiting essential DNA-dependent enzymes, namely topoisomerases I and II. With this in mind, we studied the anticancer activity of a series of carbazole derivatives against two breast cancer cell lines, namely the triple negative MDA-MB-231 and MCF-7 cells. Compounds **3** and **4** were found to be the most active towards the MDA-MB-231 cell line without interfering with the normal counterpart. Using docking simulations, we assessed the ability of these carbazole derivatives to bind human topoisomerases I and II and actin. *In vitro* specific assays confirmed that the lead compounds selectively inhibited the human topoisomerase I and interfered with the normal organization of the actin system, triggering apoptosis as a final effect. Thus, compounds **3** and **4** are strong candidates for further drug development in multi-targeted therapy for the treatment of triple negative breast cancer, for which safe therapeutic regimens are not yet available.

Keywords: human topoisomerases I/II; anticancer; docking simulation; actin dynamics

1. Introduction

One of the principal targets of pharmaceutical research is represented by the design of new and valid anticancer drugs characterized by higher selectivity on neoplastic cells and lower toxicity on normal ones. Over the years, medical and pharmaceutical researchers' attention has been focused on the carbazole scaffold, which is present in important classes of indole-containing heterocycles characterized by widespread biological activities, including anticancer, antibacterial, antiviral, antioxidant, antidiabetic, and neuroprotective ones [1]. These characteristics resulted in the extensive applications of carbazole derivatives in the field of medicinal chemistry [2,3]. The carbazole skeleton is the key structural motif of many synthetic and natural biologically effective molecules acting as DNA intercalating agents and able to interfere with the activity of crucial enzymes involved in cancer progression, such as the topoisomerases (Topos) [4]. The latter are ubiquitous enzymes, vital for gene expression, chromosome segregation, as well as DNA replication and recombination, due to their capability to solve DNA supercoiling by cutting one or both strands of the DNA duplex [5]. Based on their mechanism of action, two classes of DNA topoisomerases have been identified from eukaryotes: the DNA topoisomerase I (Topo I) which acts by making a temporary break in one strand of DNA, while the DNA topoisomerase II (Topo II) mediates the ATP-dependent DNA double-strand breaks [6]. Anticancer agents targeting both Topo I and II have been proven to be highly effective in cancer treatment [7]. Because of these interesting properties, extensive research concerning the chemistry and biological activities of carbazoles targeting topoisomerases has been reported since the characterization of the 9*H*-carbazole in 1872 by Graebe and Glaser [1,8–10]. Ellipticine (Figure 1) is one of the first studied natural occurring alkaloids with a carbazole nucleus and several studies have evidenced its wide biological effectiveness, highlighting its capability to act as DNA intercalating and Topos inhibitor [11]. However, Ellipticine is also characterized by a high toxicity. Thus, to obtain more active and safer derivatives, great efforts in the design and synthesis of carbazole analogues have been made over the last years. For instance, new pyrrolo[2,3-α]carbazole derivatives were found to significantly reduce the Topo I activity in a concentration dependent manner [12], while some 11*H*-pyrido[*a*]carbazole ones resulted good DNA intercalating and Topo II inhibitors [13]. Moreover, a new series of racemic and chiral carbazole aminoalcohols was proved to possess a potent Topo I inhibitory activity [14]. Two Ellipticine analogues acted as good inhibitors of human topoisomerase II (hTopo II) and resulted more potent than the reference molecule. In our previous works, more than one series of carbazole compounds have been designed and prepared. In particular, the new synthesized *N*-thioalkylcarbazole, *N*,*N*,*N*-trimethylethanammonium iodide alkylcarbazole, and 1,4-dimethylcarbazole derivatives exhibited exciting cytotoxic profiles on different breast cancer cell lines and inhibited the hTopo II decatenation activity [15–17]. Moreover, we demonstrated that some benzothienoquinazolinones carbazole bioisosters [18] exerted good anticancer activity on breast cancer cells due to their capability to inhibit the hTopo I supercoil relaxing activity. Beyond that, the benzothienoquinazolinones carbazole bioisoster also targeted the microtubule network involved in vital cellular functions, such as mitosis, cell migration, and cell signaling. These findings are seminal in the fight against breast cancer, which still represents the main cancer-related cause of disease for women, and its incidence and mortality have risen worldwide in recent years [19]. Over the years, numerous carbazole derivatives were designed, synthesized, and examined for their anti-breast cancer activity [20–22]. Recently, Vlaar et al. studied new analogues of EHop-016, a carbazole compound acting as a Rac1 inhibitor, implicated in the intracellular actin polymerization. Some of them exerted good antiproliferative activity on different breast cancer cells lines and inhibited the migration process in the metastatic MDA-MB-231 cells. In addition, the lead compound improved by approximately four-fold *in vitro* efficacy in inhibiting the activity of the Rho GTPase Rac1 in both MDA-MB-231 and MDA-MB-435 cell lines if compared with EHop-016 [23]. Moreover, Butler-Fernández et al. published a new series of *N*-alkyl-3,6-dibromocarbazole and *N*-alkyl-5-bromoindole derivatives, whose anticancer and anti-migratory effects in MCF-7 and MDA- MB-231 breast cancer cell lines are connected to the interference with the intracellular actin dynamics [24].

Building upon these findings, the goal of this work was to evaluate the anticancer activity of a series of five carbazole derivatives (**1–5**) (Figure 1), previously synthesized by us [25], against two human breast cancer cell lines, namely MCF-7 and MDA-MB-231. Docking studies and *in vitro* assays were performed in order to elucidate the mechanism of their action, revealing that the two individuated leads target hTopo I and actin. Finally, they can be considered promising candidates for the development of new multi-target agents in the treatment of triple negative breast cancer.

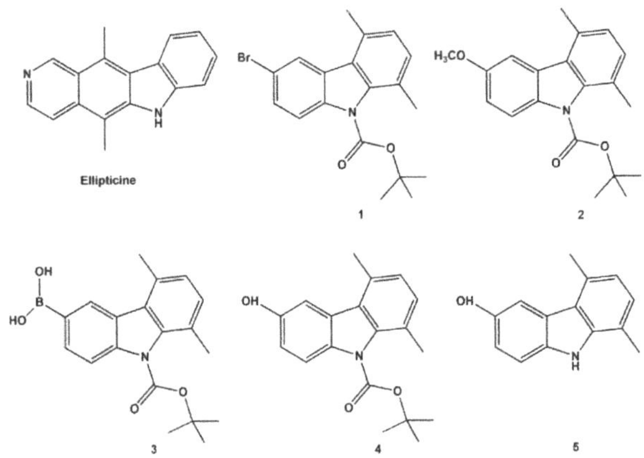

Figure 1. Structures of Ellipticine and the studied carbazole derivatives (**1–5**) previously synthesized by us [25].

2. Results

2.1. Chemistry

Carbazole derivatives (**1–4**) were synthesized as reported in the literature [25]. The synthesized molecules (**1–4**) differ by the substituent in position 6, while compound **5** is analogous to compound **4**, but does not bear the Boc group. The substituents selected for the design of the molecules were: -Br, $-OCH_3$, $-B(OH)_2$ e -OH, whose presence allowed to obtain valuable information about the correlation between the architecture of the molecules and the biological activity.

2.2. Effects on Breast Cancer and Normal Cells Viability

The inhibition of cancer cell growth exerted by the five considered compounds (**1–5**) has been evaluated *in vitro* against two models of breast cancer, namely the estrogen receptor positive (ER+) MCF-7 and the triple negative MDA-MB-231 human breast cancer cells (lacking ER, PR, and HER-2/Neu amplification), together with the normal counterpart, namely the non-malignant breast epithelial cells MCF-10A. The IC_{50} values were determined at 72 h after treatment by means of the MTT assay and are reported in Table 1. As reference molecule, we adopted Ellipticine, the parent molecule with a carbazole scaffold from which they have been derived. Our results indicated that, among the carbazole derivatives, compound **4** resulted the most active, since it drastically reduced the growth of the MDA-MB-231 cells with an IC_{50} value of 0.73 ± 0.74 μM. Then, compound **3** also exhibited good anticancer activity against the triple negative cancer cells, with an IC_{50} value of 1.44 ± 0.97 μM. A moderate activity was noticed against the MCF-7 cells, with compound **4** being most active. More interestingly, compound **4** did not affect the MCF-10A cells viability, until the concentration of 100 μM, whereas compound **3** induced the death of the half of cells at a concentration of approximately 52 μM. Compounds **5**, **2**, and **1** exhibited decreasing anticancer activity against the MDA-MB- 231 cells, with IC_{50} values equal to 6.59 ± 0.68, 8.19 ± 0.26, and 43.45 ± 1.21 μM, respectively. Moreover, no effects were recorded against the MCF-7 cells, at least until the concentration of 100 μM. These outcomes suggest that the presence of the bromine substituent on the carbazole nucleus is responsible of the net decrease of the activity, whereas the hydroxy and boron substituents seem to be determinant for the observed higher activity. It is important to highlight that compounds **3** and **4** possess a higher anticancer activity against the MDA-MB-231 cells and a better cytotoxic profile than the Ellipticine, which is not even selective amongst breast cancer and normal cells. Thus, the two lead compounds were chosen for subsequent studies aiming at individuating the molecular mechanisms underlying the observed anticancer activity.

Table 1. IC_{50} values of carbazoles derivatives and Ellipticine expressed in μM. The means ± standard deviations are shown.

	IC_{50} (μM)		
Compounds	**MDA-MB-231**	**MCF-7**	**MCF-10A**
1	43.45 ± 1.21	>100	>100
2	8.19 ± 0.26	>100	89.16 ± 0.47
3	1.44 ± 0.97	27.58 ± 0.71	51.89 ± 0.88
4	0.73 ± 0.74	19.76 ± 1.12	>100
5	6.59 ± 0.68	>100	>100
Ellipticine	1.92 ± 0.38	1.34 ± 0.40	1.12 ± 0.51

2.3. Docking Studies

Molecular docking simulations were performed in order to understand the possible binding modes of the compounds described above (Figure 1) and the target proteins, namely the actin and the human Topoisomerases I and II. We considered a "blind-docking approach" for our simulations: no "a priori" information about the binding site was provided to the system. This kind of procedure has been successfully used by our research group in several other studies (some examples in [26–29]). Using this kind of approach, we aimed both to identify the most promising candidate among our compounds and to further improve the atomic structure of our molecules in order to design and synthesize better lead compounds. Our study was based on the compound binding affinity to the two human Topoisomerases and actin (Table 2), as calculated by the program Autodock (this program calculates a binding affinity constant Ki based on the binding Energy, according to the expression Ki = exp (deltaG/(R * T)). To discriminate the successful candidates, we took into consideration the clusterization of the results from the simulations, as discussed in previous work [30]. Eventually, the obtained binding mode was examined to evaluate the quality of the protein–ligand interactions. Docking simulations (Figures 2 and S1, panels C) suggested our compounds were able to dock actin and the hTopo I and II, forming several hydrogen and hydrophobic interactions. Particularly, our molecules dock the actin in a protein cleft occupied by the Latrunculin B in the crystal structure described by Wang et al. [31]. The actin residues Asp 157, Lys 213. Glu 214, Thr 303, and Tyr 306 are involved in the hydrogen bonding to our compounds. The protein cleft is completed by the hydrophobic residues Ala 181, Leu 185, Leu 216, Cys 217, Pro 307, and Ile 309 that contribute to stabilize the compounds binding. Simulations carried over the hTopo I, as the protein target, identified three distinct possible binding sites for our compounds. Compound **5** (Figure S1, panel A) binds a first site forming hydrogen bonds with Asp 331 and hydrophobic interactions with the backbone of the loop Asn 327- Gly 339 and the side chains of residues Ala 334 and Val 338. In the same pose, compound **3** (Figure 2, panel A) forms additional hydrogen bonds with residues Lys 329, ser 342 Arg 344, and Asp 399. A second binding lays on the opposite site of the loop Asn 327- Gly 339 and hosts compounds **4** and **1** (Figures 2 and S1, panels A, respectively). These compounds are both stabilized by hydrophobic interactions with Ala 334 and Thr 337. Compound **4** forms hydrogen bonds with His 203, Lys 329, Thr 337, and Lys 368, while compound **1** loses the interaction with His 203 but adds a halogen bond between its bromide moiety and Ser 342. On the other hand, compound **2** (Figure S1, panel A) docks to a protein area that is far from the previously described sites. In this case, compound **2** is surrounded by hydrophobic residues Leu 321, the aromatic ring of Tyr 373, Val 377, Ile 464, and Val 550, forming hydrogen bonds with Asn 467 and Gln 469. The same molecules described above bind the hTopo II close to the DNA-gate (Figures 2 and S1, panels B). The residues Gln 726 and Asn 851 are hydrogen-bonded to compound **5** (Figure S1, panel A), Gln 544 and Lys 550 to compound **3** (Figure 2, panel A), and Gln 544, Lys 550, and Gln 542 to compound **4** (Figure S1, panel A). Due to the presence of a methyl group, compound **2** (Figure S1, panel A) loses the ability to form hydrogen bonds with Gln 544, Lys 550, and Gln 542 (those three molecules are almost superposed), whilst compound **1** (Figure S1, panel A) makes a halogen bond with Arg 672. The pocket is contoured by hydrophobic residues Leu 592, Pro 593 Leu 705, Leu 685, and Tyr 686.

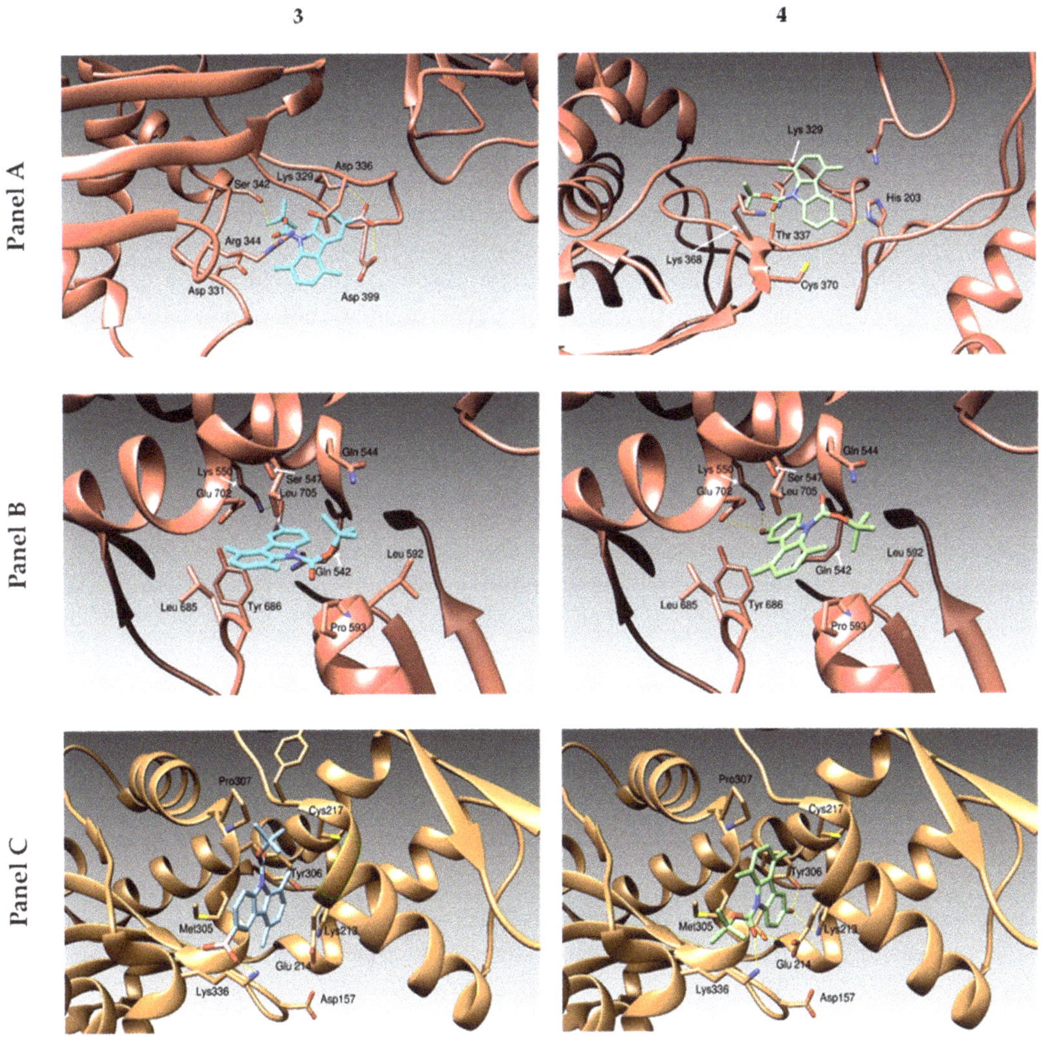

Figure 2. The three-dimensional structure of the human proteins Topoisomerase I (Panel A), Topoisomerase II (Panel B) and Actin (Panel C) bound to compounds **3** and **4** are drawn. Proteins are schematically reported as ribbons. Ligands binding poses are described as colored sticks.

Table 2. Binding free energies of carbazole derivatives against hTopo I, hTopo II and actin.

Compounds	hTopo I	hTopo II	Actin
1	−7.22	−8.13	−7.52
2	−6.46	−7.17	−7.56
3	−8.62	−7.61	−7.42
4	−7.95	−8.75	−8.0
5	−6.30	−7.75	−6.88

The binding energies are calculated using the software Autodock 4.0.2 and expressed in Kcal/mol.

2.4. Inhibition Assays on Human Topoisomerase I and II

The carbazole derivatives **1–5** were tested for their ability to inhibit the human Topoisomerase I (hTopo I) and II (hTopo II), using specific enzymatic assays, as reported in [17,18], and the obtained results are shown in Figure 3.

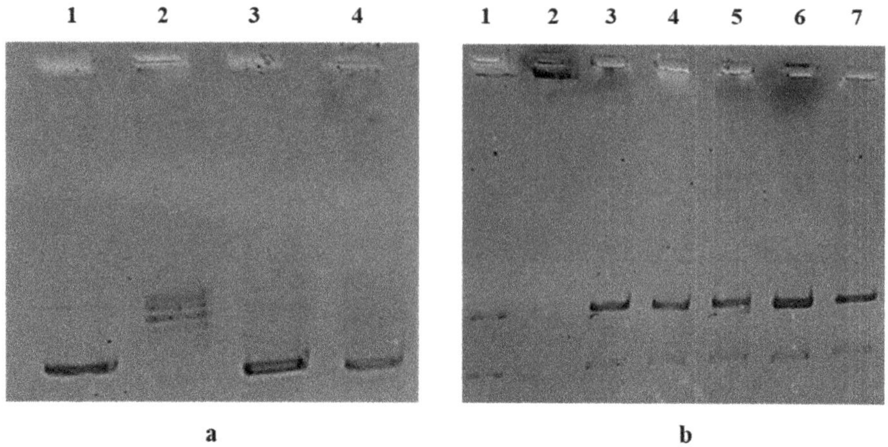

Figure 3. (**a**) hTopo I supercoiled relaxing activity. hTopo I was exposed to the vehicle alone (DMSO, lane 2) or compounds **3** and **4** at the concentration of 1 μM (lanes 3 and 4). Then, the hTopo I reaction products were visualized on agarose gel. Supercoiled DNA (plasmid pHOT1) was used as marker (lane 1). (**b**) hTopo II decatenation assay. hTopo II was exposed to the vehicle alone (DMSO, lane 3) or compounds **3** and **4** at the concentrations of 1 μM (lanes 4 and 5) and 10 μM (lanes 6 and 7). Then, the hTopo II reaction products were visualized on agarose gel. Decatenated DNA and kinetoplast DNA (kDNA) were used as markers (lanes 1 and 2).

First, only **3** and **4** totally inhibited the hTopo I supercoiling relaxing activity at the concentration of 1 μM. Indeed, as shown in Figure 3a (lanes 3 and 4), a clear band of the uncut plasmid DNA at the bottom of the gel is present. The uncut plasmid pHOT1, used as marker, is present in Figure 3a, lane 1. On the contrary, in the control reaction (only vehicle, Figure 3a, lane 2), it is possible to observe the presence of multiple bands corresponding to the relaxed plasmid DNA, used as substrate.

Next, we screened all the compounds for their activity against the hTopo II but, in this case, none of the tested compounds have been able to block the decatenation activity at the concentration of 1 μM (Figure 3b, lanes 4 and 5). Moreover, since **3** and **4** totally blocked the hTopo I, in order to exclude a dose-dependent activity, the concentration was increased to 10 μM (see Figure 3b, lanes 6 and 7). Again, no inhibitory activity was recorded. Indeed, two bands related to the DNA decatenation products are present at the bottom of the agarose gel, which denote the enzyme full activity, both in the control (only vehicle, Figure 3b, lane 3) and the tested compound reactions (Figure 3b, lanes 4–7). Thus, we can conclude that the lead compounds **3** and **4** are selective inhibitors of the hTopo I, being inactive against the hTopo II.

2.5. Influence of Compounds 3 and 4 on Actin Dynamics

In order to determine whether the lead compounds **3** and **4** may effectively regulate the actin system, we employed both the immunofluorescence and *in vitro* direct enzymatic assays. Thus, MDA-MB-231 were treated for 24 h with compounds **3** and **4** at a concentration equal to their IC_{50} values, respectively. As negative and positive controls, we used the only vehicle (DMSO) and latrunculin A (LA) at a concentration of 0.1 μM, respectively. After processing, the cells were observed under a fluorescent microscope, (see experimental section for the details). Our outcomes showed that the actin filaments are regularly organized in the cell cytoplasm in the DMSO treated cells (Figure 4, CTRL, panel B), whereas under the LA exposure (Figure 4, LA, panel B), the cells underwent an important shape change, appearing circular, because of the interference with the actin system. Moreover, the latter looked brighter and stocked in the cytoplasm in dot-like structures or thicker fibers. Again, the MDA-MB-231 cells, under exposure to compounds **3** and **4**, lost their shape (Figure 4, compounds **3** and **4**, panels B), as we observed under LA treatment, and with a very similar arrangement of the actin network. These outcomes suggest that compounds **3** and **4** interfere with the normal actin organization in MDA-MB-231 cells, with a similar behavior of LA.

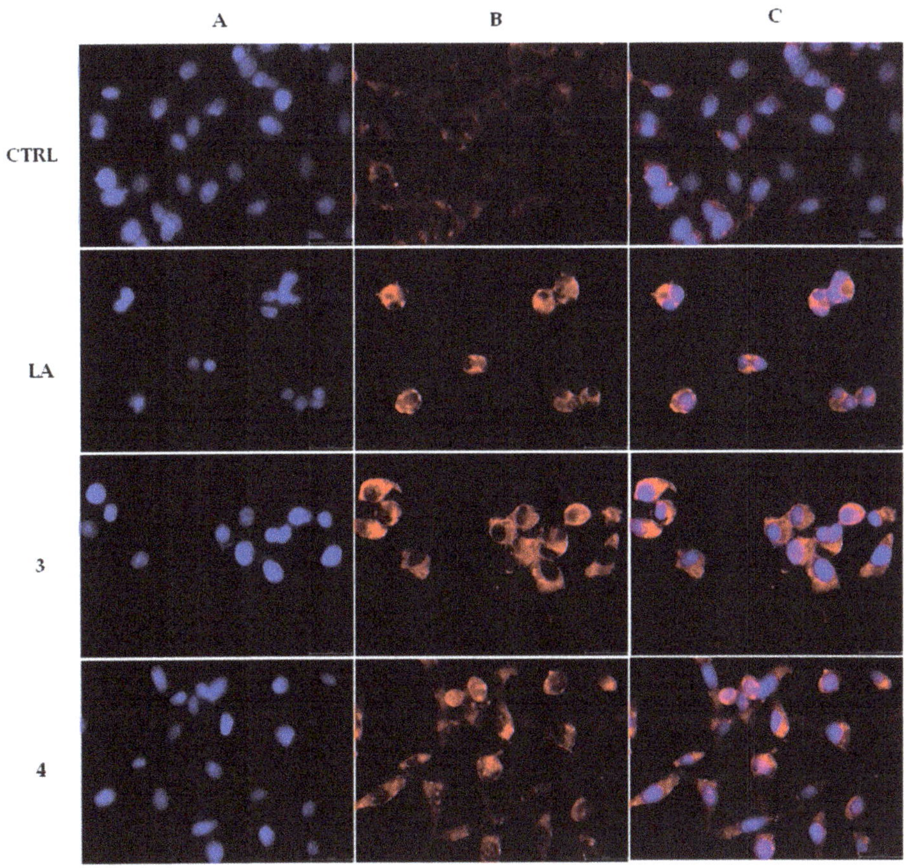

Figure 4. Actin immunofluorescence studies. MDA-MB-231 cells were exposed for 24 h to the vehicle alone (CTRL), 0.1 μM LA or carbazole derivatives **3** and **4** (used at their IC_{50} values). Then the cells were further processed, as indicated in the experimental section. The inverted fluorescence microscope was adopted to observe and image all the immunofluorescence figures (40× magnification). Panels A: nuclear stain with DAPI ($\lambda_{ex}/\lambda_{em}$ = 350/460 nm); Panels B: β-actin (Alexa Fluor® 568; $\lambda_{ex}/\lambda_{em}$ = 644/665 nm); Panels C show a merge. Representative fields are reported.

With the aim to confirm the immunofluorescence results and substantiate whether compounds **3** and **4** could inhibit the actin polymerization mechanism and/or boost the F-actin depolymerization, we employed a fluorescent-labeled purified rabbit actin. LA and cytochalasin B (CB) were used as positive controls for both the polymerization inhibition and F-actin subunits dissociation, or only the polymerization inhibition, respectively. In the negative control reaction (only vehicle), actin monomers undergo a normal polymerization, as visible in Figure 5, panel A. The reaction curve rises in approximately 5 min to a value of about 25,000 RFU (plateau) which remains almost unvaried until the end of the experiment. Contrariwise, the two reference molecules, i.e., LA and CB, used at the concentration of 5 μM, braked the actin polymerization, mostly LA rather than CB, under the adopted experimental conditions, and the LA curve dropped until a value of about 8000 RFU at the reaction end. The CB curve, similar to that of LA, showed a decrease and ended at approximately 13,000 RFU.

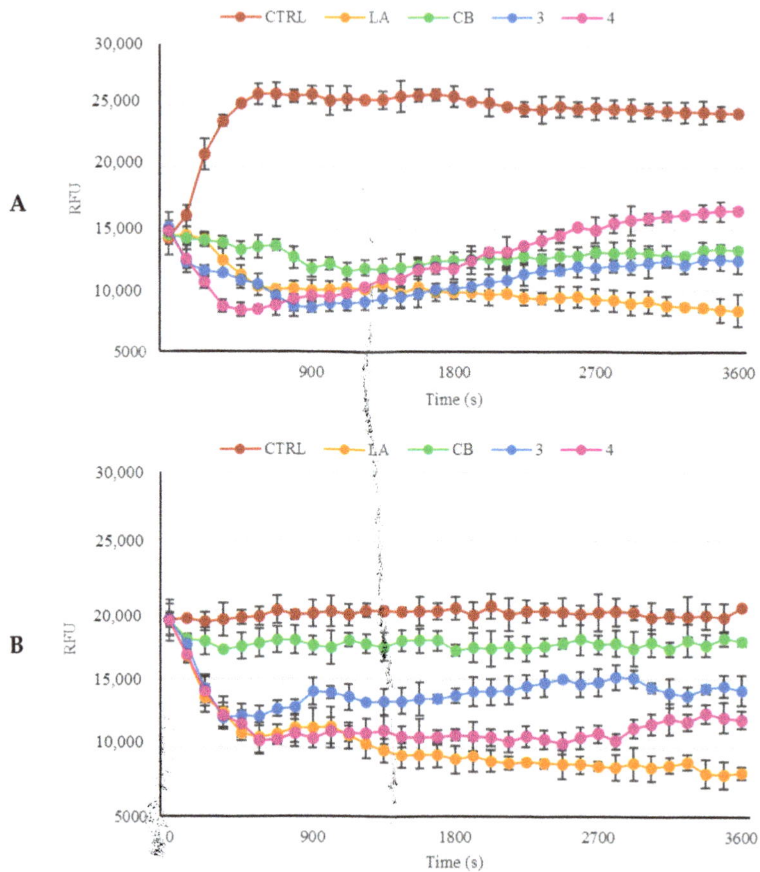

Figure 5. (**A**) *In vitro* actin polymerization assay. Compounds **3** and **4** (used at the concentration of 5 μM) were incubated with the labeled rabbit muscle actin in order to verify their ability to inhibit the protein polymerization. (**B**) *In vitro* actin depolymerization assay. After actin polymerization, compounds **3** and **4** (5 μM) were added to the reaction mixture, in order to determine their ability to act as depolymerizing agents. For both the assays, the vehicle DMSO was used as a negative control. Actin-targeting agents, LA and CB, both at the concentration of 5 μM, were used as positive controls. The assemblage of the actin filaments was established by monitoring the fluorescence ($\lambda_{Ex/Em}$= 365/410 nm) in kinetic mode for 1 h at room temperature by using a microplate reader. The graphics are representative of three separate experiments and error bars represent the standard deviations.

Lastly, compounds **3** and **4**, used at the concentration of 5 μM, both exhibited an inhibitory effect against actin polymerization, but with a lesser efficacy than LA. However, compound **3** curve was lower than that of CB and reached a final value of about 13,000 RFU, which is pretty similar to that of CB, whereas the final value of compound **4** was around 16,000 RFU. Thus, compound **3** has an inhibitory effect slightly superior to that of CB throughout the whole reaction, whereas compound **4** seemed to inhibit better the actin polymerization than CB only in the first 40 min.

Moreover, we performed the F-actin depolymerization assay to verify whether compounds **3** and **4** could induce the actin depolymerization as well as the LA does (Figure 5, panel B). For this purpose, we started the actin polymerization for one hour under the same experimental conditions used previously. Then, we added the compounds or the reference molecules at the concentration of 5 μM and monitored the reactions for an additional hour. Our results indicated that the LA induced an important fall of the curve until a final value of about 8000 RFU, indicating the F-actin depolymerization. On the contrary, the control polymerization reaction (only vehicle) exhibited an almost constant value (20,000 RFU) until the reaction end. The addition of CB was not able to produce the same effect of LA. Instead, its behavior follows that of the vehicle, suggesting that CB does not induce the F-actin depolymerization. Contrarily, both the compounds were able to induce an important depolymerizing effect on the F-actin, most evident in the first 7 min after the exposure. Particularly, compounds **3** and **4** produced final values of approximately 14,000 and 12,000 RFU, respectively. Altogether, our outcomes suggest that compounds **3** and **4** produced a similar effect to that of LA on actin, even if to a lesser extent, inducing the inhibition of the polymerization reaction and, at the same time, accelerating the dissociation of F-actin. This combined effect leads to a disorganization of the intracellular actin network.

2.6. *Compounds* **3** *and* **4** *Trigger Apoptosis in MDA-MB-231 Cells*

Having individuated two intracellular targets, we wondered whether the lead compounds could induce apoptosis in MDA-MB-231 cells. Thus, we performed a TUNEL assay. The cells were treated and processed as described in the experimental section (Section 4.2.6) and the obtained outcomes, shown in Figure 6, suggested that both the compounds are able to trigger apoptosis in MDA-MB- 231 cells. Indeed, the exposure to both compounds produced a green nuclear fluorescence, already at 24 h, in MDA-MB-231 cells (Figure 6, **3** and **4**, panels B, CF^{TM}488A) as a consequence of the damaged DNA. This event did not happen in the DMSO-treated cells (Figure 6, CTRL, panel B, CF^{TM}488A), indicating the lack of a massive DNA break. The overlay channel (Figure 6, panels C) is also shown.

2.7. *Druglike Properties, Toxicity and Drug-Likeness*

'Drug-like' molecules were evaluated *in silico* for their ADMET profile in order to rapidly screen multiple properties [32]. Compounds that have been predicted to exhibit toxicity, high blood–brain barrier permeability, low water solubility, and poor Caco2-permeability were excluded from potential hits. The server pkCSM [33] was used for this purpose. pkCSM relies on graph-based signatures. These encode distance patterns between atoms in order to represent the small molecule and to train predictive models (Table S1).

Computational studies are considered a viable approach to drug discovery, and they have several advantages over in vivo studies, especially in reducing cost, time, and animals sacrifice. These approaches, nowadays, are broadly used in studies of the physicochemical and pharmacokinetic properties of compounds in medicinal chemistry. Lipinski's rule of five (Ro5) is considered a standard for drug development [34], and violations of Ro5 are MW > 500, lipophilicity (LogPo/w) > 5, hydrogen bond determined for binding donors (HBD) < 5. Violations of these rules lead to reduced intestinal absorption, penetration, or solubility [35]. The next extension of the Lipinski Ro5 includes polar surface area (PSA <140 Å^2), which is an important predictor of drug oral bioavailability.

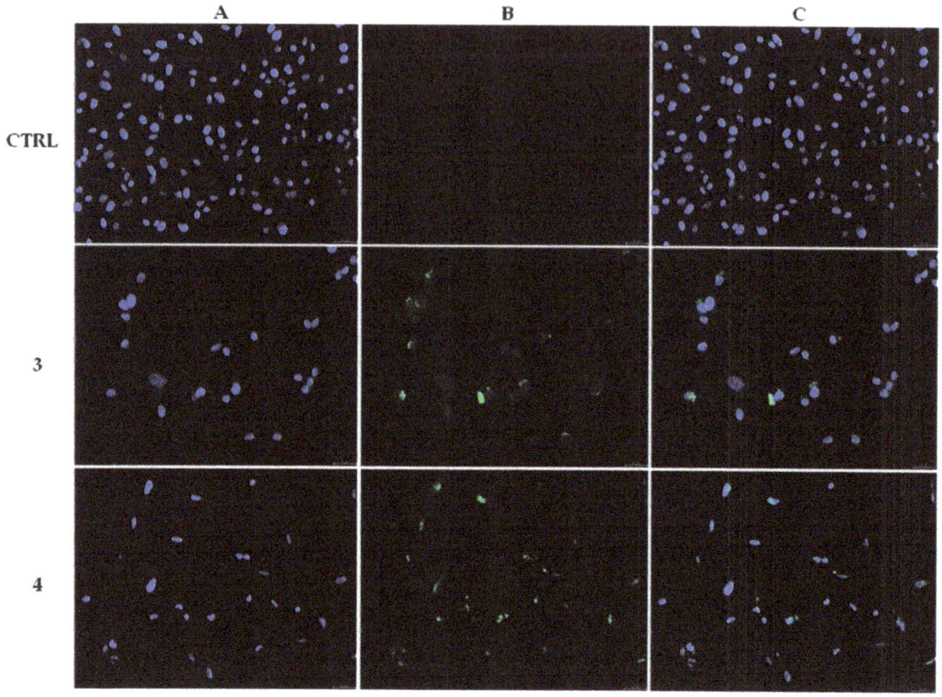

Figure 6. Tunel Assay. MDA-MB-231 breast cancer cells were treated with compounds **3** and **4** at the concentration equal to their IC_{50} values or with vehicle (CTRL) for 24 h. Then they were exposed to the TdT enzyme, further processed (see experimental section for more details) and visualized under a fluorescence microscope (20× magnification). Panels A: DAPI, $\lambda_{ex/em}$= 350 nm/460 nm. Panels B: CF^{TM}488 A, $\lambda_{ex/em}$ = 490 nm/515 nm. Panels C show the overlay channels.

The synthesized compounds have been studied *in silico* using the Swiss ADME software [36].The drug-likeness and bioavailability scores of all tested compounds are shown in Table 3. According to prediction results, the bioavailability score of all compounds was approximately 0.55. Furthermore, all compounds displayed moderate to good drug-likeness scores, ranging from −0.43 to 0.52. The best in the in-silico prediction result was achieved for the most active compounds, **3** and **4** with a drug-likeness score of 0.52 and 0.42 (Figure 7, Table 3). Moreover, these compounds showed no violation in all rules.

The BOILED-Egg allows for the evaluation of passive gastrointestinal absorption (GIA), brain penetration (BBB), and P-glycoprotein (P-gp) activity in the presence of the molecule. The white region of the "BOILED-egg" represents the high probability of passive absorption by the gastrointestinal tract, and the yellow region (yolk) the high probability of brain penetration. Moreover, the points are colored in blue if predicted as active effluxes by P-gp (PGP+) and in red if predicted as non-substrate of P-gp.

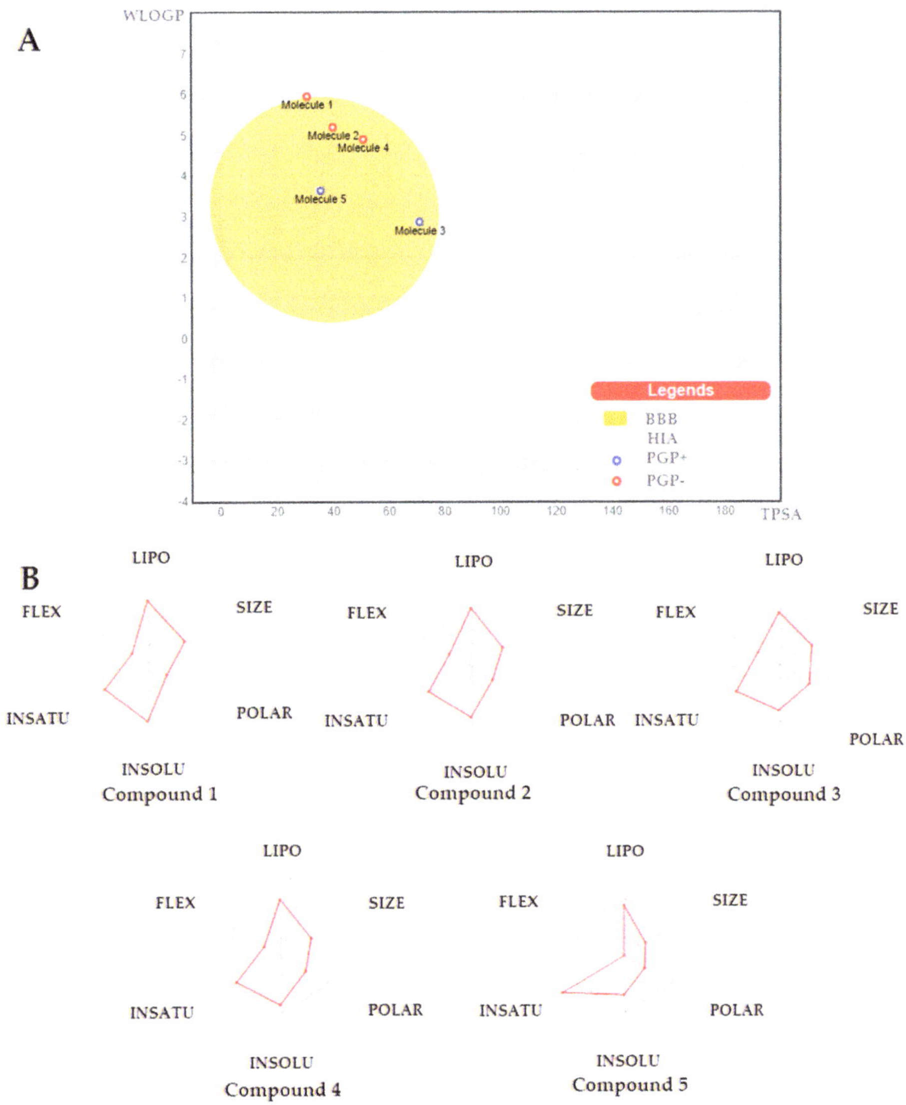

Figure 7. (**A**) BOILED-Egg diagram for all the designed compounds. (**B**) Bioavailability radar chart of all compounds. The pink area represents the optimal range for each property for oral bioavailability, Lipophilicity (LIPO): XLOGP3 between −0.7 and +5.0, Molecular weight (SIZE): MW between 150 and 500 g/mol, Polarity (POLAR) TPSA between 20 and 130 Å^2, Solubility (INSOLU): log S not higher than 6, Saturation (INSATU): fraction of carbons in the sp3 hybridization not less than 0.25, and Flexibility (FLEX): no more than 9 rotatable bonds.

Table 3. Drug likeness predictions of tested compounds.

No	MW	Number of HBA [a]	Number of HBD [b]	Log $P_{o/w}$ (iLOGP) [c]	Log S [d]	TPSA [e]	BBB Permeant [f]	Lipinski, Ghose, Veber, Egan, and Muegge Violations	Bioavailability Score	Drug-Likeness Model Score
1	374.27	2	0	3.85	Poorly soluble	31.23	No	0	0.55	−0.43
2	325.40	3	0	3.73	Moderately Soluble	40.46	Yes	0	0.55	−0.47
3	339.19	4	2	0.00	Moderately Soluble	71.69	Yes	0	0.55	0.52
4	311.37	3	1	3.11	Moderately soluble	51.46	Yes	0	0.55	0.42
5	211.26	1	2	1.91	Moderately soluble	36.02	Yes	0	0.55	−0.15

(a) number of hydrogen bond acceptors; (b) number of hydrogen bond donors; (c) lipophilicity; (d) Water solubility (SILICOS-IT [S=Soluble]); (e) topological polar surface area (Å^2); (f) Blood Brain Barrier permeant; lipinsky NorO > 10.

3. Discussion

It is noteworthy that breast cancer is the most frequently diagnosed malignant tumor and the second most deadly cancer in women. Many important developments in cancer prevention, early diagnosis, and treatment have been achieved, but the complex etiopathogenesis and the development of chemoresistance make the fight against breast cancer difficult to win. Consequently, scientific research has pursued different approaches, such as multitarget therapies able to reduce the cancer cells growth and invasion and exert very low effects on normal cells [37]. Despite the use of successful targeted and tailored therapies, many types of cancer, amongst them the triple negative breast cancers (TNBC), are still difficult to treat, mostly because of their heterogeneity and resistance onset [38]. A successful strategy in breast cancer therapy is based on the apoptosis induction in tumor cells acting on different pathways that are potential targets, with a minimal or null effect on the growth of the normal cells [39,40]. With this in mind, we studied the anticancer properties of a series of carbazole derivatives, (**1–5**), since our experience and a lot of studies from literature indicated their important antitumor activities, together with antibacterial, anti-inflammatory, and many other properties [8,20,41,42]. One of the first studied compounds with a carbazole scaffold was Ellipticine, a planar natural alkaloid, with known antitumor activities due to its ability to intercalate DNA and regulating several cell pathways with a multimodal effect [43]. Thus, we adopted two human breast cancer cell lines, namely MCF-7 and the most aggressive and metastatic MDA-MB-231, together with the normal counterpart, MCF-10A cells. The obtained viability data, resumed in Table 1, demonstrated a low to high activity in MDA-MBA cells and a moderate to null activity in MCF-7 cells, under the experimental conditions used in these assays. Particularly, the most active compounds in MDA-MB-231 cells were **3** and **4**, which also exhibited moderate activity in MCF-7 cells and a lack of cytotoxicity in the MCF-10A cells. It is worthwhile to highlight that both the compounds possess a better cytotoxic profile and selectivity with respect to the reference molecule, Ellipticine. The latter and its several synthetic derivatives possess a wide range of intracellular targets, whose exact mechanisms of action are not yet totally clear. However, our previous studies, and others from the literature, suggested a major role in DNA intercalation and inhibition of DNA topoisomerases [20,44,45]. These enzymes are implicated in the correct DNA metabolism and are the main target of numerous chemotherapeutics that produce irreversible genomic damages and cancer cell death [46]. The molecular docking approach gave us a prediction of interaction between our molecules and target proteins [47], which suggested that all the compounds could dock to the hTopo I and II in different sites and with a different mode. Particularly interesting is the case of the hTopo I, where compounds **1** and **5** dock to a loop involved into the protein dimerization processes, therefore creating some local rigidity and impairing the functional oligomerization of the complex. On the other side, compounds **3** and **4** are positioned in an area that is normally occupied by DNA, while compound **2** is located in a site that does not seem functional. Only compound **3** was found to dock to a site usually occupied by the DNA in the case of hTopo II. Next, we adopted direct inhibition assays, which indicated that only the two most active compounds, **3** and **4**, were able to block totally the hTopo I activity at a concentration of 1 μM, whereas no inhibition was recorder for the other compounds. On the contrary, neither at the same concentration nor rising up to 10 μM were compounds **3** and **4** (Figure 3), and the other ones as well, able to inhibit the hTopo II, differently from the predictive docking simulations. These results are in agreement with literature studies [12,18,48] reporting that different carbazole derivatives may block both the hTopos or, selectively, only one and induce cancer cells death. Amongst the targets of carbazole derivatives, cytoskeletal proteins have attracted the attention of many researchers, and some of them were found to provoke a net disorganization of the tubulin filaments and their accumulation around cell nuclei [22,49,50]. However, the literature is still lacking studies reporting the effects on the actin metabolism, with the exception, e.g., of a study on cell motility exerted by the carbazole derivative wiskostatin. The latter is a cell-permeable N-alkylated carbazole derivative found to be a selective inhibitor of actin filaments assembly [51]. It is known that the cell cytoskeleton is implicated in the cancer cell metastasis formation process, a dramatic phenomenon that causes many deaths for cancer. Particularly, the actin and many regulatory proteins are modified to allow the abnormal growth of cancer cells and the development of migratory properties [52,53]. Again, *in silico* and *in vitro* approaches allowed us to prove the effects on actin dynamics, which evidenced, overall, a net regulation of the actin cytoskeleton. Indeed, MDA-MB-231 cells treated with **3** or **4** change their normal morphology, because of the evident interference with intracellular actin, whose network appeared disorganized and forms bundles unevenly distributed in the cell cytoplasm. Moreover, the polymerization/depolymerization assays indicated a behavior similar to that of LA, instead of CB, which only blocks actin polymerization but not the opposite reaction, confirming what has already been observed. Finally, as expected, MDA-MB-231 cells, under exposure to compounds **3** or **4**, underwent cell death by apoptosis, recorded by means of the TUNEL assay. This ultimate effect is due to the observed hTopo I and actin filament formation inhibition necessary to sustain the uncontrolled cancer cells growth and progression. Drug-likeness has greatly impacted the most recent medicinal chemistry, which considers different molecular properties, such as hydrophobicity, size, flexibility, presence of various pharmacophores features, bioavailability, transport properties, and so on. Regarding this, the obtained scores indicated that the most active derivatives **3** and **4** do not violate any rules and are predicted to be orally active, making them the most promising compounds to be further developed.

4. Materials and Methods

4.1. Docking Studies

We built the three dimensional models of hTopo I and II, as previously described [30], using as templates the crystal structures of the hTopo I in covalent and noncovalent complexes with DNA (PDB code 1A35) [54] and of Topo IIα in complex with a short DNA fragment and etoposide (PDB Code 5gwk) [31]. The crystal structure of the complex formed between the Beta/Gamma-Actin with Profilin and the acetyltransferase AnCoA-NAA80 [55] (PDB code 6nbw) was also used as a target for the docking simulations. The structures of the tested compounds have been built and energy minimized using the program MarvinSketch (ChemAxon ltd, Budapest, Hungary). Autodock v.4.2.2. program suite [56] was employed to evaluate the possible binding modes and the binding energies of our compounds to the above mentioned proteins. We chose to adopt a "blind docking" strategy for our simulations: the docking of the compounds to the different targets were done without any a priori knowledge of the binding site by the system. All the simulations were performed adopting the standard program default values. The protein and the ligands were prepared using the ADT graphical interface [57]. For each protein, polar hydrogens were added, Kollman charges assigned, and solvation parameters calculated. While the ligands were considered as fully flexible objects, each protein was considered as full rigid. To properly calculate affinity maps, a searching grid was extended all over the protein and the search was carried out using a Lamarckian genetic algorithm. Using this protocol, a population of 100 individuals with a mutation rate of 0.02 was evolved for 100 generations and the final evaluation of the results was conducted, listing the different poses of each molecule accordingly to its predicted binding energy. Further on, an analysis cluster based on root mean squares deviation (RMSD) values of each pose with respect to the starting geometry was performed. The lowest energetic conformation of the most populated cluster was considered as the best candidate. In case two or more clusters were almost equipopulated and their energy distribution was spread, the corresponding were considered as bad ligands [26]. The docking poses resulting from our simulations were ranked in order of their binding energy values and clustered on the basis of a RMSD cut-off value of 2.0 Å. From the structural analysis of the lowest energy solutions of each cluster, we could spot the protein binding site. Figures were drawn using the program Chimera [58].

4.2. Biology

4.2.1. Cell Cultures

The used cell lines (MCF-7, MDA-MB-231 and MCF-10A) were obtained from American Type Culture Collection (ATCC, Manassas, VA, USA) and cultured as already indicated [59].

4.2.2. MTT Assay

MTT assays (Sigma Aldrich (St.Louis, MO, USA)) were employed to evaluate the *in vitro* anticancer activities of all the studied compounds, as previously described [59]. The compounds were tested at different concentrations (0.1-1-10-25-50-100 μM) for 72 h. The IC_{50} values were calculated from the percent (%) of control using GraphPad Prism 9 (GraphPad Software, La Jolla, CA, USA).

4.2.3. hTopo I Relaxation Assay and hTopo II Decatenation Assay

hTopo I relaxation assays were performed as indicated in the manufacturer's protocol (TopoGEN, Port Orange, FL, USA) with some revisions [15]. hTopo I relaxation assays were performed in a final volume of 20 μL: 0.25 μg of supercoiled pHOT1 in TE buffer [TE: 10 mM Tris-HCl (pH 7.5), 1 mM EDTA] was added to a solution containing water, 1× assay buffer (10 mM Tris-HCl (pH 7.9), 1 mM EDTA, 0.5 mM NaCl, 0.1% bovine serum albumin, 0.1 mM spermidine and 5% glycerol) and the tested compounds. The mix was incubated for 15 min at 37 °C. Then, the reaction was initiated by addition of recombinant hTopo I (2 U), incubated at 37 °C for 1 h and terminated by the addition of 5× stop buffer (5% sarkosyl, 25% glycerol, 0.125% bromophenol blue). The aqueous phase was loaded onto a 1% agarose gel containing 1× TAE buffer (diluted from50× buffer containing 242 g Tris base, 57.1 mL glacial acetic acid and 100 mL of 0.5 M EDTA) without ethidium bromide (EB). At the end, 1× TAE buffer containing EB (0.5 μg/mL) was used to stain agarose gel for 30 min and after washing with distilled water for 15 min, it was visualized using a UV transilluminator. Similarly, hTopo II decatenation assays were carried out, as indicated in the manufacturer's procedures (TopoGEN, Port Orange, FL, USA) with some revisions [15].

hTopo II decatenation assays were performed in a final volume of 20 μL: 0.3 μg of kinetoplast DNA (kDNA) was added to a solution containing 1× assay buffer (50 mM Tris-HCl, pH of 8, 150 mM NaCl, 10 mM MgCl2, 0.5 mM Dithiothreitol (DTT), 30 μg/mL bovine serum albumin (BSA), 1 mM ATP) and the tested compounds. The mix was incubated for 15 min at 37 °C. Then, the reaction was started by adding 3 U of hTopo II and incubating at 37 °C for 1 h. Then, 5× stop buffer was added and the samples were treated as described in the previous paragraph. The aqueous phase was loaded on a 1% agarose gel containing 1× TAE buffer with EB (0.5 μg/mL) and visualized using an UV transilluminator.

4.2.4. Immunofluorescence Analysis

The cells were plated and further processed, as previously indicated [30]. Specifically, the rabbit anti-β-Actin (Santa Cruz Biotechnology, Dallas, TX, USA), diluted 1:100 in bovine serum albumin (BSA) 2%, was used as primary antibody and incubated overnight at 4 °C. The secondary antibody, Alexa Fluor® 488 conjugate goat-anti-rabbit, was diluted 1:500 and incubated for 2 h at 37 °C. DAPI (4′,6-diamidino-2-phenylindole, Sigma Aldrich, Milan, Italy) 0.2 μg/mL was used for nuclei staining. A fluorescence microscope (Leica DM 6000) was utilized for fluorescence detection (40× magnification). LAS-X software allowed acquiring and processing all the fluorescence images, which are representative of three separate experiments.

4.2.5. Actin Polymerization/Depolymerization Assay

An Actin Polymerization/Depolymerization Assay Kit, purchased from Abcam, was employed to assess the ability of compounds **3** and **4** to interfere with the actin polymerization and depolymerization processes. To perform the essays, the manufacturer's instructions were followed with some modifications [60]. In particular, polymerization assay was carried out incubating in a white 96-well plate the reconstituted actin with supplemented Buffer G, compounds **3** and **4** and then Buffer P was added in order to induce actin polymerization. For the Actin Depolymerization Assay, the actin polymerization was first induced, incubating supplemented Buffers P and G at room temperature for one hour. Then, compounds **3** and **4** were added at a concentration of 5 μM. Latrunculin A (LA) and Cytochalasin B (CB) were utilized as positive control at a concentration of 5 μM. For both the assays, the assemblage of the actin filaments was defined by measuring the fluorescence (Ex/Em: 365/410 nm) in kinetic mode for 1 h at room temperature in a microplate reader.

4.2.6. Tunel Assay

TUNEL assay was employed to assess the cells apoptosis, following the manufacturer's protocols (CF™488A TUNEL Assay Apoptosis Detection Kit, Biotium, Hayward, CA, USA) with few revisions. Briefly, the cells were plated and then additional processed as previously described [30]. DAPI (0.2 μg/mL, Sigma Aldrich, Milan, Italy) was used for nuclei staining. A fluorescence microscope (Leica DM 6000) was used for fluorescence detection (20× magnification). LAS-X software allowed acquiring and processing all the fluorescence images, which are representative of three separate experiments.

4.3. In-Silico Predictive Studies

The targeted molecules were appraised for predicting the drug-likeness based on 5 separate filters, namely Lipinski, Ghose, Veber, Egan, and Muegge [61–66] rules, accompanying bioavailability and drug-likeness scores obtained using the Molsoft software and SwissADME program (http://swissadme.ch, The access date was 10 January 2022) using the ChemAxon's Marvin JS structure drawing tool.

5. Conclusions

Since the discovery of Ellipticine, carbazole derivatives have attracted the interest of the scientific world because of their versatility and wide range of applications. Herein, we described the interesting anticancer properties of a small series of carbazole derivatives observed using *in silico* and *in vitro* studies. The most active compounds **3** and **4** were found to be particularly active against the highly aggressive and metastatic MDA-MB-231 cells, without cytotoxicity on the normal counterpart. Further studies proved that they are able to selectively inhibit the hTopo I and actin polymerization, promoting, at the same time, the F-actin depolymerization. As a final and combined effect, both compounds induced apoptosis in the MDA-MB-231 breast cancer cells. Finally, these compounds deserve to be further developed as new multi-target agents in the treatment of triple negative breast cancer, currently characterized by a poor prognosis and for which few valid therapeutic options are available.

Supplementary Materials: The following supporting information can be downloaded at: https://www.mdpi.com/article/10.3390/ph16030353/s1, Figure S1: The three-dimensional structure of the human proteins Topoisomerase I (Panel A), Topoisomerase II (Panel B) and Actin (Panel C) bound to compounds **1**, **2** and **5** are drawn. Proteins are schematically reported as ribbons. Ligands binding poses are described as colored sticks.; Table S1: The main calculated pharmacokinetic descriptors studied on pkCSM predictive models.

Author Contributions: Conceptualization, M.S.S. and C.R.; methodology, P.L. and C.S.; software, C.R. and A.P.; validation, A.G.; formal analysis, A.C. (Anna Caruso); investigation, J.C. and D.I.; resources, P.L. and M.S.S.; data curation, C.S. and A.G.; writing—original draft preparation, J.C. and D.I.; writing—review and editing, D.I., A.M. and A.C. (Alessia Catalano); visualization, A.C. (Alessia Catalano); supervision, M.S.S. All authors have read and agreed to the published version of the manuscript.

Funding: This work was supported by the Ministero della Salute, Fondi Ricerca Corrente 2022 (C.R.). M.S.S. was supported by PON "R&I" 2014–2020—project from Area di Specializzazione

"Salute", ARS01_00568 titled "SI.F.I.PA.CRO.DE.—Sviluppo e industrializzazione farmaci innovativi per terapia molecolare personalizzata PA.CRO.DE." for providing lab tools.

Institutional Review Board Statement: Not applicable.

Informed Consent Statement: Not applicable.

Data Availability Statement: Not applicable.

Conflicts of Interest: The authors declare no conflict of interest.

References

1. Gluszynska, A. Biological potential of carbazole derivatives. *Eur. J. Med. Chem.* **2015**, *94*, 405–426. [CrossRef] [PubMed]
2. Caruso, A.; Chimento, A.; El-Kashef, H.; Lancelot, J.-C.; Panno, A.; Pezzi, V.; Saturnino, C.; Sinicropi, M.S.; Sirianni, R.; Rault, S. Antiproliferative activity of some 1,4-dimethylcarbazoles on cells that express estrogen receptors: Part I. *J. Enzym. Inhib. Med. Chem.* **2012**, *27*, 609–613. [CrossRef] [PubMed]
3. Caruso, A.; Voisin-Chiret, A.S.; Lancelot, J.C.; Sinicropi, M.S.; Garofalo, A.; Rault, S. Efficient and simple synthesis of 6-aryl-1,4-dimethyl-9H-carbazoles. *Molecules* **2008**, *13*, 1312–1320. [CrossRef] [PubMed]
4. Asche, C.; Demeunynck, M. Antitumor carbazoles. *Anticancer Agents Med. Chem.* **2007**, *7*, 247–267. [CrossRef] [PubMed]
5. Bjornsti, M.A.; Kaufmann, S.H. Topoisomerases and cancer chemotherapy: Recent advances and unanswered questions. *F1000Research* **2019**, *8*, F1000 Faculty Rev-1704. [CrossRef] [PubMed]
6. Salerno, S.; Da Settimo, F.; Taliani, S.; Simorini, F.; La Motta, C.; Fornaciari, G.; Marini, A.M. Recent advances in the development of dual topoisomerase I and II inhibitors as anticancer drugs. *Curr. Med. Chem.* **2010**, *17*, 4270–4290. [CrossRef]
7. You, F.; Gao, C. Topoisomerase Inhibitors and Targeted Delivery in Cancer Therapy. *Curr. Top. Med. Chem.* **2019**, *19*, 713–729. [CrossRef] [PubMed]
8. Bashir, M.; Bano, A.; Ijaz, A.S.; Chaudhary, B.A. Recent Developments and Biological Activities of N-Substituted Carbazole Derivatives: A Review. *Molecules* **2015**, *20*, 13496–13517. [CrossRef]
9. Caruso, A.; Lancelot, J.-C.; El-Kashef, H.; Sinicropi, M.S.; Legay, R.; Lesnard, A.; Rault, S. A rapid and versatile synthesis of novel pyrimido [5,4-b] carbazoles. *Tetrahedron* **2009**, *65*, 10400–10405. [CrossRef]
10. Saini, S.; Kumar, K.; Meena, S.; Dandia, A.; Ameta, K.L.; Parewa, V. An Insight into the Synthesis and Pharmacological Activities of Indoles, Isoindoles and Carbazoles. In *N-Heterocycles: Synthesis and Biological Evaluation*; Ameta, K.L., Kant, R., Penoni, A., Maspero, A., Scapinello, L., Eds.; Springer Nature Singapore: Singapore, 2022; pp. 395–413.
11. Kizek, R.; Adam, V.; Hrabeta, J.; Eckschlager, T.; Smutny, S.; Burda, J.V.; Frei, E.; Stiborova, M. Anthracyclines and ellipticines as DNA-damaging anticancer drugs: Recent advances. *Pharmacol. Ther.* **2012**, *133*, 26–39. [CrossRef]
12. Lampropoulou, E.; Manioudaki, M.; Fousteris, M.; Koutsourea, A.; Nikolaropoulos, S.; Papadimitriou, E. Pyrrolo [2,3-alpha]carbazole derivatives as topoisomerase I inhibitors that affect viability of glioma and endothelial cells in vitro and angiogenesis in vivo. *Biomed. Pharmacother.* **2011**, *65*, 142–150. [CrossRef] [PubMed]
13. Ferlin, M.G.; Gia, O.; Dalla Via, L. Synthesis and antiproliferative activity of some ellipticine-like 11H-pyrido[a]carbazole derivatives. *ChemMedChem* **2011**, *6*, 1872–1883. [CrossRef] [PubMed]
14. Chen, Z.; Yang, T.; Wang, W.; Yao, J.; Han, S.; Tao, Y.; Wang, R.; Duan, L. Synthesis and Biological Evaluation of Carbazole Aminoalcohols as Antitumor Agents. *Chem.* **2018**, *3*, 12630–12638. [CrossRef]
15. Iacopetta, D.; Rosano, C.; Puoci, F.; Parisi, O.I.; Saturnino, C.; Caruso, A.; Longo, P.; Ceramella, J.; Malzert-Freon, A.; Dallemagne, P.; et al. Multifaceted properties of 1,4-dimethylcarbazoles: Focus on trimethoxybenzamide and trimethoxyphenylurea derivatives as novel human topoisomerase II inhibitors. *Eur. J. Pharm. Sci.* **2017**, *96*, 263–272. [CrossRef] [PubMed]
16. Saturnino, C.; Caruso, A.; Iacopetta, D.; Rosano, C.; Ceramella, J.; Muia, N.; Mariconda, A.; Bonomo, M.G.; Ponassi, M.; Rosace, G.; et al. Inhibition of Human Topoisomerase II by N,N,N-Trimethylethanammonium Iodide Alkylcarbazole Derivatives. *ChemMedChem* **2018**, *13*, 2635–2643. [CrossRef]
17. Sinicropi, M.S.; Iacopetta, D.; Rosano, C.; Randino, R.; Caruso, A.; Saturnino, C.; Muia, N.; Ceramella, J.; Puoci, F.; Rodriquez, M.; et al. N-thioalkylcarbazoles derivatives as new anti-proliferative agents: Synthesis, characterisation and molecular mechanism evaluation. *J. Enzym. Inhib. Med. Chem.* **2018**, *33*, 434–444. [CrossRef]
18. Ceramella, J.; Caruso, A.; Occhiuzzi, M.A.; Iacopetta, D.; Barbarossa, A.; Rizzuti, B.; Dallemagne, P.; Rault, S.; El-Kashef, H.; Saturnino, C.; et al. Benzothienoquinazolinones as new multi-target scaffolds: Dual inhibition of human Topoisomerase I and tubulin polymerization. *Eur. J. Med. Chem.* **2019**, *181*, 111583. [CrossRef]
19. Lacey, J.V., Jr.; Devesa, S.S.; Brinton, L.A. Recent trends in breast cancer incidence and mortality. *Environ. Mol. Mutagen.* **2002**, *39*, 82–88. [CrossRef]
20. Caruso, A.; Iacopetta, D.; Puoci, F.; Cappello, A.R.; Saturnino, C.; Sinicropi, M.S. Carbazole derivatives: A promising scenario for breast cancer treatment. *Mini Rev. Med. Chem.* **2016**, *16*, 630–643. [CrossRef]
21. Issa, S.; Prandina, A.; Bedel, N.; Rongved, P.; Yous, S.; Le Borgne, M.; Bouaziz, Z. Carbazole scaffolds in cancer therapy: A review from 2012 to 2018. *J. Enzym. Inhib. Med. Chem.* **2019**, *34*, 1321–1346. [CrossRef] [PubMed]
22. Sinicropi, M.S.; Tavani, C.; Rosano, C.; Ceramella, J.; Iacopetta, D.; Barbarossa, A.; Bianchi, L.; Benzi, A.; Maccagno, M.; Ponassi, M.; et al. A Nitrocarbazole as a New Microtubule-Targeting Agent in Breast Cancer Treatment. *Appl. Sci.* **2021**, *11*, 9139. [CrossRef]

23. Vlaar, C.P.; Castillo-Pichardo, L.; Medina, J.I.; Marrero-Serra, C.M.; Velez, E.; Ramos, Z.; Hernandez, E. Design, synthesis and biological evaluation of new carbazole derivatives as anti-cancer and anti-migratory agents. *Bioorg. Med. Chem.* **2018**, *26*, 884–890. [CrossRef]
24. Butler-Fernandez, K.M.; Ramos, Z.; Francis-Malave, A.M.; Bloom, J.; Dharmawardhane, S.; Hernandez, E. Synthesis, Anti-Cancer and Anti-Migratory Evaluation of 3,6-Dibromocarbazole and 5-Bromoindole Derivatives. *Molecules* **2019**, *24*, 2686. [CrossRef]
25. Caruso, A.; Voisin-Chiret, A.S.; Lancelot, J.C.; Sinicropi, M.S.; Garofalo, A.; Rault, S. Novel and Efficient Synthesis of 5,8-Dimethyl-9H-carbazol-3-ol via a Hydroxydeboronation Reaction. *Heterocycles* **2007**, *71*, 2203–2210. [CrossRef]
26. Rosano, C.; Lappano, R.; Santolla, M.F.; Ponassi, M.; Donadini, A.; Maggiolini, M. Recent advances in the rationale design of GPER ligands. *Curr. Med. Chem.* **2012**, *19*, 6199–6206. [CrossRef]
27. Viale, M.; Cordazzo, C.; de Totero, D.; Budriesi, R.; Rosano, C.; Leoni, A.; Ioan, P.; Aiello, C.; Croce, M.; Andreani, A.; et al. Inhibition of MDR1 activity and induction of apoptosis by analogues of nifedipine and diltiazem: An in vitro analysis. *Investig. New Drugs* **2011**, *29*, 98–109. [CrossRef]
28. Santolla, M.F.; De Francesco, E.M.; Lappano, R.; Rosano, C.; Abonante, S.; Maggiolini, M. Niacin activates the G protein estrogen receptor (GPER)-mediated signalling. *Cell Signal* **2014**, *26*, 1466–1475. [CrossRef]
29. Sanchez-Carranza, J.N.; Gonzalez-Maya, L.; Razo-Hernandez, R.S.; Salas-Vidal, E.; Nolasco-Quintana, N.Y.; Clemente-Soto, A.F.; Garcia-Arizmendi, L.; Sanchez-Ramos, M.; Marquina, S.; Alvarez, L. Achillin Increases Chemosensitivity to Paclitaxel, Overcoming Resistance and Enhancing Apoptosis in Human Hepatocellular Carcinoma Cell Line Resistant to Paclitaxel (Hep3B/PTX). *Pharmaceutics* **2019**, *11*, 512. [CrossRef] [PubMed]
30. Iacopetta, D.; Mariconda, A.; Saturnino, C.; Caruso, A.; Palma, G.; Ceramella, J.; Muia, N.; Perri, M.; Sinicropi, M.S.; Caroleo, M.C.; et al. Novel Gold and Silver Carbene Complexes Exert Antitumor Effects Triggering the Reactive Oxygen Species Dependent Intrinsic Apoptotic Pathway. *ChemMedChem* **2017**, *12*, 2054–2065. [CrossRef] [PubMed]
31. Wang, Y.R.; Chen, S.F.; Wu, C.C.; Liao, Y.W.; Lin, T.S.; Liu, K.T.; Chen, Y.S.; Li, T.K.; Chien, T.C.; Chan, N.L. Producing irreversible topoisomerase II-mediated DNA breaks by site-specific Pt(II)-methionine coordination chemistry. *Nucleic Acids Res.* **2017**, *45*, 10861–10871. [CrossRef]
32. Arne, K.; Gonçalves, M.V.; Carsten, W.; Thales, K. ADME Profiling in Drug Discovery and a New Path Paved on Silica. In *Drug Discovery and Development*; Vishwanath, G., Partha, K., Ashit, T., Eds.; IntechOpen: Rijeka, Croatia, 2019; p. 6.
33. Pires, D.E.; Blundell, T.L.; Ascher, D.B. pkCSM: Predicting Small-Molecule Pharmacokinetic and Toxicity Properties Using Graph-Based Signatures. *J. Med. Chem.* **2015**, *58*, 4066–4072. [CrossRef]
34. Benet, L.Z.; Hosey, C.M.; Ursu, O.; Oprea, T.I. BDDCS, the Rule of 5 and drugability. *Adv. Drug Deliv. Rev.* **2016**, *101*, 89–98. [CrossRef] [PubMed]
35. Jagannathan, R. Characterization of Drug-like Chemical Space for Cytotoxic Marine Metabolites Using Multivariate Methods. *ACS Omega* **2019**, *4*, 5402–5411. [CrossRef] [PubMed]
36. Daina, A.; Michielin, O.; Zoete, V. SwissADME: A free web tool to evaluate pharmacokinetics, drug-likeness and medicinal chemistry friendliness of small molecules. *Sci. Rep.* **2017**, *7*, 42717. [CrossRef] [PubMed]
37. Xie, L.; Bourne, P.E. Developing multi-target therapeutics to fine-tune the evolutionary dynamics of the cancer ecosystem. *Front. Pharmacol.* **2015**, *6*, 209. [CrossRef]
38. Catalano, A.; Iacopetta, D.; Ceramella, J.; Mariconda, A.; Rosano, C.; Scumaci, D.; Saturnino, C.; Longo, P.; Sinicropi, M.S. New Achievements for the Treatment of Triple-Negative Breast Cancer. *Appl. Sci.* **2022**, *12*, 5554. [CrossRef]
39. Iacopetta, D. Special Issue on "Anticancer Drugs Activity and Underlying Mechanisms". *Appl. Sci.* **2021**, *11*, 8169. [CrossRef]
40. Sellers, W.R.; Fisher, D.E. Apoptosis and cancer drug targeting. *J. Clin. Investig.* **1999**, *104*, 1655–1661. [CrossRef]
41. Ceramella, J.; Iacopetta, D.; Barbarossa, A.; Caruso, A.; Grande, F.; Bonomo, M.G.; Mariconda, A.; Longo, P.; Carmela, S.; Sinicropi, M.S. Carbazole Derivatives as Kinase-Targeting Inhibitors for Cancer Treatment. *Mini Rev. Med. Chem.* **2020**, *20*, 444–465. [CrossRef]
42. Wang, G.; Sun, S.; Guo, H. Current status of carbazole hybrids as anticancer agents. *Eur. J. Med. Chem.* **2022**, *229*, 113999. [CrossRef]
43. Mazumder, K.; Aktar, A.; Roy, P.; Biswas, B.; Hossain, M.E.; Sarkar, K.K.; Bachar, S.C.; Ahmed, F.; Monjur-Al-Hossain, A.S.M.; Fukase, K. A Review on Mechanistic Insight of Plant Derived Anticancer Bioactive Phytocompounds and Their Structure Activity Relationship. *Molecules* **2022**, *27*, 3036. [CrossRef] [PubMed]
44. Denny, W.A.; Baguley, B.C. Dual topoisomerase I/II inhibitors in cancer therapy. *Curr. Top. Med. Chem.* **2003**, *3*, 339–353. [CrossRef] [PubMed]
45. Stiborova, M.; Frei, E. Ellipticines as DNA-targeted chemotherapeutics. *Curr. Med. Chem.* **2014**, *21*, 575–591. [CrossRef]
46. McClendon, A.K.; Osheroff, N. DNA topoisomerase II, genotoxicity, and cancer. *Mutat. Res.* **2007**, *623*, 83–97. [CrossRef]
47. Meng, X.Y.; Zhang, H.X.; Mezei, M.; Cui, M. Molecular docking: A powerful approach for structure-based drug discovery. *Curr. Comput. Aided Drug Des.* **2011**, *7*, 146–157. [CrossRef]
48. Facompre, M.; Carrasco, C.; Colson, P.; Houssier, C.; Chisholm, J.D.; Van Vranken, D.L.; Bailly, C. DNA binding and topoisomerase I poisoning activities of novel disaccharide indolocarbazoles. *Mol. Pharmacol.* **2002**, *62*, 1215–1227. [CrossRef] [PubMed]
49. Naret, T.; Khelifi, I.; Provot, O.; Bignon, J.; Levaique, H.; Dubois, J.; Souce, M.; Kasselouri, A.; Deroussent, A.; Paci, A.; et al. 1,1-Diheterocyclic Ethylenes Derived from Quinaldine and Carbazole as New Tubulin-Polymerization Inhibitors: Synthesis, Metabolism, and Biological Evaluation. *J. Med. Chem.* **2019**, *62*, 1902–1916. [CrossRef]

50. Padmaja, P.; Rao, G.K.; Indrasena, A.; Reddy, B.V.S.; Patel, N.; Shaik, A.B.; Reddy, N.; Dubey, P.K.; Bhadra, M.P. Synthesis and biological evaluation of novel pyrano[3,2-c]carbazole derivatives as anti-tumor agents inducing apoptosis via tubulin polymerization inhibition. *Org. Biomol. Chem.* **2015**, *13*, 1404–1414. [CrossRef] [PubMed]
51. Pfannes, E.K.; Theves, M.; Wegner, C.; Beta, C. Impact of the carbazole derivative wiskostatin on mechanical stability and dynamics of motile cells. *J. Muscle Res. Cell Motil.* **2012**, *33*, 95–106. [CrossRef]
52. Aseervatham, J. Cytoskeletal Remodeling in Cancer. *Biology* **2020**, *9*, 385. [CrossRef]
53. Datta, A.; Deng, S.; Gopal, V.; Yap, K.C.; Halim, C.E.; Lye, M.L.; Ong, M.S.; Tan, T.Z.; Sethi, G.; Hooi, S.C.; et al. Cytoskeletal Dynamics in Epithelial-Mesenchymal Transition: Insights into Therapeutic Targets for Cancer Metastasis. *Cancers* **2021**, *13*, 1882. [CrossRef]
54. Redinbo, M.R.; Stewart, L.; Kuhn, P.; Champoux, J.J.; Hol, W.G. Crystal structures of human topoisomerase I in covalent and noncovalent complexes with DNA. *Science* **1998**, *279*, 1504–1513. [CrossRef]
55. Rebowski, G.; Boczkowska, M.; Drazic, A.; Ree, R.; Goris, M.; Arnesen, T.; Dominguez, R. Mechanism of actin N-terminal acetylation. *Sci. Adv.* **2020**, *6*, eaay8793. [CrossRef]
56. Morris, G.M.; Huey, R.; Lindstrom, W.; Sanner, M.F.; Belew, R.K.; Goodsell, D.S.; Olson, A.J. AutoDock4 and AutoDockTools4: Automated docking with selective receptor flexibility. *J. Comput. Chem.* **2009**, *30*, 2785–2791. [CrossRef]
57. Sanner, M.F.; Duncan, B.S.; Carrillo, C.J.; Olson, A.J. Integrating computation and visualization for biomolecular analysis: An example using python and AVS. *Pac. Symp. Biocomput.* **1999**, 401–412.
58. Pettersen, E.F.; Goddard, T.D.; Huang, C.C.; Couch, G.S.; Greenblatt, D.M.; Meng, E.C.; Ferrin, T.E. UCSF Chimera–a visualization system for exploratory research and analysis. *J. Comput. Chem.* **2004**, *25*, 1605–1612. [CrossRef]
59. Iacopetta, D.; Rosano, C.; Sirignano, M.; Mariconda, A.; Ceramella, J.; Ponassi, M.; Saturnino, C.; Sinicropi, M.S.; Longo, P. Is the Way to Fight Cancer Paved with Gold? Metal-Based Carbene Complexes with Multiple and Fascinating Biological Features. *Pharmaceuticals* **2020**, *13*, 91. [CrossRef]
60. Ceramella, J.; Mariconda, A.; Sirignano, M.; Iacopetta, D.; Rosano, C.; Catalano, A.; Saturnino, C.; Sinicropi, M.S.; Longo, P. Novel Au Carbene Complexes as Promising Multi-Target Agents in Breast Cancer Treatment. *Pharmaceuticals* **2022**, *15*, 507. [CrossRef]
61. Kanamori, K.; Roberts, J.D. Nitrogen-15 nuclear magnetic resonance study of benzenesulfonamide and cyanate binding to carbonic anhydrase. *Biochemistry* **1983**, *22*, 2658–2664. [CrossRef]
62. Lipinski, C.A. Lead- and drug-like compounds: The rule-of-five revolution. *Drug Discov. Today Technol.* **2004**, *1*, 337–341. [CrossRef]
63. Egan, W.J.; Merz, K.M., Jr.; Baldwin, J.J. Prediction of drug absorption using multivariate statistics. *J. Med. Chem.* **2000**, *43*, 3867–3877. [CrossRef]
64. Ghose, A.K.; Viswanadhan, V.N.; Wendoloski, J.J. A knowledge-based approach in designing combinatorial or medicinal chemistry libraries for drug discovery. 1. A qualitative and quantitative characterization of known drug databases. *J. Comb. Chem.* **1999**, *1*, 55–68. [CrossRef] [PubMed]
65. Muegge, I.; Heald, S.L.; Brittelli, D. Simple selection criteria for drug-like chemical matter. *J. Med. Chem.* **2001**, *44*, 1841–1846. [CrossRef] [PubMed]
66. Veber, D.F.; Johnson, S.R.; Cheng, H.Y.; Smith, B.R.; Ward, K.W.; Kopple, K.D. Molecular properties that influence the oral bioavailability of drug candidates. *J. Med. Chem.* **2002**, *45*, 2615–2623. [CrossRef] [PubMed]

pharmaceuticals

MDPI

Article

Phenotypic Discovery of Thiocarbohydrazone with Anticancer Properties and Catalytic Inhibition of Human DNA Topoisomerase IIα

Ilija N. Cvijetić [1,2], Barbara Herlah [2,3], Aleksandar Marinković [4], Andrej Perdih [2,3,*] and Snežana K. Bjelogrlić [5]

1 Faculty of Chemistry, University of Belgrade, Studentski trg 12-16, 11000 Belgrade, Serbia
2 National Institute of Chemistry, Hajdrihova 19, SI 1000 Ljubljana, Slovenia
3 Faculty of Pharmacy, University of Ljubljana, Aškerčeva 7, SI 1000 Ljubljana, Slovenia
4 Faculty of Technology and Metallurgy, University of Belgrade, Karnegijeva 4, 11120 Belgrade, Serbia
5 National Cancer Research Center, Pasterova 14, 11000 Belgrade, Serbia
* Correspondence: andrej.perdih@ki.si; Tel.: +386-1-4760-376

Abstract: Phenotypic screening of α-substituted thiocarbohydrazones revealed promising activity of 1,5-bis(salicylidene)thiocarbohydrazide against leukemia and breast cancer cells. Supplementary cell-based studies indicated an impairment of DNA replication via the ROS-independent pathway. The structural similarity of α-substituted thiocarbohydrazone to previously published thiosemicarbazone catalytic inhibitors targeting the ATP-binding site of human DNA topoisomerase IIα prompted us to investigate the inhibition activity on this target. Thiocarbohydrazone acted as a catalytic inhibitor and did not intercalate the DNA molecule, which validated their engagement with this cancer target. A comprehensive computational assessment of molecular recognition for a selected thiosemicarbazone and thiocarbohydrazone provided useful information for further optimization of this discovered lead compound for chemotherapeutic anticancer drug discovery.

Keywords: thiocarbohydrazones; human DNA topoisomerase IIα; catalytic inhibitors; dynophores; molecular dynamics; cancer research; phenotypic screening

1. Introduction

Cancer is the second leading cause of death worldwide, and, according to the World Health Organization (WHO), it was responsible for nearly 10 million deaths in 2020, or nearly one in six deaths [1]. It comprises a diverse group of diseases that result from abnormal cell growth and can potentially invade and metastasize to other parts of the body. One of the hallmarks of cancer is rapid, uncontrolled cell proliferation [2,3], and the inhibition of this process has been the focus of cancer research since its inception, resulting in many efficient chemotherapy regimens [4].

Rapidly dividing cancer cells require the enhanced activity of a family of DNA topoisomerases, efficient biological nanomachines that catalyze formation of either transient single-strand breaks (type I topoisomerases, topo I) or double-strand breaks (type II topoisomerases, topo II) and regulate the topological changes of the DNA molecule. An important member of this family is the human topoisomerase IIα, an ATP-dependent enzyme [5] that exists in α and β isoforms [6] and shares approximately 70% sequence similarity but is differentially regulated during cell growth. Topo IIα is elevated in proliferating cells, whereas topo IIβ is present in proliferating as well as postmitotic cells. Human topo IIα represents a main target of the type II family for cancer therapies and is targeted by many established anticancer drugs such as etoposide, doxorubicin, daunorubicin, and mitoxantrone [7,8]. These compounds are classified as topoisomerase poisons because they exert anticancer activity by stabilizing the transient covalent DNA–topo II complex, which blocks DNA replication and transcription and promotes cell apoptosis. Although they are clinically highly efficient, the use of topo II poisons is limited by rapidly developing cancer resistance and severe side effects such as the induction of secondary malignancies and cardiotoxicity [7,9–11]. To avoid these side effects, an emerging group of catalytic topo II inhibitors was extensively studied [12]. These compounds interfere with the catalytic cycle of topo II without inducing DNA damage, acting through several mechanisms such as preventing DNA cleavage, competing with ATP for the same binding site [13–15] to prevent ATP hydrolysis, and interfering with DNA–topo II binding [16].

The design of dual-target or multitarget inhibitors is a promising approach to overcome cancer cell resistance and the side effects of topoisomerase II inhibitors while reducing the pharmacokinetic issues associated with combinatorial therapy. One of the proposed strategies is to target proteins that are structurally related to topo II, such as Hsp90, and kinases that have similar ATP-binding domains [17]. Alternatively, molecular hybridization of two pharmacophores resulting in dual-binding inhibitors, such as the daunorubicin-suberoylanilide hydroxamic acid (SAHA) hybrid that targets topo II and histone deacetylases (HDACs), also presents itself as a viable new strategy [18].

Intracellular targets as well as the polypharmacological profiles of the newly studied anticancer agents are often discovered retrospectively. In silico tools that analyze pharmacophoric similarity with drugs with known mechanisms of action are often very useful for this task [19]. Moreover, the use of molecular dynamics (MD) simulations in combination with pharmacophore modeling is a successful approach in the development of potent telomerase inhibitors as anticancer drugs [20].

In the last 10 years, there has been a resurgence of interest in phenotypic drug discovery (PDD) in both academic research and the pharmaceutical industry [21]. PDD screening yields hit compounds with diverse mechanisms of action, but the further development of optimized compounds is complicated by additional variables such as cell permeability, potential polypharmacology, binding to transport proteins, and metabolic stability. Although there are successful examples of hit-to-lead optimization using ligand-based structure–activity relationships (SAR), lack of target knowledge is considered a major risk for clinical development and regulatory approval [22].

Thiosemicarbazones (TSCs) are a well-known class of molecules with anticancer properties. Among them, the α-(N)-heterocyclic TSCs are of particular interest due to their ability to also chelate metal ions. These ligands bind strongly to Fe ions and inhibit metal-dependent enzymes such as ribonucleotide reductase, which is essential for DNA biosynthesis. The best-known example of this class of compounds is 3-aminopyridine-2-carboxaldehyde TSC (3-AP, triapine), a ribonucleotide reductase inhibitor that was tested in more than 30 Phase I and Phase II trials and is currently in Phase III clinical trials for radiotherapy in combination with cisplatin [23]. The inhibitory effect of triapine on topo II was investigated. Yalowich et. al. reported that triapine and some other TSCs did not induce cleavage of plasmid DNA or inhibit topo IIα decatenation [24]. On the other hand, Huang et. al. reported a series of α-(N)-heterocyclic TSC derivatives (Figure 1), such as compound **TSC24,** that act as catalytic inhibitors of topo IIα and bind to the ATPase domain where the ATP binding site is located [25].

Thiocarbohydrazones (TCHs) are higher homologs of TSCs with an additional N atom that can act as a metal-coordinating center. Compared to TSCs, the reports on the anticancer activity and mechanistic studies of TCHs are scarce. In a previous study, a series of mono- and bis-TCHs displayed a polypharmacological profile of anticancer activity with strong indications of a multitarget mechanism of action [26]. Moreover, salicylaldehyde monothiocarbohydrazone was reported to act as a copper ion ionophore and an antiproliferative agent against breast cancer and human prostate adenocarcinoma cell lines with low toxicity to normal human keratinocytes [27]. It was reported that the addition of Cu^{2+} increased the antiproliferative activity of salicylaldehyde mono-TCH. In addition, the complex of 1,5-bis(salicylidene)thiocarbohydrazide (compound **2**, Figure 1) and Cu^{2+} formed in situ efficiently cleaved DNA via oxidative and hydrolytic pathways, resulting in significant antiproliferative activity against HeLa and MCF-7 cancer cell lines [28]. However, no studies on the activity of TCHs on topo IIα have been reported so far.

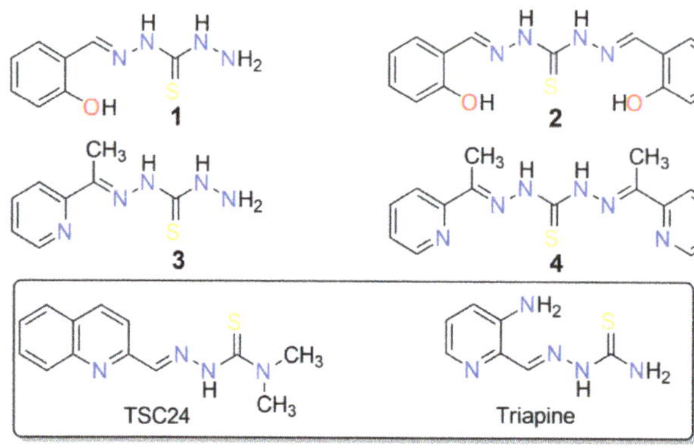

Figure 1. Structures of thiocarbohydrazones **1–4** investigated in this study, along with structurally similar thiosemicarbazones (triapine) that possess anticancer properties, and **TSC24,** which additionally inhibits the human DNA topoisomerase IIα.

Here, we report the phenotypic screening of a small series of mono- and bis-TCHs (salicylaldehyde or 2-acetylpyridine) bearing an α-(N)-atom or an α-hydroxyl group as metal chelating centers and radical scavenging groups (Figure 1) against acute monocytic leukemia (THP-1), breast adenocarcinoma (MCF-7), and pancreatic adenocarcinoma (AsPC-1) cell lines. Encouraged by the observed activity of these compounds and their similarity to the TSC-based compound **TSC24,** a known catalytic topo IIα inhibitor [25], we then investigated their topo IIα inhibitory activity. For the most promising compound (**2**), we performed additional biochemical assays to further investigate its mechanism of action. Computational study of the binding properties of the active compounds **2** and **TSC24** in the ATP binding site of topo IIα, with molecular simulations, dynamic pharmacophore models, and MM/GBSA binding free energy calculations enabled a deeper insight into molecular recognition and provided information for further optimization. We also performed target fishing and molecular docking to outline other plausible targets and evaluated TCH's drug-like properties.

2. Results and Discussion

2.1. Phenotypic Screening of Thiocarbohydrazones on Human Cancer Cell Lines

First, we tested the proapoptotic effect of thiocarbohydrazones **1–4** on leukemia (THP-1), breast cancer (MCF-7), and prostate adenocarcinoma (AsPC-1) cancer cell lines. Apoptosis is the main type of cell death in THP-1 and MCF-7, whereas AsPC-1 proved to be a highly resistant cancer stem cell, with its cell death events being exclusively necrotic [29, 30]. The obtained results shown in Table S1 demonstrate that compound **4** induced cell death in three cell lines and suggest a higher efficacy of bis-TCH derivatives compared to mono-TCHs.

The apoptotic response was concentration-dependent for all compounds except for compound **2**, 1,5-bis(salicylidene)thiocarbohydrazide (Figure S1), for which a non-standard biphasic curve with two exponential phases and a middle plateau situated between 10 and 50 μM was observed (Figure S2). These changes strongly correlate with the distribution of THP-1 cells within the phases of mitotic division (Figure 2A). The suspended increase in the percentage of apoptotic cells detected at 10 and 50 μM of compound **2** coincides with the cell cycle arrest at the G2 checkpoint. A burst in the incidence of cells in the advanced phases of apoptosis at 75 μM of compound **2** was accompanied by a decreased frequency of cells at the G2/M phase and an apparently restored G1 phase, whereas compound **2** at 100 μM stimulated arrest at the G1-to-S transition point. The described fluctuations in apoptotic response and cell cycle distribution imply that the DNA repairing machinery is initiated and operates in THP-1 cells treated with compound **2** within 10–50 μM concentration range, which strongly indicates that compound **2** interferes with chromosomal replication. At this point, it remains unclear whether compound **2** at concentrations greater than 50 μM affects the THP-1 cells via the same mechanism, or if another one emerges.

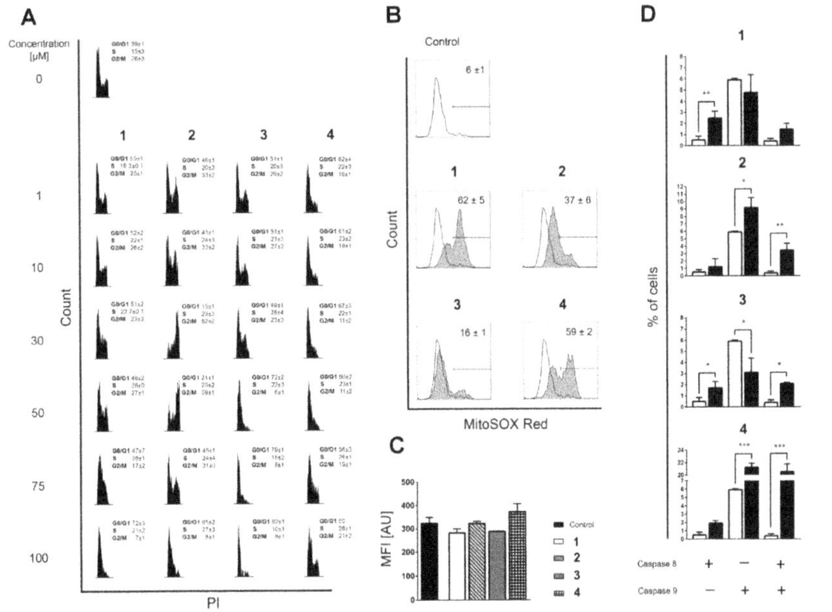

Figure 2. (**A**) Distribution of THP-1 cells within phases of mitotic division assessed on the remaining cells after Annexin V/PI readouts. Incidences of cells in phases G0/G1, S, and G2/M were determined according to the distribution of cells in nontreated populations. All results are expressed as mean ± SD of two replicates from independent experiments. (**B**) Mitochondrial superoxide ($O_2^{\bullet-}$) generation in nontreated and cells treated with the investigated compounds, determined after 6 h of incubation by using MitoSOX Red staining. Results represent mean ± SD percentage of cells positive for mitochondrial $O_2^{\bullet-}$ production in three replicates from independent experiments. (**C**) Mean fluorescent intensity (MFI) computed for $O_2^{\bullet-}$-positive subpopulations, expressed in arbitrary units (AU). Results are presented as the mean ± SD of three replicates from independent experiments. The Kruskal–Wallis test showed no statistical difference in mitochondrial $O_2^{\bullet-}$ accumulation between the analyzed samples. (**D**) Percentages of cells positive for activated caspase-8, caspase-9, or both caspases acquired in nontreated samples (open bars) or in samples treated for 6 h with the investigated compounds (closed bars). Bars represent the mean ± SD of three replicates from independent experiments. Statistical evaluation was performed using an unpaired *t*-test with Welch's correction comparing treated to non-treated populations. Significant differences are marked with *, **, and ***.

Although compounds **1**, **3**, and **4** also arrested the THP-1 cells at the G0/G1 phase or at the G1-to-S checkpoint, none of these changes provide an indication that DNA repair is ongoing at any of the concentrations tested (Figure 2A). DNA replication is a very complex process that may be interfered with in various ways [31,32]. Therefore, it was interesting to determine whether the currently investigated compounds stimulate the production of reactive oxygen species (ROS) that may damage DNA integrity and compromise its replication [33]. Therefore, we assessed the impact of compounds **1–4** on the mitochondrial superoxide anion ($O_2^{\bullet-}$) production in mitochondria, which is the cellular organelle known as the key source of ROS generation. The compounds display quite different pro-oxidant activity regarding the percentage of cells positive for $O_2^{\bullet-}$ (Figure 2B), but the values of mean fluorescence intensity (MFI) in all treated samples are on the level of non-treated controls (Figure 2C). The MFI results indicate that $O_2^{\bullet-}$ does not accumulate but is most likely rapidly neutralized by the mitochondrial superoxide dismutase [34], implying that DNA damage by ROS is unlikely to be the reason why DNA repair machinery was launched after the treatment of THP-1 cells with compound **2**.

2.2. Similarity-Based Quest for Possible Targets: Human DNA Topoisomerase IIα

In search of the possible molecular targets that could rationalize the observations obtained in the phenotypic screening, we came across a structurally similar thiosemicarbazone derivative called compound **TSC24** (Figure 1). To quantify its similarity with our series, we calculated the Tanimoto maximum common structure similarity (MCS) index for compounds **2** and **3** with **TSC24** and obtained values of 0.60 and 0.38, respectively [35,36]. This compound acted as a human topo IIα catalytic inhibitor and potentially targeted the ATP binding site on the enzyme's ATPase domain. To test whether the thiocarbohydrazones target the same enzyme, we first screened compounds **1–4** and the etoposide, a known topoisomerase poison, at a concentration of 50 μM with HTS topo IIα relaxation assay [37]. The assay was performed as previously described [15]. The selected screening concentration corresponded to the concentration that induced the cell cycle arrest at the G2 phase in THP-1 cells treated with compound **2** (Figure 2A). The percentage of topo IIα inhibition for etoposide was in line with the literature [13], and the results (Table S2) showed first indications that the α-hydroxyphenyl-thiocarbohydrazones (TCHs) can inhibit topo IIα. The most promising compound was 1,5-bis(salicylidene)thiocarbohydrazide (compound **2**) with a higher percentage of inhibition than etoposide (83% vs. 70%) at 50 μM. This highlights that compound **2** may be a potential inhibitor of topo IIα, and it allows speculation on the underlying cause of the arrest of THP-1 cells at the G2 checkpoint after treatment with 50 μM of compound **2** (Figure 2A). Moreover, the activation of the intrinsic apoptotic pathway for bis-TCH analog compound **2** suggests that the damaged cellular homeostasis is responsible for its pro-apoptotic activity (Figure 2D) [38,39].

To study the mechanism of topo IIα inhibition in detail, additional assays are required, as topo II operates as a complex molecular machine [40]. Thus, we focused on the most promising thiocarbohydrazone, compound **2**, from the initial HTS assay and performed a topo IIα-mediated decatenation assay together with etoposide as a reference molecule. The results of this assay allow direct observation of the inhibition process on the gel. Analysis of the data obtained for etoposide showed that the first traces of decatenated kinetoplast DNA (kDNA) appeared at 125 μM and became more intense at 31.5 μM, which is consistent with the literature (Figure 3A). Furthermore, compound **2** inhibited the decatenation of kDNA in a concentration-dependent manner, with 56% inhibition occurring at 50 μM (Figure 3A) (see also Figure S3A and Table S3 for more details).

To distinguish the catalytic inhibitors from the topoisomerase poisons, we performed a topo IIα-mediated cleavage assay. The topoisomerase II poisons act by stabilizing the short-lived DNA-enzyme cleavage complex, which prevents the relegation of the plasmid and leads to the accumulation of the linear, cleaved plasmid. The results shown in Figure 3B (and in Figure S3B and Table S4) confirm that etoposide gradually increases the amount of linear plasmid. In the case of compound **2**, there was no significant increase in linear plasmid at any concentration tested, indicating that compound **2** acts as a catalytic inhibitor of topo IIα.

In addition, we examined the activity of compound **2** as a DNA intercalator using the unwinding assay with mAMSA as a positive control. The results in Figures 3C and S3C show that the intercalation of mAMSA was completed at 62.5 μM, whereas no unwinding was observed even at 250 μM of compound **2** with supercoiled or relaxed substrate. Therefore, compound **2** is not a DNA intercalator. To obtain further direct evidence of topo IIα inhibition, we also performed a topo IIα-mediated relaxation assay on the gel. The results shown in Figure 3D (and in Figure S3D and Table S5) independently confirmed the results of the decatenation assay, i.e., that compound **2** inhibits topo II-mediated DNA topology changes in a concentration-dependent manner (e.g., 61% inhibition at 62.5 μM). Taken together, the performed biochemical assays strongly imply that compound **2** acts as a catalytic inhibitor of human topo IIα.

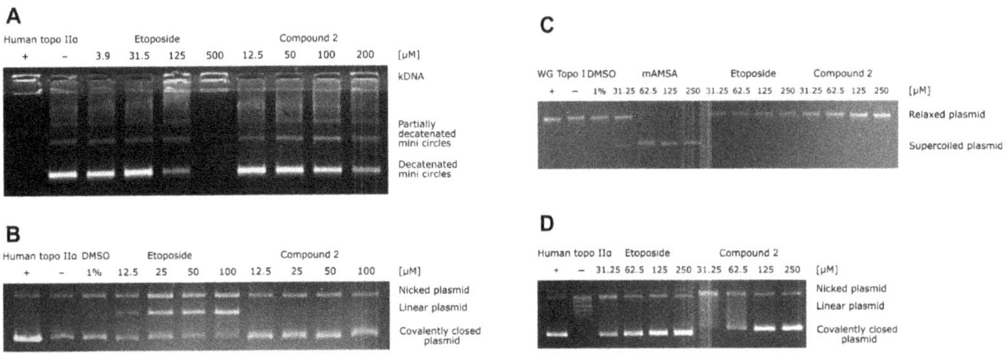

Figure 3. (**A**) Results of the human topo IIα-mediated decatenation assay. The assay was performed for four different concentrations of compound **2** (12.5, 50, 100, and 200 μM) and for etoposide as a positive control (3.9, 31.5, 125, and 500 μM). (**B**) Human topo IIα-mediated cleavage assay. The assay was performed for four different concentrations of compound **2** (12.5, 25, 50, and 100 μM), and etoposide was used as a positive control (12.5, 25, 50, and 100 μM). (**C**) Unwinding assay using supercoiled substrate. The assay was performed for four different concentrations of compound **2** and etoposide (31.5, 62.5, 125, and 250 μM) and of the positive control intercalator mAMSA (31.5, 62.5, 125, and 250 μM). (**D**) Results of the topoisomerase IIα-mediated relaxation assay. The assay was performed for four different concentrations of compound **2** (31.5, 62.5, 125, and 250 μM) and etoposide as positive control (31.5, 62.5, 125, and 250 μM).

2.3. Thiosemicarbazide and Thiocarbohydrazone Binding to Topo IIα ATPase Domain Investigated with Molecular Simulations

The structural resemblance between thiosemicarbazone **TSC24** and thiocarbohydrazone **2** further implies that both compounds could target the ATP binding site located on the topo IIα ATPase domain. Currently, there is no available crystal structure of human topoisomerase IIα in complex with a catalytic inhibitor bound to the ATP binding site; thus, the binding mode of **TSC24** was proposed with molecular docking using the available structure of the ATPase domain of human topo IIα, as in the original study [25]. The quinoline moiety of **TSC24** was placed deep in the binding pocket, forming mainly hydrophobic interactions with Ile125, Ile141, and Phe142, as well as with Asn95 and Asn120 residues located within the adenine sub-pocket (Figure 4A). This binding mode differs from the previously published mode [25], wherein the α-N atom of the quinoline moiety was reported to interact with the hydroxyl group of Ser149. This placed the **TSC24** ligand in the middle of the adenine binding pocket near the sugar-binding region. This difference in binding poses likely originates from the inclusion of a Mg^{2+} ion in the definition of the active site [25]. As Mg^{2+} generally binds to the enzymes in a complex with ATP [41], we omitted this ion from the active site in our docking experiments, as such a binding site topology could be considered more realistic [13–15,40].

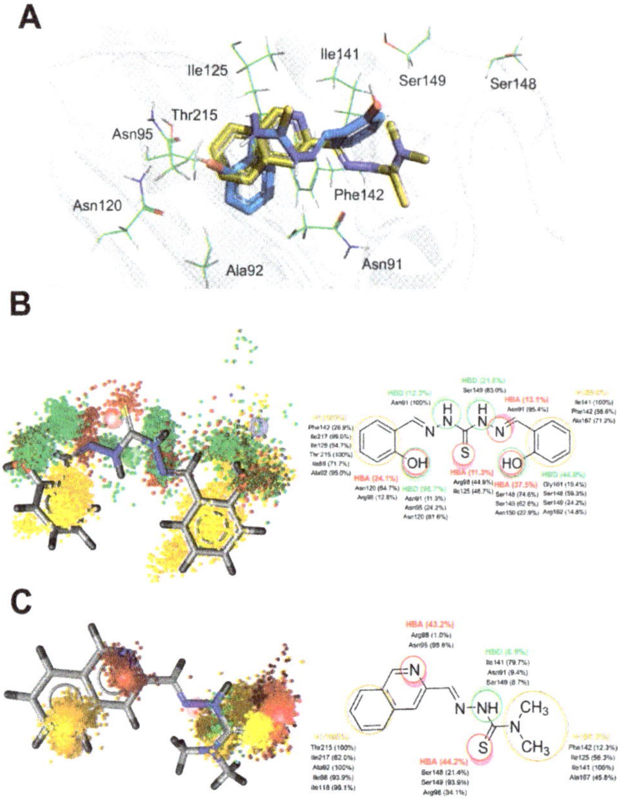

Figure 4. (**A**) Representative binding poses of compounds **2** and **TSC24** within the ATPase domain of human topo IIα obtained via molecular docking (PDB:1ZXM); dynophore models represented as superfeature clouds (left) and the contribution of individual amino acid residues to each interaction (right) for compounds **2** (**B**) and **TSC24** (**C**).

We also docked thiocarbohydrazone **2**, and one of its phenolic moieties was positioned within the adenine part of the ATP pocket, where it interacted with residues Asn120 and Asn95. The hydrazide NH atoms were located near Asn91, a part of the phosphate-binding region of the ATPase domain. Both aromatic rings of compound **2** were oriented towards the hydrophobic pocket and comprised residues Ile125, Ile141, and Phe142 (Figure 4A).

Starting from the docking poses of ligands **TSC24** and **2** in the ATP binding site, we then performed 200 ns of unrestrained MD simulations of a fully solvated system. For compound **2**, we observed in the performed replica 1 (R1) simulation that the interactions with residues defining the adenine subpocket (e.g., Asn120 and Asn95) as well as the hydrophobic subpocket (Ile125, Ile141, and Phe142) were mostly conserved during the entire trajectory. On the other hand, the thiocarbohydrazide linker and the second phenolic group adopted two orientations in the binding site, which we labeled as CF1 and CF2 (Figure S4). To better capture the dynamics of these interactions, we performed two additional simulations termed R2 and R3, each with a length of 200 ns that started from the different initial structures generated by prolonging the NPT equilibration stage. The results indicate that the CF1 and CF2 orientations were exchanged throughout the entire trajectory. The root mean square deviations (RMSD) of the ligand **2** for three replicas of the MD simulation ranged from 1.28 Å for R1 to 1.75 Å for R3 (Figure S5A and Table S6). The bound conformation of the ligand **TSC24** displayed less conformational variation within the ATP binding site during the simulations. Lower average RMSD values of the ligand, between 0.92 Å for the first simulation (R1) and 0.95 Å for the second replica (R2), suggested general conformational confinement to the docking mode (Figure S5B and Table S7). Because the **TSC24** binding conformation was strongly preserved and stable in both replicas R1 and R2, we did not perform a third simulation.

To complement the geometric analysis of the binding modes of thiocarbohydrazone **2** and the compound **TSC24**, we also calculated the dynamical pharmacophores (dynophores) [42] for all performed simulations. With this approach, one can more effectively investigate the hydrophobic interactions as well as H-bonds between the ligand and the binding site. The dynophore models for the R2 simulation of compound **2** and R2 of **TSC24** are shown in Figure 4B,C. The remaining models and their corresponding animations can be found in the Supplementary Information (Figures S6–S8 and Videos S1–S5).

Dynophores pinpointed the favorable hydrophobic interactions between the first phenolic moiety of compound **2** and residues Ile125, Ile217, Ile88, Phe142, Thr215, and Ala92, which contribute to the stability within the ATP binding site. Furthermore, the hydrogen bonds of the phenolic OH of compound **2** with Asn91, Asn95, Arg98, and particularly Asn120 were conserved during the MD trajectories, providing the additional stabilization of the ligand in the binding pocket. The conformational mobility of the thiocarbohydrazide linker and the second phenolic group of compound **2** observed in the simulation, oscillating between the orientations CF1 and CF2, resulted in a scattered pattern of pharmacophore super-features around the amino acids responsible for sugar and phosphate binding of the ATP molecule. More precisely, the thiocarbohydrazide linker established hydrogen bond donor (HBD) interactions with Asn91, Ser149, Asn150, and Arg162 in approximately 30% of the simulation time. The hydrogen bond acceptor (HBA) interactions of the thioketo sulfur atom with Arg98 and Ile125 were observed in 11% of the simulation time. The second phenolic group, located in the outer part of the binding site, was bound to Ser148, Ser149 and Asn150 via HBA interactions. This part of the molecule was also stabilized by hydrophobic interactions with Ile141, Phe142, and Ala167 (Figure 4B). Conformational analysis of the second replica of the MD simulation (Supplementary Video S2) shows that ligand **2** preferred the bent conformation (CF1). The driving forces for this conformational change of the ligand to its linear conformation CF2 appear to be interactions with residues of the ATP-sugar binding region as well as the ATP-binding loop, i.e., residues Ser148, Ser149, Asn150, Arg162, Asn163, Gly164, and Ala167.

Compared to a larger conformational space that bound compound **2** exhibits during the simulations, the ligand **TSC24** remains predominantly in a single conformation. The quinoline moiety of the ligand is located in the adenine binding pocket of the binding site and forms hydrophobic interactions with Ile88, Ala92, Ile118, Thr215, and Ile217 and HBA interactions with Asn95 and Arg98. Two interactions that stabilize compound **2** within the adenine subpocket were not found for quinoline ring of **TSC24**; hydrogen bond interaction with Asn120 and hydrophobic interactions with the hydrophobic pocket comprised Ile125, Ile141, and Phe142. Two methyl groups of the thiosemicarbazone moiety established stable hydrophobic interactions with this pocket and with Ala167. This ligand was additionally stabilized by HBD interactions with the main chain of Ile141 and HBA bonds with Arg98, Ser148, and Ser149 (Figure 4C).

Overall, the bound conformation of **TSC24** exhibited a more conformationally restricted and stable pattern of interactions than compound **2**, where determined interactions were more dispersed. Decatenation and relaxation experiments demonstrated that at the 25 μM concentration of **TSC24**, topo IIα was completely inhibited [25], and it thus exhibited better inhibition properties than compound **2**. The obtained simulation data imply that optimization of the flexible thiocarbohydrazide linker of this lead compound by either structure rigidification and/or more efficient exploitation of the interactions revealed from the dynophore models could pave the way to more potent and selective topo IIα-acting compounds.

To complement the geometry-based information with energetic data, we performed Molecular Mechanics/Generalized Born Surface Area (MM/GBSA) free energy calculations.

This approach does not consider explicit water molecules that might be involved in hydrogen bonding between the protein and the ligand. Instead, the explicit solvent is removed and replaced with an implicit continuum solvent to significantly speed up computation time. Interestingly, this calculation suggested that compound **2** binds more strongly to the ATPase domain compared to **TSC24** ($\Delta G = -34.92 \pm 6.14$ vs. -29.71 ± 3.78 kcal/mol, respectively (Figure S9)). Although the average binding affinity of compound **2** was higher overall for topo IIα, it was also associated with a significantly higher standard deviation of ΔG compared with conformationally restrained **TSC24**, consistent with the more stable binding position observed for **TSC24**.

To evaluate the importance of the observed interactions, we analyzed the energy contributions of the individual residues that were shown by the dynophore models to stabilize the ligands in a particular conformation (Table 1). The results show that the hydrogen bond interactions with Asn91, Asn95, and Asn120 contributed most strongly to the stability of a phenolic moiety of compound **2** bound in the adenine binding pocket. Hydrophobic interactions with Ile125, Ile141, Phe142 also stabilized compound **2** in CF1 and CF2. The main differences between the two stable conformations CF1 and CF2 are the smaller contribution of Phe142 in the linear conformation CF2 as the second phenol ring shifts away, weaker interactions of CF2 with the adenine-binding part, and stronger interactions with residues Arg98, Asn163, and Gly164 of the ATP-binding loop.

Table 1. The results of MM/GBSA binding free energy analysis for the selected timeframes in which ligands **2** and **TSC24** adopted a particular conformation. All residues with binding free energy contributions above 0.5 kcal/mol are listed and ordered by descending contributions.

	Replica, Timeframe	Conformation	Main Amino Acid Residues (Total Energy Decomposition ΔG in kcal/mol)
Compound 2	R1, 40–60 ns	CF1, bent	Asn95 (−3.21), Asn91 (−2.97), Asn120 (−1.79), Phe142 (−1.43), Ile141 (−1.29), Ile 217 (−0.96), Thr215 (−0.92), Gly166 (−0.85), Ala167 (−0.85), Ile125 (−0.84)
	R1, 100–120 ns	CF2, linear	Asn91 (−1.85), Asn95 (−1.79), Ile141 (−1.78), Ile125 (−1.65), Asn120 (−1.47), Ile217 (−0.90), Arg98 (−0.83), Asn163 (−0.83), Phe142 (−0.79), Gly164 (−0.72), Thr215 (−0.67)
TSC24	40–60 ns	restrained	Ile141 (−2.17), Phe142 (−1.82), Asn91 (−1.06), Ala92 (−0.94), Ile125 (−0.94), Ser148 (−0.86), Thr215 (−0.85), Ile217 (−0.84), Gly164 (−0.78), Ile88 (−0.77)
	160–180 ns	restrained	Ala92 (−1.86), Ile217 (−1.79), Ile125 (−1.49), Gly164 (−1.09), Phe142 (−0.99), Ile141 (−0.98), Thr215 (−0.92), Asn91 (−0.80), Ile88 (−0.68), Asn120 (−0.67)

The reference ligand **TSC24** forms hydrophobic interactions with Ala92, Ile217, Ile125, Ile141, and Phe142 of the adenine binding region (Table 1). MM/GBSA analysis of the two selected 20 ns timeframes of the trajectory revealed that Asn95 and Arg98 are not important for molecular recognition between **TSC24** and the topo IIα catalytic domain in contrast to the dynophore analysis (Figure 4). The interactions of the aliphatic part of **TSC24** with Ser148 and Gly164 of the phosphate binding part of the ATPase binding site were also suggested to be important for the stability of the protein/ligand complex by analyzing the MM/GBSA calculations.

The conformational behavior of the ATPase domain of topo IIα with both bound compounds was also investigated. The average RMSD of the protein backbone for the topo IIα-**TSC24** complex was between 1.66 and 1.70 Å, whereas for the three replicas of topo IIα with bound compound **2**, the averages were 2.01, 2.15, and 2.49 Å (as shown in Tables S6 and S7). The temporal RMSD plots show that complexes reached equilibrium after about 20–30 ns of simulation. To further pinpoint the main source of flexibility, we performed root-mean-square fluctuation (RMSF) calculations [43], where the most profound fluctuations were associated with the movement of residues in the transducer domain of topo IIα, comprising residues 265–405 (Figure 5A), which is similar to observations in our previous studies [5]. The protein flexibility induced by the binding of the two catalytic inhibitors is similar, and two conformations of compound **2** revealed by MD (Figure S4) did not influence protein flexibility considerably (Table S6). The adenine and sugar-binding moiety of the ATP binding site were essentially unchanged during the simulations, whereas some structural fluctuations were observed in the ATP binding loop and the hydrophobic pocket above the ATP binding site.

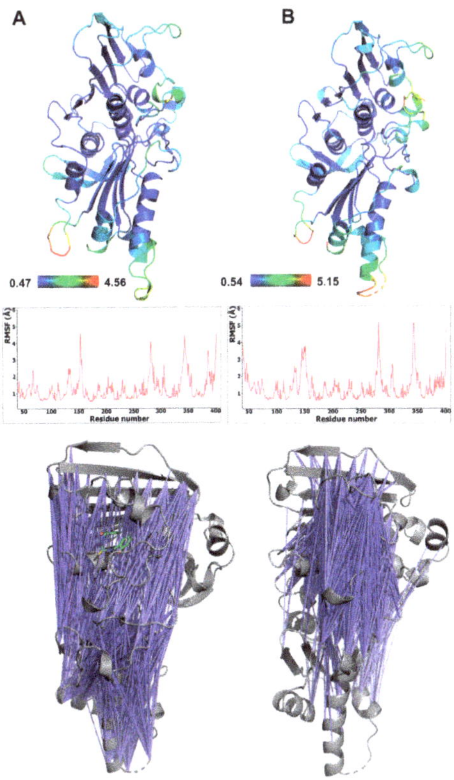

Figure 5. Flexibility of the ATPase domain of human topo IIα, depicted as a B-factor representation colored by RMSF values (top) and also shown as RMSF graphs s (middle) for complexes with compound **2** (**A**) and **TSC24** (**B**). The bottom panel shows pairs of residues of topo IIα ATPase domain displaying anticorrelation movements (correlation coefficients between −0.4 and −0.6) (left), topo IIα complex with compound **2**, and (right) topo IIα complex with compound **TSC24**.

The calculated cross-correlation matrices of the protein residues revealed significant anticorrelation movements of the ATP binding site residues located on the GHKL domain and the residues of the transducer domain. These pairs of residues in complex with bound **TSC24** displayed more pronounced and focused anticorrelations around the ATP binding site compared to complexes with compound **2** (Figure 5), which is in line with its superior topo IIα inhibition [25]. The binding of both compounds results in enhanced and focused movement compared to the apo structure of topo IIα we simulated in our previous study, which showed considerably more randomly distributed anticorrelation movements [44]. This indicated that the ligands can mimic some of the interactions of the native ATP ligand, which was also able to enhance the focused movement compared to the apo structure [44].

2.4. *In Silico Target Fishing of Thiocarbohydrazones and Assessment of Drug-Likeness*

Phenotypic screening (Figures 2A and S2) indicated that the cytotoxic activity of compound **2** is probably associated with the activity on multiple targets. In this study, we evaluated that compound **2** is a catalytic inhibitor of human DNA topoisomerase IIα. To gain further insights and deconvolute the results of the phenotypic screening, we searched for additional potential macromolecular targets using the PharmMapper web server [45]. This program analyzes the pharmacophoric similarity between the multiple conformations of the input structure and several thousand pharmacophore models mapping the active sites of different protein targets. The results indicated that glutathione S-transferase P, carbonic anhydrase 12, and cell division protein kinase 2 were the three best-matched proteins associated with cancer (Table S8).

Next, we docked compound **2** into the active site of these three proteins, relaxed the obtained complexes using a short MD simulation to remove steric hindrances, and rescored binding poses using the ChemPLP, PLP, and PLP95 scoring functions [46]. The docking scores highlighted cell division protein kinase 2 (CDPK2) as a promising target (Table S9). The binding mode of compound **2** in the active site of CDPK2 shows that two phenolic moieties establish a rich pattern of hydrophobic, HBD, and HBA interactions within the binding pocket (Figure S10). These data provide a starting point for future studies to further evaluate the mode of action of these compounds.

According to the Biopharmaceutics Classification System (BCS) and the estimated ADMET properties (Table S10), compound **2** belongs to Class I, i.e., compounds with high permeability and high solubility. It follows Lipinski's rule of five [47] and shows promising bioavailability; thus, it is likely to be well absorbed upon oral administration. Moreover, compound **2** has low blood–brain barrier (BBB) permeability and is not a substrate for P-glycoprotein. In terms of potential toxicity, compound **2** is predicted as non-hepatotoxic but could inhibit the hERG II ion channel. In addition, the hydrazone phenol moiety of compound **2** is recognized as a potential pan-assay interference structure (PAINS), which urges caution in further development. Nevertheless, some drugs used in the clinic such as the anticancer agent mitoguazone, the antibiotic agent nifuroxazide and the antihypertensive agent dihydralazine possess a hydrazone moiety [48]. Compound **2** thus possesses lead-like properties; however, future modifications should aim to reduce potential cardiotoxicity and carefully validate the PAINS potential or, preferably, eliminate it completely.

3. Materials and Methods

3.1. *Chemistry*

The synthesis, characterization, and isomerism of thiocarbohydrazones **1–4** were reported in our previous paper [49].

3.2. *Biology*

3.2.1. Cell Culture and Preparation of Solutions

Human acute monocytic leukemia cell line (THP-1, ATCC® TIB-202) was maintained in RPMI-1640 (Life Technologies, Paisley, UK, Cat. No. 11875-093), supplemented with 10% (*v*/*v*) heat inactivated fetal bovine serum (FBS, Life Technologies, Paisley, UK, Cat No 10270-106) and 1% (*v*/*v*) penicillin-streptomycin (10,000 units/mL and 10,000 μg/mL, Life Technologies, Paisley, UK, Cat No 15140-122). Cells were kept at 37 °C in a humidified atmosphere containing 5% (*v*/*v*) CO_2 during their exponential growing phase and during incubation with investigated compounds.

Investigated compounds **1–4** were dissolved in DMSO to the stock concentration of 20 mM. Further dilutions to the experimental concentrations applied on the cells were done with RPMI-1640 or DMEM media immediately before each experiment; thus, the final concentration of DMSO on cells treated with the highest applied concentration of an investigated compound was 0.5% (*v*/*v*).

3.2.2. Annexin V and Propidium Iodide Staining

Cells were seeded in 96 flat-bottom well plates (Corning® Costar®, Cat. No. CLS3596) in a volume of 0.1 mL at a density of 10,000 per well. Investigated compounds **1–4** were added in a range of six concentrations 24 h after cell seeding. As controls, non-treated cells, cells treated with 0.5% DMSO, and cells treated with Celastrol (Enzo Life Sciences, Lausen, Switzerland, Cat. No. ALX-350-332-M025) at 50 µM concentration were used. After 24 h of treatment, Annexin V-FITC (Immuno Tools, Friesoythe, Germany, Cat No 31490013) and propidium iodide (PI, Miltenyl Biotec Inc., Auburn, AB, USA, Cat No 130-093-233) were added to wells in a volume of 3 µL each. Plates were analyzed after 30 min of incubation in the dark with a Guava® easyCyte 12HT Benchtop flow microcapillary cytometer (Millipore, Merck, Darmstadt, Germany) using the dedicated InCyte™ 3.1 software package. Cells were classified according to Annexin V-FITC (green fluorescence) and PI (red fluorescence) labeling on viable cells (double negative), pre-apoptotic cells (Annexin V-FITC single-stained cells), necrotic cells (PI single-stained cells), and cells in advanced phases of apoptosis (double-stained cells).

3.2.3. Concentration–Response Curve Plotting

Percentages of Annexin V single-stained and double-stained cells were summarized for each concentration of investigated compound and plotted against corresponding concentrations. Concentration–response curves were drawn using the sigmoidal asymmetric five-parameter logistic equation, or the biphasic model, for the hill-shaped curve in GraphPad Prism 6 software (GraphPad Software, Inc., Boston, MA, United States).

3.2.4. Cell Cycle Analysis

The distribution of cells within phases of mitotic division was evaluated on the remaining cells after Annexin V/PI analysis, which, right after the readout was finished, were fixed in ethanol overnight at 4 °C. Before reading, plates were centrifuged on 450× *g* for 10 min, ethanol was discarded, and PBS was added in a volume of 100 µL per well. Cells were stained with 50 µL of FxCycle™ PI/RNAse Staining solution (Molecular Probes, Eugene, OR, United States; Thermo Fisher Scientific, Waltham, MA, United States, Cat. No. F10797) and incubated at 37 °C for 30 min in the dark. Plates were analyzed with a Guava® easyCyte 12HT Benchtop flow microcapillary cytometer using the dedicated InCyte™ 3.1 software package.

3.2.5. Caspase-8 and Caspase-9 Activities

Cells were treated with investigated compounds at 50 µM concentration for 6 h; afterwards, the activity of caspase-8 and caspase-9 were assayed by means of Guava Caspase 9 SR and Caspase 8 FAM kits (EMD Millipore, Merck, Darmstadt, Germany, Cat. No. 4500-0640) by following the manufacturer's instructions. Cells were analyzed with a Guava® easyCyte 12HT Benchtop flow microcapillary cytometer using the dedicated InCyte™ 3.1 software package. Acquired data cells were discriminated according to their expression of caspase-8 (Grn-B fluorescence) and caspase-9 (Yel-B fluorescence).

3.2.6. Generation of Radical Oxygen Species in Mitochondria

Cells were treated over 6 h with investigated compounds in a concentration of 50 µM; afterwards, they were stained with MitoSox Red (Molecular Probes, Cat. No. M36008) according to the manufacturer's recommendations. Analysis was performed with a Guava® easyCyte 12HT Benchtop flow microcapillary cytometer using the dedicated InCyte™ 3.1 software package. The generation of $O_2^{\bullet-}$ was evaluated by means of two parameters: percentage of $O_2^{\bullet-}$-generating cells, and mean fluorescence intensity (MFI) expressed in arbitrary units (AU). The MFI was computed for the $O_2^{\bullet-}$-positive subpopulation, and it indicated the average quantity of $O_2^{\bullet-}$ per cell.

3.2.7. Human DNA Topoisomerase IIα-Mediated Decatenation Assay

This topo IIα assay, along with additional assays described in Section 2.2, were performed in collaboration with Inspiralis (Norwich, UK). One U of topo II was incubated with 200 ng kDNA in a 30 μL reaction at 37 °C for 30 min under the following conditions: 50 mM Tris HCl (pH 7.5), 125 mM NaCl, 10 mM $MgCl_2$, 5 mM DTT, 0.5 mM EDTA, 0.1 mg/mL bovine serum albumin (BSA), and 1 mM ATP. The reaction was then stopped by the addition of 30 μL chloroform/isoamyl alcohol (26:1) and 30 μL Stop Dye (40% sucrose (*w*/*v*), 100 mM Tris. HCl (pH 7.5), 10 mM EDTA, 0.5 μg/mL bromophenol blue) before being loaded on a 1% TAE gel run at 85 V for 90 min.

Bands were visualized via ethidium bromide staining for 15 min and destained for 10 min. Gels were scanned using documentation equipment (GeneGenius, Syngene, Cambridge, UK), and inhibition levels were calculated from the band data obtained with the gel scanning software (GeneTools, Syngene, Cambridge, UK). Assays were performed for active compound **2** at concentrations of 12.5, 50, 100, and 200 μM and for etoposide standard at concentrations of 3.9, 31.5, 125, and 500 μM.

3.2.8. Human DNA Topoisomerase IIα-Mediated Relaxation Assay

The activity of the enzyme was determined prior to the testing of the compounds, and 1 U was defined as the amount of enzyme required to fully relax the substrate. Compound **2** was tested at 31.25 μM, 62.5 μM, 125 μM, and 250 μM, and these volumes were added to the reaction before the addition of the enzyme. Final DMSO concentration in the assays was 1% (*v*/*v*). One U of human topo IIα was incubated with 500 ng supercoiled pBR322 in a 30 μL reaction at 37 °C for 30 min under the following conditions: 50 mM Tris HCl (pH 7.5), 125 mM NaCl, 10 mM $MgCl_2$, 5 mM DTT, 0.5 mM EDTA, 0.1 mg/mL bovine serum albumin (BSA), and 1 mM ATP. Each reaction was stopped by the addition of 30 μL chloroform/isoamyl alcohol (24:1) and 30 μL Stop Dye before being loaded on a 1.0% TAE gel run at 90 V for 90 min. Bands were visualized via ethidium staining for 10 min, destained for 10 min in water, analyzed by using gel documentation equipment (Syngene, Cambridge, UK), and quantified using Syngene Gene Tools software. Raw gel data (fluorescent band volumes) collected from Syngene, GeneTools gel analysis software were calculated as a percentage of the 100% control (the fully supercoiled DNA band) and converted to percent inhibition.

3.2.9. Human DNA Topoisomerase IIα Cleavage Assay

The activity of the human topoisomerase IIα was determined prior to testing of the compounds, and 1 unit (U) was defined as the amount of enzyme required to reach the maximum cleavage of the substrate. The final DMSO concentration in all reactions was 1% (*v*/*v*). The compound was serially diluted in 100% DMSO and added to the reaction before the addition of the enzyme. The control compound for all assays was etoposide. One U of the human topo IIα was incubated with 0.5 μg supercoiled plasmid DNA (pBR322) in a 30 μL reaction at 37 °C for 30 min under the following conditions: 20 mM Tris HCl (pH 7.5), 200 mM NaCl, 0.25 mM EDTA, and 5% glycerol. The reaction was then incubated for a further 30 min with 0.2% SDS and 0.5 μg/μL proteinase K. The reaction was then stopped by the addition of 30 μL chloroform/isoamyl alcohol (26:1) and 30 μL Stop Dye (40% sucrose (*w*/*v*), 100 mM Tris HCl (pH 7.5), 10 mM EDTA, 0.5 μg/mL bromophenol blue) before being loaded on a 1% TAE gel run at 80 V for 2 h. Assays were performed for compound **2** at concentrations of 12.5, 25, 50, and 100 μM and for etoposide (control) at concentrations of 12.5, 25, 50 and 100 μM. Bands were visualized as scanned and as described in the decatenation assay.

3.2.10. Wheatgerm Topo I Unwinding ASSAY

A volume of 1 U of wheatgerm topo I was incubated with 0.5 μg supercoiled or relaxed plasmid DNA (pBR322) in a 30 μL reaction at 37 °C for 30 min under the following conditions: 50 mM Tris HCl (pH 7.9), 50 mM NaCl, 1.0 mM EDTA, 1.0 mM DTT, and 20% glycerol. Each reaction was stopped, and the compounds were removed, prior to the running of the gels by the addition of 50 μL butanol and 30 μL of water. The samples were vortexed, and the aqueous layer was removed before the addition of 30 μL chloroform/iso-amyl alcohol (24:1) and 30 μL Stop Dye. These were then loaded on a 1.0% TAE gel run at 90 V for 2 h. Bands were visualized via ethidium bromide staining for 15 min and destaining for 10 min. Gels were scanned using documentation equipment (GeneGenius, Syngene, Cambridge, UK). The assay was performed at four different concentrations of compound **2** (31.5, 62.5, 125, and 250 μM) and of the positive control, intercalator mAMSA (31.5, 62.5, 125, and 250 μM).

3.3. *Molecular Modeling*

3.3.1. Molecular Docking Calculations

The generated conformations of compounds **2** and **TSC24** were optimized with a MMFF94 force field [50], with the obtained conformations additionally optimized by using the semiempirical PM7 method [51] implemented in MOPAC2016 [52]. This approach generated structures with precise bond lengths and optimal conformational properties to provide a good starting geometry for the molecular docking. The active site of one protomer of human topo IIα ATPase domain (PDB code:1ZXM [53]) was defined as all residues up to 10 Å away from the co-crystallized ligand AMP-PNP. The ligand, water, and metal ions were removed, and the hydrogen atoms were added to resemble the protonation state of the protein at pH 7.4 as predicted by PROPKA [54]. The .pdbqt files of compound **2**, **TSC24**, and the receptor were prepared in Vega ZZ 3.2.0 [55,56]. AutoDockVina 1.1 [57] was used for docking. Exhaustiveness was set to 100, and 20 binding poses were stored. All other settings were kept at default values.

3.3.2. Molecular Dynamics Simulations

The best docking poses of compounds **2** and **TSC24** were taken as starting points for the protein–ligand complexes used for molecular dynamics simulations. We first determined the force field parameters for both ligands. The molecular geometry and electronic structure of compounds **2** and **TSC24** were first optimized at the Hartree–Fock level using the 6–31 G* basis set. The partial charges of the ligands were obtained by performing a Merz–Kollman population analysis. Quantum mechanical calculations were carried out in Gaussian 16, revision C.01 [58]. The RESP charges were then fitted using the Antechamber module of Amber18 [59]. The bond distances, bond angles, and dihedrals of the ligands were obtained from the optimized geometries using Antechamber. The force field parameters of the ligands were represented in the General Amber Force Field of second generation (gaff2). The partial atomic charges of the ligands **2** and **TSC24** and their atom types are provided in the Supplementary Materials, Tables S11 and S12. The monomer of human topo IIα ATPase domain (chain A, residues 39–345 and 350–405) comprised the simulated protein molecule.

Both complexes were solvated using a cubic box of TIP3P-type water molecules [60] with at least 10 Å distance between the edges of the box and the protein. Three chloride ions were added to neutralize the system, and the final system consisted of approximately 137,000 atoms. The Amber14SB force field was used to describe the protein [61], and gaff2 was used for the ligand atoms [62]. The energies of the solvated protein–ligand complexes were minimized by applying 10,000 steps of steepest descent minimization followed by 20,000 steps of conjugate gradient minimization. Next, we performed four NVT equilibration runs in which the system was gradually heated to 300 K using the Langevin thermostat. In each run, 10,000 steps with a time step of 2 fs were applied by gradually releasing the applied constraints on the protein from 100 kcal mol^{-1} Å^{-2} (first run) over 60 kcal mol^{-1} Å^{-2} (second) and 30 kcal mol^{-1} Å^{-2} (third run) to the fourth run, which had no restraints. Afterward, one run of 200 ps NPT equilibration with 2 fs time step was performed, applying 20 kcal mol^{-1} Å^{-2} force constants on the protein. The final run of NPT equilibration was done without applying any constraints. The pressure was maintained at 1 bar using the Berendsen thermostat [63].

The initial configurations of the three replicas of the MD simulation of the topo IIα-**2** complex were generated by varying the equilibration time of unconstrained NPT equilibration to 0.2, 0.6, and 1.0 ns. For the topo IIα-**TSC24** complex, 0.6 ns and 1.0 ns of the final NPT equilibration were taken for generating input structures of two replicas. The simulations were performed by applying periodic boundary conditions, with long-range electrostatics treated using the Particle Mesh Evald method [64] with the cut-off value of 10 Å. The lengths of all bonds involving hydrogen atoms were constrained using the SHAKE algorithm to achieve the time step of 2 fs [65]. Each of five MD replicas was 0.2 μs long with the production simulations done in the Amber18 pmemd.cuda program [66]. Molecular simulations were performed using the computational resources of the Azman high-performance computing (HPC) center at the National Institute of Chemistry in Slovenia.

3.3.3. Analysis of the MD Trajectories

Molecular trajectories obtained during the production stage of the MD simulations were analyzed by determining the root-mean-square deviations (RMSD) of the protein backbone Cα and the ligand, which were calculated in the VMD software where the trajectories were also visualized [67]. The root-mean-square fluctuation (RMSF) values of the protein were calculated using the Cpptraj module of Amber 18 [59]. The initial structure of the protein–ligand complex was used as a reference frame for the RMSD and RMSF calculations. Dynamic pharmacophores were calculated using DynophoreApp software and using 400 equidistant frames for each simulated system [42].

The cross-correlation maps determining the extents of the pairwise residual correlations were calculated with the Bio3D package [68] and using its dccm function, which derived the covariation matrices and calculated the Pearson's correlation coefficients (C_{ij}) on the Cα atom pairs, i and j. The anticorrelated movements of pairs of residues were visualized in PyMOL [69].

The binding free energy calculations of the protein–ligand complexes were performed using the Molecular Mechanics/Generalized Born Surface Area (MM/GBSA) method implemented in Amber Tools 20 [70,71]. Calculations were performed on 400 equidistant frames using the Generalized Born IGB method 5 and 0.15 M salt concentration. Moreover, we performed per-residue decomposition analysis to evaluate the contributions of individual amino acid residues to overall binding.

3.3.4. Pharmacophore Similarity Search and ADME-Tox Predictions

The PharmMapper web server was used for the identification of other potential molecular targets for active compound **2** [45]. All protein targets from the v2010 database (7302 items) were matched against the database of pharmacophores generated for 300 conformers of compound **2** and ranked according to normalized Fit scores and Z-scores. The ADME-Tox properties of compound **2** were predicted using the Swiss-ADME [72] and pkCSM [73] online tools. The SMILES string of the compound was used as input.

4. Conclusions

Phenotypic screening of a small series of mono- and bis-substituted thiocarbohydrazones against three cancer cell lines revealed that 1,5-bis(salicylidene)thiocarbohydrazide **2** induces the apoptosis and activates the DNA repair machinery via an ROS-independent pathway. Structurally related thiosemicarbazones are known catalytic topo IIα inhibitors, and compound **2** exhibited a similar catalytic inhibition of action and did not act as a DNA intercalator. The molecular simulations of complexes of compounds **2** and **TSC24** with the ATPase domain of topo IIα, which were followed by dynamic pharmacophore and

MM/GBSA analyses, revealed differences in the interaction patterns and conformational stability of the two ligands. Thiocarbohydrazide **2** was more flexible in the targeted ATP binding site compared with **TSC24**. Furthermore, the model of molecular recognition suggested that rigidifying the structure and optimizing the interactions with the sugar- and phosphate-binding portions of the topo IIα ATP site could result in more potent and selective derivatives.

Human DNA topoisomerase IIα is a validated anticancer drug target, but rapidly evolving cancer resistance and severe side effects associated with the use of topoisomerase poisons such as etoposide and doxorubicin limit its applicability. The design and development of catalytic topo IIα inhibitors is a promising approach to overcome these issues. This study highlights 1,5-bis(salicylidene)thiocarbohydrazide as a potential novel lead compound with cytotoxic and topo II inhibition properties useful for chemotherapeutic anticancer drug discovery.

Supplementary Materials: The following supporting information can be downloaded at: https://www.mdpi.com/article/10.3390/ph16030341/s1, Table S1: Anticancer activity of thiocarbohydrazones **1–4**; Figure S1: Types of cell death in THP-1 cells treated with investigated compounds, determined by means of Annexin V/PI dual staining assay after 24 h incubation; Figure S2: Concentration-response curve for compound **2** on THP-1 cells after 24 h treatment; Table S2: The results of HTS human topo IIα relaxation assay; Figure S3: The results of second run of human topo IIα decatenation assay, human topo IIα cleavage assay, unwinding assay, and topoisomerase IIα relaxation assay; Table S3: Results of decatenation assay (in duplicate) for thiocarbohydrazone **2** and etoposide at different concentrations represented as the % of the decatenated kDNA; Table S4: Percentage of linear DNA, determined in cleavage assay (in duplicate) for etoposide and thiocarbohydrazone **2** at four concentrations; Table S5: Results of relaxation assay (in duplicate) for thiocarbohydrazone **2** and etoposide at different concentrations represented as the % of topo IIα inhibition; Figure S4: Two conformations of compound **2** observed in three replicas of MD simulations; Figure S5: RMSD of the ligand heavy atoms for each replica of MD simulation of **2** (A) and **TSC24** (B) bound within the ATPase domain of human topo IIα; Table S6: Average RMSD of the ligand **2** and topo IIα protein for the three replicas of MD simulation; Table S7: Average RMSD of the **TSC24** ligand and topo IIα protein for two replicas of MD simulation; Figure S6: Dynophore model for the first replica (R1) of MD simulation of the ligand **2** in complex with ATPase domain of human topoisomerase IIα; Figure S7: Dynophore model for the third replica (R3) of MD simulation of the ligand **2** in complex with ATPase domain of human topoisomerase IIα; Figure S8: Dynophore model for the first replica (R1) of MD simulation of the ligand **TSC24** in complex with ATPase domain of human topoisomerase IIα; Figure S9: The MM/GBSA binding free energy (ΔG) for the binding of compounds **2** and **TSC24** to topo IIα; Table S8: Cancer-related drug target candidates for thiocarbohydrazone **2** identified through PharmMapper search; Table S9: ChemPLP, PLP, and PLP95 docking scores of the minimized complexes of compound **2** and the best three protein targets identified via pharmacophore similarity search; Figure S10: Binding mode and ligand interaction diagram for the binding of compound **2** to cell division protein kinase 2; Table S10: The pharmacokinetics and drug-likeness of compound **2** predicted using pkCSM and SwissADME; Table S11: The partial atomic charges of the ligand **2**; Table S12: The partial atomic charges of the ligand **TSC24**; Video S1: Dynophores_R1_compound2; Video S2: Dynophores_R2_compound2; Video S3: Dynophores_R3_compound2; Video S4: TSC24_R1_Dynophores; Video S5: TSC24_R2_Dynophores.

Author Contributions: Conceptualization, I.N.C. and A.P.; methodology, S.K.B., I.N.C. and A.P.; software, B.H., I.N.C. and A.P.; validation, S.K.B., A.M. and A.P.; formal analysis, I.N.C.; investigation, I.N.C., B.H. and S.K.B.; resources, A.M. and A.P.; data curation, I.N.C., S.K.B. and B.H.; writing—original draft preparation, I.N.C., A.P. and S.K.B.; writing—review and editing, I.N.C., B.H. and A.P.; visualization, I.N.C. and B.H.; supervision, A.P.; project administration, I.N.C. and A.P.; funding acquisition, I.N.C. and A.P. All authors have read and agreed to the published version of the manuscript.

Funding: This work was supported by the Slovenian Research Agency with research project J1-4402 (A.P.) and research program P1-0012 (A.P.) as well as by a young researcher grant (B.H.) and the Ministry of Education, Science and Technological Development of Republic of Serbia, Contract number: 451-03-68/2022-14/200168 (I.C.).

Institutional Review Board Statement: Not applicable.

Informed Consent Statement: Not applicable.

Data Availability Statement: Data are contained within the article and Supplementary Materials.

Acknowledgments: Alison Howells and Nicolas Burton (Inspiralis, Norwich, UK) are acknowledged for performing the human DNA topoisomerase IIα decatenation, relaxation, cleavage, and unwinding assays. Kaja Bergant Loboda from the National Institute of Chemistry, Slovenia is thanked for technical assistance in the initial stage of this work. We also acknowledge Gerhard Wolber for providing us access to dynophore calculations at the Freie Universität Berlin, Germany and the Azman high-performance computing (HPC) center at the National Institute of Chemistry in Ljubljana for computational resources.

Conflicts of Interest: The authors declare no conflict of interest.

References

1. WHO. Available online: https://www.who.int/news-room/fact-sheets/detail/cancer (accessed on 9 December 2022).
2. Hanahan, D.; Weinberg, R.A. The Hallmarks of Cancer. *Cell* **2000**, *100*, 57–70. [CrossRef]
3. Hanahan, D.; Weinberg, R.A. Hallmarks of Cancer: The Next Generation. *Cell* **2011**, *144*, 646–674. [CrossRef]
4. Carter, S.K. The Search for Therapeutic Cell Controls by the Chemotherapy Program of the National Cancer Institute. *J. Investig. Dermatol.* **1972**, *59*, 128–138. [CrossRef]
5. Ogrizek, M.; Janežič, M.; Valjavec, K.; Perdih, A. Catalytic Mechanism of ATP Hydrolysis in the ATPase Domain of Human DNA Topoisomerase IIα. *J. Chem. Inf. Model.* **2022**, *62*, 3896–3909. [CrossRef] [PubMed]
6. Drake, F.H.; Zimmerman, J.P.; McCabe, F.L.; Bartus, H.F.; Per, S.R.; Sullivan, D.M.; Ross, W.E.; Mattern, M.R.; Johnson, R.K.; Crooke, S.T. Purification of Topoisomerase II from Amsacrine-Resistant P388 Leukemia Cells. Evidence for Two Forms of the Enzyme. *J. Biol. Chem.* **1987**, *262*, 16739–16747. [CrossRef] [PubMed]
7. Nitiss, J.L. Targeting DNA Topoisomerase II in Cancer Chemotherapy. *Nat. Rev. Cancer* **2009**, *9*, 338–350. [CrossRef] [PubMed]
8. Walker, J.V.; Nitiss, J.L. DNA Topoisomerase II as a Target for Cancer Chemotherapy. *Cancer Investig.* **2002**, *20*, 570–589. [CrossRef]
9. Felix, C.A. Secondary Leukemias Induced by Topoisomerase-Targeted Drugs. *BBA Gene Struct. Expr.* **1998**, *1400*, 233–255. [CrossRef] [PubMed]
10. Minotti, G.; Menna, P.; Salvatorelli, E.; Cairo, G.; Gianni, L. Anthracyclines: Molecular Advances and Pharmacologic Developments in Antitumor Activity and Cardiotoxicity. *Pharmacol. Rev.* **2004**, *56*, 185. [CrossRef] [PubMed]
11. Pogorelčnik, B.; Perdih, A.; Solmajer, T. Recent Developments of DNA Poisons—Human DNA Topoisomerase IIα Inhibitor-as Anticancer Agents. *Curr. Pharm. Des.* **2013**, *19*, 2474–2488. [CrossRef]
12. Pogorelcnik, B.; Perdih, A.; Solmajer, T. Recent Advances in the Development of Catalytic Inhibitors of Human DNA Topoisomerase IIα As Novel Anticancer Agents. *Curr. Med. Chem.* **2013**, *20*, 694–709. [CrossRef]
13. Bergant, K.; Janežič, M.; Valjavec, K.; Sosič, I.; Pajk, S.; Štampar, M.; Žegura, B.; Gobec, S.; Filipič, M.; Perdih, A. Structure-Guided Optimization of 4,6-Substituted-1,3,5-Triazin-2(1H)-Ones as Catalytic Inhibitors of Human DNA Topoisomerase IIα. *Eur. J. Med. Chem.* **2019**, *175*, 330–348. [CrossRef] [PubMed]
14. Bergant Loboda, K.; Janežič, M.; Štampar, M.; Žegura, B.; Filipič, M.; Perdih, A. Substituted 4,5′-Bithiazoles as Catalytic Inhibitors of Human DNA Topoisomerase IIα. *J. Chem. Inf. Model.* **2020**, *60*, 3662–3678. [CrossRef] [PubMed]
15. Loboda, K.B.; Valjavec, K.; Štampar, M.; Wolber, G.; Žegura, B.; Filipič, M.; Dolenc, M.S.; Perdih, A. Design and Synthesis of 3,5-Substituted 1,2,4-Oxadiazoles as Catalytic Inhibitors of Human DNA Topoisomerase IIα. *Bioorg. Chem.* **2020**, *99*, 103828. [CrossRef] [PubMed]
16. Larsen, A.K.; Escargueil, A.E.; Skladanowski, A. Catalytic Topoisomerase II Inhibitors in Cancer Therapy. *Pharmacol. Ther.* **2003**, *99*, 167–181. [CrossRef]
17. Skok, Ž.; Zidar, N.; Kikelj, D.; Ilaš, J. Dual Inhibitors of Human DNA Topoisomerase II and Other Cancer-Related Targets. *J. Med. Chem.* **2020**, *63*, 884–904. [CrossRef]
18. Guerrant, W.; Patil, V.; Canzoneri, J.C.; Oyelere, A.K. Dual Targeting of Histone Deacetylase and Topoisomerase II with Novel Bifunctional Inhibitors. *J. Med. Chem.* **2012**, *55*, 1465–1477. [CrossRef]
19. Proschak, E.; Stark, H.; Merk, D. Polypharmacology by Design: A Medicinal Chemist's Perspective on Multitargeting Compounds. *J. Med. Chem.* **2019**, *62*, 420–444. [CrossRef]
20. Culletta, G.; Allegra, M.; Almerico, A.M.; Restivo, I.; Tutone, M. In Silico Design, Synthesis, and Biological Evaluation of Anticancer Arylsulfonamide Endowed with Anti-Telomerase Activity. *Pharmaceuticals* **2022**, *15*, 82. [CrossRef]

21. Moffat, J.G.; Vincent, F.; Lee, J.A.; Eder, J.; Prunotto, M. Opportunities and Challenges in Phenotypic Drug Discovery: An Industry Perspective. *Nat. Rev. Drug Discov.* **2017**, *16*, 531–543. [CrossRef]
22. Moffat, J.G.; Rudolph, J.; Bailey, D. Phenotypic Screening in Cancer Drug Discovery—Past, Present and Future. *Nat. Rev. Drug Discov.* **2014**, *13*, 588–602. [CrossRef] [PubMed]
23. Merlot, A.M.; Kalinowski, D.S.; Richardson, D.R. Novel Chelators for Cancer Treatment: Where Are We Now? *Antioxid. Redox Signal.* **2012**, *18*, 973–1006. [CrossRef] [PubMed]
24. Yalowich, J.C.; Wu, X.; Zhang, R.; Kanagasabai, R.; Hornbaker, M.; Hasinoff, B.B. The Anticancer Thiosemicarbazones Dp44mT and Triapine Lack Inhibitory Effects as Catalytic Inhibitors or Poisons of DNA Topoisomerase IIα. *Biochem. Pharmacol.* **2012**, *84*, 52–58. [CrossRef]
25. Huang, H.; Chen, Q.; Ku, X.; Meng, L.; Lin, L.; Wang, X.; Zhu, C.; Wang, Y.; Chen, Z.; Li, M.; et al. A Series of α-Heterocyclic Carboxaldehyde Thiosemicarbazones Inhibit Topoisomerase IIα Catalytic Activity. *J. Med. Chem.* **2010**, *53*, 3048–3064. [CrossRef] [PubMed]
26. Božić, A.; Marinković, A.; Bjelogrlić, S.; Todorović, T.R.; Cvijetić, I.N.; Novaković, I.; Muller, C.D.; Filipović, N.R. Quinoline Based Mono- and Bis-(Thio)Carbohydrazones: Synthesis, Anticancer Activity in 2D and 3D Cancer and Cancer Stem Cell Models. *RSC Adv.* **2016**, *6*, 104763–104781. [CrossRef]
27. Bonaccorso, C.; Grasso, G.; Musso, N.; Barresi, V.; Condorelli, D.F.; La Mendola, D.; Rizzarelli, E. Water Soluble Glucose Derivative of Thiocarbohydrazone Acts as Ionophore with Cytotoxic Effects on Tumor Cells. *J. Inorg. Biochem.* **2018**, *182*, 92–102. [CrossRef]
28. Parsekar, S.U.; Haldar, P.; Antharjanam, P.K.S.; Kumar, M.; Koley, A.P. Synthesis, Characterization, Crystal Structure, DNA and Human Serum Albumin Interactions, as Well as Antiproliferative Activity of a Cu(II) Complex Containing a Schiff Base Ligand Formed in Situ from the Cu(II)-Induced Cyclization of 1,5-Bis(Salicylidene)Thiocarbohydrazide. *Appl. Organomet. Chem.* **2021**, *35*, e6152. [CrossRef]
29. Patel, P.; Chen, E.I. Cancer Stem Cells, Tumor Dormancy, and Metastasis. *Front. Endocrinol.* **2012**, *3*, 125. [CrossRef]
30. Sarkar, F.H.; Li, Y.; Wang, Z.; Kong, D. Pancreatic Cancer Stem Cells and EMT in Drug Resistance and Metastasis. *Minerva Chir.* **2009**, *64*, 489–500.
31. Burgers, P.M.J.; Kunkel, T.A. Eukaryotic DNA Replication Fork. *Annu. Rev. Biochem.* **2017**, *86*, 417–438. [CrossRef]
32. Jones, R.M.; Petermann, E. Replication Fork Dynamics and the DNA Damage Response. *Biochem. J.* **2012**, *443*, 13–26. [CrossRef]
33. Srinivas, U.S.; Tan, B.W.Q.; Vellayappan, B.A.; Jeyasekharan, A.D. ROS and the DNA Damage Response in Cancer. *Redox Biol.* **2019**, *25*, 101084. [CrossRef]
34. Wang, Y.; Branicky, R.; Noë, A.; Hekimi, S. Superoxide Dismutases: Dual Roles in Controlling ROS Damage and Regulating ROS Signaling. *J. Cell Biol.* **2018**, *217*, 1915–1928. [CrossRef] [PubMed]
35. Backman, T.W.H.; Cao, Y.; Girke, T. ChemMine Tools: An Online Service for Analyzing and Clustering Small Molecules. *Nucleic Acids Res.* **2011**, *39*, W486-491. [CrossRef] [PubMed]
36. Lima, Â.C.d.O.; Conceição, R.S.; Freitas, L.S.; de Carvalho, C.A.L.; Conceição, A.L.d.S.; Freitas, H.F.; Pita, S.S.d.R.; Ifa, D.R.; Pinheiro, A.M.; Branco, A. Hydroxycinnamic Acid-Spermidine Amides from Tetragonisca Angustula Honey as Anti-Neospora Caninum: In Vitro and in Silico Studies. *Chem. Biol. Drug Des.* **2021**, *98*, 1104–1115. [CrossRef] [PubMed]
37. Montecucco, A.; Zanetta, F.; Biamonti, G. Molecular Mechanisms of Etoposide. *EXCLI J.* **2015**, *14*, 95–108. [CrossRef]
38. Avrutsky, M.I.; Troy, C.M. Caspase-9: A Multimodal Therapeutic Target With Diverse Cellular Expression in Human Disease. *Front. Pharmacol.* **2021**, *12*, 701301. [CrossRef]
39. Li, P.; Zhou, L.; Zhao, T.; Liu, X.; Zhang, P.; Liu, Y.; Zheng, X.; Li, Q. Caspase-9: Structure, Mechanisms and Clinical Application. *Oncotarget* **2017**, *8*, 23996–24008. [CrossRef]
40. Bergant, K.; Janezic, M.; Perdih, A. Bioassays and In Silico Methods in the Identification of Human DNA Topoisomerase IIα Inhibitors. *Curr. Med. Chem.* **2018**, *25*, 3286–3318. [CrossRef]
41. Pilotelle-Bunner, A.; Cornelius, F.; Sebban, P.; Kuchel, P.W.; Clarke, R.J. Mechanism of Mg^{2+} Binding in the Na^+,K^+-ATPase. *Biophys. J.* **2009**, *96*, 3753–3761. [CrossRef]
42. Schaller, D.; Šribar, D.; Noonan, T.; Deng, L.; Nguyen, T.N.; Pach, S.; Machalz, D.; Bermudez, M.; Wolber, G. Next Generation 3D Pharmacophore Modeling. *Wiley Interdiscip. Rev. Comput. Mol. Sci.* **2020**, *10*, e1468. [CrossRef]
43. Lima, L.R.; Bastos, R.S.; Ferreira, E.F.B.; Leão, R.P.; Araújo, P.H.F.; Pita, S.S.d.R.; De Freitas, H.F.; Espejo-Román, J.M.; Dos Santos, E.L.V.S.; Ramos, R.d.S.; et al. Identification of Potential New Aedes Aegypti Juvenile Hormone Inhibitors from N-Acyl Piperidine Derivatives: A Bioinformatics Approach. *Int. J. Mol. Sci.* **2022**, *23*, 9927. [CrossRef] [PubMed]
44. Janežič, M.; Valjavec, K.; Loboda, K.B.; Herlah, B.; Ogris, I.; Kozorog, M.; Podobnik, M.; Grdadolnik, S.G.; Wolber, G.; Perdih, A. Dynophore-Based Approach in Virtual Screening: A Case of Human DNA Topoisomerase IIα. *Int. J. Mol. Sci.* **2021**, *22*, 13474. [CrossRef] [PubMed]
45. Liu, X.; Ouyang, S.; Yu, B.; Liu, Y.; Huang, K.; Gong, J.; Zheng, S.; Li, Z.; Li, H.; Jiang, H. PharmMapper Server: A Web Server for Potential Drug Target Identification Using Pharmacophore Mapping Approach. *Nucleic Acids Res.* **2010**, *38*, W609–W614. [CrossRef]
46. Korb, O.; Stützle, T.; Exner, T.E. Empirical Scoring Functions for Advanced Protein−Ligand Docking with PLANTS. *J. Chem. Inf. Model.* **2009**, *49*, 84–96. [CrossRef]
47. Lipinski, C.A.; Lombardo, F.; Dominy, B.W.; Feeney, P.J. Experimental and Computational Approaches to Estimate Solubility and Permeability in Drug Discovery and Development Settings. *Adv. Drug Deliv. Rev.* **1997**, *23*, 3–25. [CrossRef]

48. Amine Khodja, I.; Boulebd, H. Synthesis, Biological Evaluation, Theoretical Investigations, Docking Study and ADME Parameters of Some 1,4-Bisphenylhydrazone Derivatives as Potent Antioxidant Agents and Acetylcholinesterase Inhibitors. *Mol. Divers.* **2021**, *25*, 279–290. [CrossRef]
49. Assaleh, M.H.; Božić, A.R.; Bjelogrlić, S.; Milošević, M.; Simić, M.; Marinković, A.D.; Cvijetić, I.N. Water-Induced Isomerism of Salicylaldehyde and 2-Acetylpyridine Mono-and Bis-(Thiocarbohydrazones) Improves the Antioxidant Activity: Spectroscopic and DFT Study. *Struct. Chem.* **2019**, *30*, 2447–2457. [CrossRef]
50. Halgren, T.A. MMFF VI. MMFF94s Option for Energy Minimization Studies. *J. Comput. Chem.* **1999**, *20*, 720–729. [CrossRef]
51. Stewart, J.J.P. Optimization of Parameters for Semiempirical Methods VI: More Modifications to the NDDO Approximations and Re-Optimization of Parameters. *J. Mol. Model.* **2013**, *19*, 1–32. [CrossRef]
52. Stewart, J.J.P. MOPAC: A Semiempirical Molecular Orbital Program. *J. Comput. Aided Mol. Des.* **1990**, *4*, 1–105. [CrossRef]
53. Wei, H.; Ruthenburg, A.J.; Bechis, S.K.; Verdine, G.L. Nucleotide-Dependent Domain Movement in the ATPase Domain of a Human Type IIA DNA Topoisomerase. *J. Biol. Chem.* **2005**, *280*, 37041–37047. [CrossRef]
54. Olsson, M.H.M.; Søndergaard, C.R.; Rostkowski, M.; Jensen, J.H. PROPKA3: Consistent Treatment of Internal and Surface Residues in Empirical PKa Predictions. *J. Chem. Theory Comput.* **2011**, *7*, 525–537. [CrossRef]
55. Pedretti, A.; Villa, L.; Vistoli, G. VEGA—An Open Platform to Develop Chemo-Bio-Informatics Applications, Using Plug-in Architecture and Script Programming. *J. Comput. Aided Mol. Des.* **2004**, *18*, 167–173. [CrossRef]
56. Pedretti, A.; Mazzolari, A.; Gervasoni, S.; Fumagalli, L.; Vistoli, G. The VEGA Suite of Programs: An Versatile Platform for Cheminformatics and Drug Design Projects. *Bioinformatics* **2021**, *37*, 1174–1175. [CrossRef]
57. Trott, O.; Olson, A.J. AutoDock Vina: Improving the Speed and Accuracy of Docking with a New Scoring Function, Efficient Optimization, and Multithreading. *J. Comput. Chem.* **2010**, *31*, 455–461. [CrossRef]
58. Frisch, M.; Trucks, G.; Schlegel, H.; Scuseria, G.; Robb, M.; Cheeseman, J.; Scalmani, G.; Barone, V.; Petersson, G.; Nakatsuji, H.; et al. *Gaussian 16, Revision C. 01*; Gaussian, Inc.: Wallingford, CT, USA, 2016.
59. Case, D.; Ben-Shalom, I.; Brozell, S.; Cerutti, D.; Cheatham III, T.; Cruzeiro, V.; Darden, T.; Duke, R.; Ghoreishi, D.; Gilson, M.; et al. *AMBER 2018*; University of California: San Francisco, CA, USA, 2018.
60. Jorgensen, W.L.; Chandrasekhar, J.; Madura, J.D.; Impey, R.W.; Klein, M.L. Comparison of Simple Potential Functions for Simulating Liquid Water. *J. Chem. Phys.* **1983**, *79*, 926–935. [CrossRef]
61. Maier, J.A.; Martinez, C.; Kasavajhala, K.; Wickstrom, L.; Hauser, K.E.; Simmerling, C. Ff14SB: Improving the Accuracy of Protein Side Chain and Backbone Parameters from Ff99SB. *J. Chem. Theory Comput.* **2015**, *11*, 3696–3713. [CrossRef] [PubMed]
62. Wang, J.; Wolf, R.M.; Caldwell, J.W.; Kollman, P.A.; Case, D.A. Development and Testing of a General Amber Force Field. *J. Comput. Chem.* **2004**, *25*, 1157–1174. [CrossRef] [PubMed]
63. Berendsen, H.J.C.; Postma, J.P.M.; van Gunsteren, W.F.; DiNola, A.; Haak, J.R. Molecular Dynamics with Coupling to an External Bath. *J. Chem. Phys.* **1984**, *81*, 3684–3690. [CrossRef]
64. Darden, T.; York, D.; Pedersen, L. Particle Mesh Ewald: An N·Log (N) Method for Ewald Sums in Large Systems. *J. Chem. Phys.* **1993**, *98*, 10089–10092. [CrossRef]
65. Ryckaert, J.-P.; Ciccotti, G.; Berendsen, H.J. Numerical Integration of the Cartesian Equations of Motion of a System with Constraints: Molecular Dynamics of n-Alkanes. *J. Comput. Phys.* **1977**, *23*, 327–341. [CrossRef]
66. Götz, A.W.; Williamson, M.J.; Xu, D.; Poole, D.; Le Grand, S.; Walker, R.C. Routine Microsecond Molecular Dynamics Simulations with AMBER on GPUs. 1. Generalized Born. *J. Chem. Theory Comput.* **2012**, *8*, 1542–1555. [CrossRef] [PubMed]
67. Humphrey, W.; Dalke, A.; Schulten, K. VMD: Visual Molecular Dynamics. *J. Mol. Graph.* **1996**, *14*, 33–38. [CrossRef] [PubMed]
68. Grant, B.J.; Rodrigues, A.P.; ElSawy, K.M.; McCammon, J.A.; Caves, L.S. Bio3d: An R Package for the Comparative Analysis of Protein Structures. *Bioinformatics* **2006**, *22*, 2695–2696. [CrossRef]
69. The PyMOL Molecular Graphics System. Version 2.0. Schrodinger, LCC: New York, NY, USA, 2015.
70. Wang, C.; Greene, D.; Xiao, L.; Qi, R.; Luo, R. Recent Developments and Applications of the MMPBSA Method. *Front. Mol. Biosci.* **2018**, *4*, 87. [CrossRef] [PubMed]
71. Case, D.; Belfon, K.; Ben-Shalom, I.; Brozell, S.; Cerutti, D.; Cheatham III, T.; Cruzeiro, V.; Darden, T.; Duke, R.; Giambasu, G.; et al. *AMBER 2020*; University of California: San Francisco, CA, USA, 2020.
72. Daina, A.; Michielin, O.; Zoete, V. SwissADME: A Free Web Tool to Evaluate Pharmacokinetics, Drug-Likeness and Medicinal Chemistry Friendliness of Small Molecules. *Sci. Rep.* **2017**, *7*, 42717. [CrossRef]
73. Pires, D.E.V.; Blundell, T.L.; Ascher, D.B. PkCSM: Predicting Small-Molecule Pharmacokinetic and Toxicity Properties Using Graph-Based Signatures. *J. Med. Chem.* **2015**, *58*, 4066–4072. [CrossRef]

pharmaceuticals

MDPI

Article

In Silico and In Vitro Investigation of Cytotoxicity and Apoptosis of Acridine/Sulfonamide Hybrids Targeting Topoisomerases I and II

Mohamed Badr [1], Elshaymaa I. Elmongy [2,*], Doaa Elkhateeb [3,4], Yasmine S. Moemen [5], Ashraf Khalil [4], Hadeer Ali [3], Reem Binsuwaidan [6], Feby Awadallah [3,7] and Ibrahim El Tantawy El Sayed [3]

1 Department of Biochemistry, Faculty of Pharmacy, Menoufia University, Shebin El-Kom 6131567, Egypt; mohamed.badr@phrm.menofia.edu.eg
2 Department of Pharmaceutical Chemistry, Faculty of Pharmacy, Helwan University, Ain Helwan, Cairo P.O. Box 11795, Egypt
3 Chemistry Department, Faculty of Science, Menoufia University, Shebin El-Kom 32511, Egypt; elkhateebdoaaa@gmail.com (D.E.); yohader1@gmail.com (H.A.); feby_awadallah@yahoo.com (F.A.); ibrahimtantawy@science.menofia.edu.eg (I.E.T.E.S.)
4 Department of Clinical Biochemistry and Molecular Diagnostics, National Liver Institute, Menoufia University, Shebin El-Kom 32511, Egypt; ashkalil2010@gmail.com
5 Clinical Pathology Department, National Liver Institute, Menoufia University, Shebin El-Kom 32511, Egypt; yasmine.moemen@gmail.com
6 Department of Pharmaceutical Sciences, College of Pharmacy, Princess Nourah bint Abdulrahman University, P.O. Box 84428, Riyadh 11671, Saudi Arabia; rabinsuwaidan@pnu.edu.sa
7 Clinical pathology Department, Menoufia University Hospital, Shebin El-Kom 32511, Egypt
* Correspondence: shaymaa.taha@pharm.helwan.edu.eg

Abstract: Background: Sulfonamide acridine derivatives have garnered significant attention from medicinal chemists due to their diverse range of biological activities. Methods: In this study, eleven compounds were synthesized according to the literature, and their impact on cell growth inhibition, induction of apoptosis, and cell cycle distribution were assessed in three different cell lines. Their inhibitory effects on the topoisomerase (Topo) I and II were investigated in vitro. Molecular docking studies were conducted to predict the binding affinities of these compounds for crystallized downloaded topoisomerases. Results: The compounds were examined in vitro for their anticancer activity against human hepatic (HepG2) colon (HCT-8) and breast (MCF-7) carcinoma cell lines. Compound **8b** was the most active against HepG2, HCT-116, and MCF-7 with IC_{50} 14.51, 9.39, and 8.83 µM, respectively, compared to Doxorubicin as reference. In addition, it demonstrated the highest potency among the tested compounds against Topo-I, with an IC_{50} value of 3.41 µg/mL compared to the control camptothecin (IC_{50} of 1.46 µM). Compound **7c** displayed a significant inhibitory effect on Topo-II, with an IC_{50} of 7.33 µM, compared to an IC_{50} value of 6.49 µM via Doxorubicin, the control. Compounds **7c** and **8b** were assessed against topoisomerases showing induction of apoptosis and a reduction in the S phase of the cell cycle. Molecular docking demonstrated interaction with the active site as with those exhibited by the co-crystallized ligands of the crystallized proteins in both topoisomerases. Conclusion: Compounds **7c** and **8b** hold promise as potential anticancer drugs due to their anti-proliferative and proapoptotic effects, which are mediated by their action on the topoisomerase enzyme, particularly Topo II.

Keywords: acridines; cytotoxicity; topoisomerase; molecular docking

1. Introduction

Topoisomerases are a crucial group of enzymes that play a vital role in various DNA-related processes, such as replication and transcription. They intricately regulate DNA structure, enabling the smooth unwinding, separation, and rejoining of strands throughout cellular processes [1]. These enzymes cleave and reconnect the DNA strands by forming a covalent bond with the DNA phosphorus group. Topoisomerases are categorized into two groups according to their mechanism of action: topoisomerases I (Topo I) and topoisomerases II (Topo-II), which are then subdivided into five subfamilies [2]. Topo-I generate temporary single-stranded DNA nicks, whereas (Topo IIα and Topo IIβ) induce temporary double-stranded DNA breaks. Cancer chemotherapy targets nuclear Topo I and Topo II, while therapeutic procedures include Top inhibitors [3–5]. The use of cytostatic drugs, which inhibit enzyme activity, lead to the irreversible breakage of DNA strands (via the formation of a stable DNA-topoisomerase complex), which leads to cell death. The heightened level of topoisomerase activity observed in several types of cancer has generated significant interest in the use of topoisomerase inhibitors [6]. Over the past decade, a wide variety of bioactive compounds have been produced and subjected to biological evaluation by various scientists, with the specific aim of targeting topoisomerase. These molecules can be classified into nine categories based on their structural features, and compounds that include N-heterocycles are one of these categories. Acridine derivatives are a class of heterocyclic compounds that are currently being extensively studied for their potential as anticancer drugs. Acridine derivatives, including Amsacrine (Figure 1), are part of an extensively researched group of anticancer medicines that mainly function by interfering with DNA synthesis and inhibiting or poisoning Topo- II [2,7–14]. The acridine ring is responsible for intercalating into DNA, whereas the non-interactive head group offers selectivity for the DNA–Topo cleavage complex and enhances its activity. In 1984, acridines (specifically Amsacrine) were discovered as topoisomerase II inhibitors. Since then, numerous derivatives of acridine with various substituents on the acridine chromophore have been extensively investigated as topoisomerase inhibitors [15]. Also, during the early 1980s, cytotoxic alkaloid camptothecin (CPT) was identified as a topoisomerase I inhibitor [16].

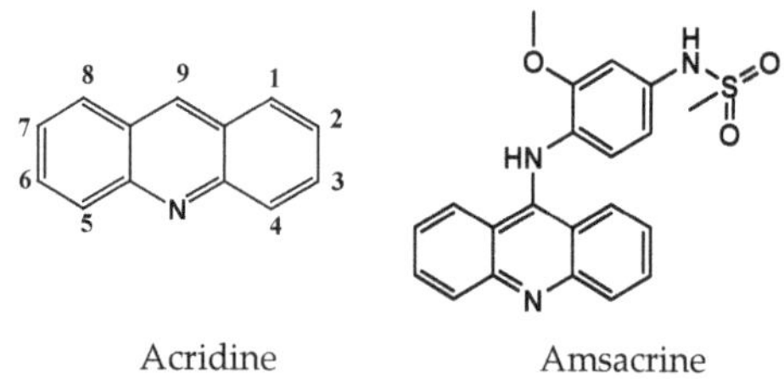

Figure 1. Structure of acridine-based Amsacrine (m-AMSA 8).

The academic literature has recorded the successful use of the hybrid pharmacophore concept, which involves combining heterocycles with established active groups such as sulfonamides, in the development of pharmaceutical drugs. Furthermore, the hybrid of thiourea and sulfonamide compounds has garnered significant interest because of their diverse range of biological effects, such as their ability to inhibit the growth of cancer cells, viruses, bacteria, parasites, and fungi [17–22]. Furthermore, the goal is to reduce drug resistance development in hybrid compounds while improving their effectiveness [23]. The combination of two bioactive chemicals via hybridization typically leads to increased activity as a result of the synergistic effects. The presence of a significant number of intramolecular hydrogen bonds between the thiourea group and the receptor pocket contributes to the enhancement of drug bioactivities [24]. In our previous study, we aimed to discover new acridine derivatives that have strong anticancer properties but cause minimal side effects. To achieve this, we synthesized and evaluated several acridine analogs with sulfonamide and thioureide side chains at the C-9 position. These analogs were prepared using nucleophilic aromatic substitution (SNAr) to replace the chlorine atom in 9-chloroacridine with either 4,4′-diaminodiphenylmethane or phenylenediamine. To make thiourea and acridine sulfonamide analogs, the 9-amino derivatives were mixed with phenyl isothiocyanate or aryl–sulfonyl chlorides. By incorporating basic side chains with para-phenylenediamine into the acridine structure at the C-9 position, we observed a significant increase in the anticancer activity of these compounds in vitro against human HepG2, colon (HCT-116), and breast (MCF-7) carcinoma cells [25]. Accordingly, this study aimed to evaluate both in silico and in vitro the inhibitory effects of the synthesized acridine analogs hybrids on both topoisomerase I and topoisomerase II, along with their antiproliferative activity on hepatocellular carcinoma cell lines.

2. Results

2.1. The Antiproliferative Activity Against HepG2, MCF7, and HCT Cell Lines

As reported in our previous study the antiproliferative activity of synthesized compounds were assessed on three cancerous cell lines, HepG2, MCF7, and HCT, using an MTT assay. Following 48 h exposure to varying concentrations of each compound, growth inhibitory curves were employed to determine the IC_{50} for each compound. These IC_{50} values were then compared to those of Doxorubicin (DOX), a known agent with powerful anti-proliferative and cytotoxic effects on these cell lines [25]. The resulting IC_{50} values for the tested compounds are tabulated below; Table 1 indicates that compounds **5b**, **8b**, **6b**, and **8a** exhibited significant anticancer activity against all three cancer cell lines. It is worth noting that **5b** and **8b** are the most active against HepG2, HCT-116, and MCF-7 with IC_{50} 8.30, 8.93, 5.88 and 14.51, 9.39, 8.83 μM, respectively, when compared to reference drug DOX used in the same assay. Dose responsive curves of antiproliferative activity for the synthesized acridine derivatives are illustrated in Figure S1.

Table 1. In vitro cytotoxicity of the tested hybrids against three cancer cell lines.

Compounds	IC_{50} (μM)		
	HepG-2	HCT-116	MCF-7
5b	8.30 ± 0.7	8.93 ± 0.8	5.88 ± 0.4
6a	36.71 ± 3.5	47.21 ± 3.9	40.93 ± 3.8
6b	17.98 ± 1.7	13.07 ± 1.2	16.47 ± 1.5
7a	62.32 ± 4.4	72.16 ± 4.7	88.36 ± 5.1
7b	93.60 ± 5.3	88.53 ± 4.8	73.81 ± 4.5
7c	31.07 ± 3.1	40.61 ± 3.7	29.08 ± 2.7
7d	45.49 ± 3.9	57.96 ± 4.3	33.65 ± 3.3
7e	55.35 ± 4.2	66.67 ± 4.5	50.76 ± 4.0
8a	20.63 ± 1.8	15.18 ± 1.6	10.54 ± 1.0
8b	14.51 ± 1.4	9.39 ± 0.9	8.83 ± 0.9
8c	22.38 ± 2.0	35.69 ± 3.4	25.47 ± 2.4
8d	23.13 ± 2.2	29.85 ± 2.8	19.69 ± 1.7
DOX	4.50 ± 0.2	5.23 ± 0.3	4.17 ± 0.2

2.2. Enzyme Inhibitory Assay

Based on the above-mentioned results and aiming to explore the specific inhibitory activity of the synthesized compounds, they were assessed for their potential inhibitory effect on both Topo-I and Topo-II isomerase enzymes. Subsequently, these compounds were employed to evaluate their capacity to inhibit topoisomerase activity at the established IC_{50}, using supercoiled DNA as a substrate. The resulting DNA products were separated via 1% gel electrophoresis and stained with ethidium bromide. As shown in Table 2, in comparison to the control (camptothecin), which displayed an IC_{50} of 1.46 μM, compounds **6b**, **8b**, and **7d** exhibited IC_{50} values of 4.17, 3.41, and 7.67 μM, respectively, effectively reducing Topo-I activity. Similarly, compounds **7a** and **7c** displayed IC_{50} values of 8.07 and

7.33 μM, respectively, in the Topo-II assay, effectively reducing Topo-II activity, however, the reference Doxorubicin, demonstrated an IC_{50} value of 6.49 (Figure S2a,b).

Table 2. Potential inhibitory effect of compounds (**6a–8d**) on both Topo I and Topo II activities reflected by their IC_{50} μM.

Compound	Topo-I (IC_{50}) μM	Topo-II (IC_{50}) μM
6a	27.6 ± 1.5	101 ± 5.51
6b	4.17 ± 0.23	24.5 ± 1.34
7a	18.4 ± 1	8.07 ± 0.44
7b	82.5 ± 4.49	234 ± 12.7
7c	16.6 ± 0.9	7.33 ± 0.4
7d	7.67 ± 0.42	21.6 ± 1.18
7e	42.2 ± 2.29	38.7 ± 2.1
8a	15 ± 0.81	10.9 ± 0.59
8b	3.41 ± 0.19	16.2 ± 0.88
8c	21.4 ± 1.17	22 ± 1.2
8d	11.9 ± 0.65	32.7 ± 1.78
Cisplatin	5.71 ± 0.31	11.3 ± 0.62
Doxorubicin	3.36± 0.18	6.49 ± 0.35
Camptothecin	1.64 ± 0.09	-

2.3. *Cell Cycle Analysis in HepG2 Cells*

Compounds **8b** and **7c** were evaluated on the hepatocellular carcinoma cell line (HepG2). Compound **8b** exhibited significant topoisomerase I inhibitory activity with an IC_{50} of 3.41, while compound **7c** displayed remarkable topoisomerase II inhibitory activity with an IC_{50} of 7.33. These compounds were selected for further investigation to assess their impact on apoptosis and cell cycle arrest. As tabulated below in Table 3, the percentage of the cells in the S phase decreased from 46.39% to 39.15% when exposed to compound **8b**. Additionally, compound **7c** induced a slight reduction in the cell population in the G0–G1 phase, from 47.02% to 42.61%, and in the S phase, from 46.39% to 29.27%. This outcome suggests that compound **7c** led to cell cycle arrest in the S phase. These results exhibit a significant resemblance to the effects of the well-established anticancer drug SAHA, highlighting the substantial efficacy of the synthesized compounds.

Table 3. Cell cycle analysis revealing DNA content percentage by control, compound **7c** and compound **8b**.

Code	G0–G1	%S	%G2/M	%Pre-G1
7c/HepG2	42.61	29.27	28.12	19.61
8b/HepG2	51.19	39.15	9.66	26.95
SAHA/HepG2	47.02	46.39	6.59	34.12
cont. HepG2	44.06	35.11	20.83	1.84

2.4. *Flow Cytometry for Compounds* **8b** *and* **7c** *in HepG2 Cells*

An Annexin V-FITC assay was utilized to detect apoptosis in HepG2 cells. The cells were exposed to compounds **8b** (3.4 μM) and **7c** (7.3 μM) for 24 h, with untreated cells and cells treated with SAHA serving as negative and positive controls, respectively. After staining with propidium iodide (PI) and Annexin V-FITC, cell apoptosis was assessed via flow cytometry. The percentages of cells in the early and late apoptotic stages, as well as those undergoing necrosis, were presented in Figure 2.

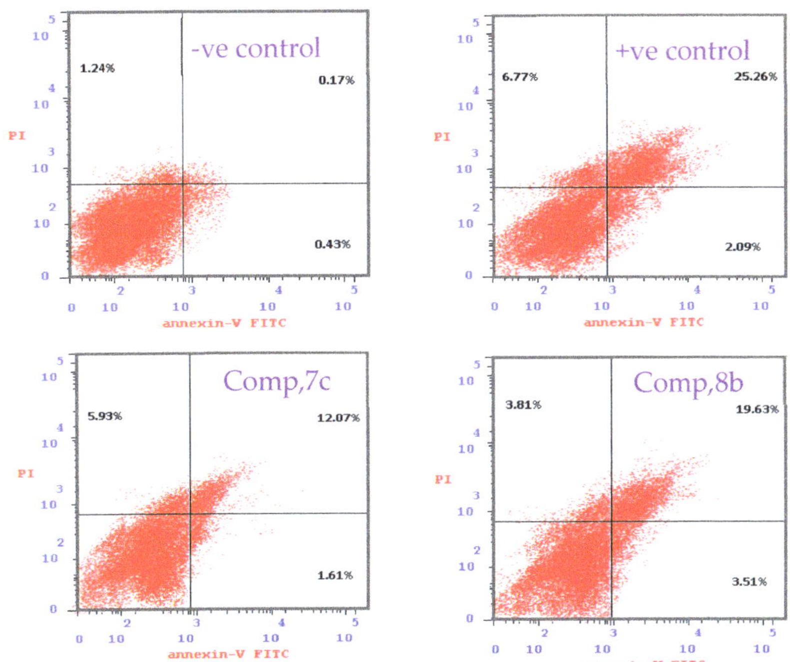

Figure 2. Impact of compounds **7c** and **8b** on apoptosis in HepG2 cells. HepG2 cells were treated with compounds **7c** and **8b** at the indicated concentrations for 72 h. Cells were collected, treated with Annexin V-FITC and PI, and subjected to flow cytometry as described in the text. Percentages of early and late apoptotic cells are indicated in the lower right and upper right corners, respectively.

Compound **7c** exhibited a 19.6% population of apoptotic cells (comprising 1.6% in the early stage and 12% in the late stage), along with 5.9% in the necrotic stage. In contrast, compound **8b** demonstrated 26.9% of the apoptotic cells (consisting of 3.5% in the early stage and 19.6% in the late stage), with 3.1% in the necrotic stage. Compound **7c** induced apoptosis to the extent of 55% and necrosis to 88% of the levels observed with SAHA, while compound **8b** induced apoptosis to the extent of 79% and necrosis to 46% of that of SAHA.

2.5. Selectivity Index

The higher the selectivity index of a compound to cells, the more selective the compounds are to killing or inhibiting the growth of a cancer cell compared to normal cells. Compounds with high SI values offer the potential for safer and more effective therapy in cancer therapy [26].

Accordingly, in an attempt to explore the selectivity of the prepared acridine-based derivatives to cancer cells over normal cells, both compounds were assessed for their cytotoxicity against normal cell line (THLE-2) in order to calculate the SI. Both compounds **7c** and **8b** exhibited relative cellular cytotoxicity against normal cell line THLE-2 with IC_{50} of 104 and 55.5 μM, respectively, which is better than that caused by the control anticancer drug SAHA (IC_{50} = 33.6 μM), as shown in Table 4.

Table 4. In vitro cytotoxicity of compounds **7c** and **8b** against THLE-2 cell line.

Compound	IC_{50} μM
7c	104 ± 5.86
8b	55.5 ± 3.13
SAHA	33.6 ± 1.89

Then, the selectivity index was calculated, and the results are tabulated below, in Table 5. Based on the test results, compound **7c** was found to be more selective for HePG2 and MCF-7 cells with SI > 3 than for HCT-116 cells. Meanwhile, compound **8b** was found to be selective for all the three tested cancer cells HePG2, HCT-116, and MCF-7 with SI > 3. These findings show that the synthetic compounds have the ability to target cancer cells specifically, while causing the least amount of damage to healthy cells, which is a desirable effect in chemotherapeutic treatments.

Table 5. Selectivity index of compounds **7c** and **8b** on HePG2, HCT-116, and MCF-7 cell lines.

Compound	HePG2		HCT-116		MCF-7		THLE-2
	IC_{50} μM	SI	IC_{50} μM	SI	IC_{50} μM	SI	IC_{50} μM
7c	31.07 ± 3.1	3.35	40.61 ± 3.7	2.56	29.08 ± 2.7	3.57	104 ± 5.86
8b	14.51 ± 1.4	3.82	9.39 ± 0.9	5.91	8.83 ± 0.9	6.28	55.5 ± 3.13

2.6. Molecular Docking

Two proteins were downloaded from the protein data bank with pdb ID: 1K4T and 5GWK for topoisomerases-I and II, respectively. Regarding topoisomerase-I (pdb ID: 1K4T), the co-crystalized ligand downloaded with topoisomerase I—"2-(1-dimethylaminomethyl-2-hydroxy-8-hydroxymethyl-9-oxo-9,11-dihydro-indolizino [1,2-b]quinolin-7-yl)-2-hydroxy-butyric acid", referred to as "TGG"—defined the residues in the binding pockets [27] Asp-533, Asn-722, Lys-532,Phe-361,Gly-363, and Arg-364, as shown in Figure 3. TGG was stabilized by stacking interaction with the DNA in addition to a hydrogen bond with Asp-533 and Asn-722. Residues Asp-533, Asn-722, Lys-532, and Arg-364 contributed to direct interaction with the drug. Phe-361, Arg-362, and Gly-363 were not essentially important to interact with the drug directly but were stabilizing the intercalation-binding pocket. Re-docking of the co-crystallized ligand was performed for validation and the recorded RMSD value was 0.94 Å. Analyzing the resulting data upon docking compounds **6a–8d** revealed similar interactions to those reported by TTG, where the main amino acids involved in most of the interactions were Lys-374, Lys-532, Arg-364, Asp-533, Arg-362, and Asn-722. The binding affinities were −8.145 to −6.274 kcal/mol with perfect fitting inside the binding site, as shown in Table 6. The acridine–sulfonamide derivative **8b** recorded the least IC_{50} value against topoisomerase I (3.41 ± 0.19)μM. Analyzing the recorded scores, compound **8b** demonstrated a binding energy of −6.739 kcal/mol and the amino acids residues involved in interaction with the receptor pocket were the main residues required for direct interaction with the co-crystallized ligand including Lys-532, Asp-533, and Arg-364, in addition to Phe-361, which is important for stabilizing the intercalation binding pocket, as seen in Figure 4. There was also hydrogen bonding at the DNA binding region at TGP11 and DG12, in addition to another hydrophobic one with DG12. The docked compounds showed high affinities; the various interactions with the mentioned main residues were consistent with the recorded low IC_{50} value.

Regarding topoisomerase-IIα (pdb ID: 5GWK), the redocking of the co-crystallized ligand was performed at the ligand binding site and showed a RMSD of 1.71 Å. Upon docking the synthesized compounds at the defined ligand site, the proposed binding mode of the substituted tricyclic planar structure of the docked acridines showed affinity values ranging from −7.508 to −5.285 kcal/mol. Analyzing the docking results of the best inhibitor **7c**, which recorded the lowest IC_{50} score 7.33 ± 0.4 μM against topoisomerase II, it can be seen that it has occupied the same pocket as the co-crystallized ligand, as shown in Figure 5, forming four interactions with the pocket residues, two of which were with the terminal sulfonamide moiety, which showed hydrogen bond interaction with Arg-804 and DC11, while the planar aromatic system of substituted acridine showed hydrophobic π-π stacked and π-π T-shaped stacked interactions with DA12 and DG 13, respectively, at the DNA hinge region, as seen in Figure 5. The docking results including residues involved in the interactions as well as the types of bonding are tabulated in Table 6.

Table 6. Docking results for compounds **6a–8d** at the downloaded proteins topoisomerase I and II.

Compound	Energy (kcal/mol)	Amino Acids (Ligand Interaction Type)	Energy (kcal/mol)	Aminoacids (Ligand Interaction Type)
6a	−7.90	Lys-374 (H-bond acceptor) Lys-532 (H-bond acceptor) Arg-364 (pi-cation) Arg-364 (pi-cation)	−6.45	His-759 (H-bond acceptor) Lys-489 (pi-alkyl) DA 12 (pi-pi) DA 12 (pi-pi)
6b	−6.52	Arg-364 (H-bond acceptor) TGP 11 (pi-pi)	−5.99	DG 13 (H-bond donor) DC 11 (pi-alkyl) DA 12 (pi-pi)
7a	−7.21	Asp-533 (H-bond donor) Arg-364 (H-bond acceptor) Arg-364 (H-bond acceptor) Arg-364 (H-bond acceptor)	−6.43	Met-766 (H-bond donor) DA 12 (pi-pi)
7b	−6.76	Thr-718 (H-bond donor) Lys-532 (H-bond acceptor)	−6.46	Met-766 (H-bond donor) DA 12 (H-pi)
7c	−8.14	Lys-532 (H-bond acceptor) Arg-364 (pi-cation)	−6.51	DC11 (H-bond acceptor) Arg-804(H-bond acceptor) DG12 (pi-pi stacked) DG 13 (pi-pi)
7d	−7.84	Asn-722 (H-bond acceptor) Lys-425 (pi-alkyl) Lys-425 (pi-alkyl)	−6.96	DC 14 (H-bond donor) Glu-461(H-bond donor) DC 14 (H-bond acceptor)
7e	−8.15	Arg-362 (pi-alkyl) Lys-425 (pi-alkyl) DG12 (H-bond acceptor)	−7.37	-
8a	−6.27	Asn-722 (H-bond acceptor)	−5.75	Met-766 (H-bond donor) Met-765 (H-bond donor) DA 12 (pi-pi)
8b	−6.74	Lys-532 (H-bond acceptor) Arg-364 (pi-cation) TGP 11 (H-bond acceptor) DG 12 (H-bond acceptor) DG 12 (pi-alkyl) Asp-533 (pi-anion) Phe-361 (pi-alkyl)	−5.29	Asp-463 (H-bond acceptor) Ser-464 (H-bond acceptor) DA 12 (pi-pi) DA 12 (pi-pi)
8c	−6.88	Arg-364 (H-bond acceptor)	−6.78	DA 12 (H-bond donor) DA 12 (pi-alkyl)
8d	−7.49	Lys-532 (H-bond acceptor) TGP11 (pi-alkyl)	−7.51	Met-762 (H-bond donor) DG 13 (H-bond donor) DC 11 (H-bond acceptor) DA 12 (H-bond acceptor)

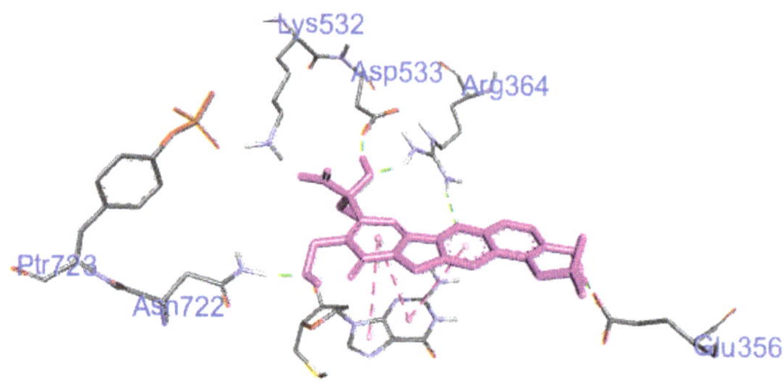

Figure 3. Site view of the co-crystallized ligand (represented as pink sticks) of Topo-I showing the most important active site residues, which define the receptor pocket.

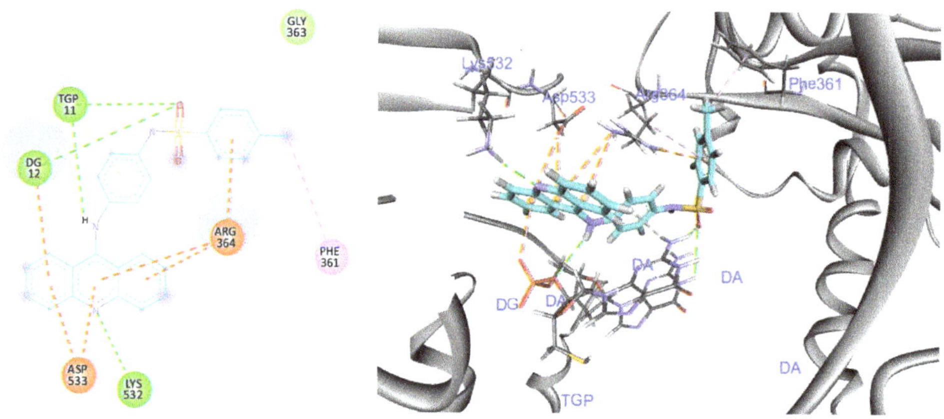

Figure 4. The acridine derivative **8b** docked in Topo-I ligand site (2D and 3D) showing the hydrogen bonds in a green dashed line, the hydrophobic interactions in a purple dashed line, and the pi-ion bonds in an orange dashed line.

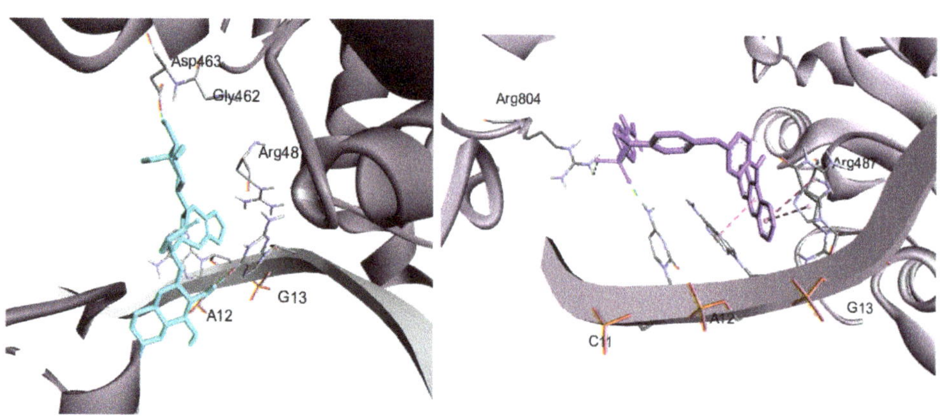

Figure 5. The co-crystallized ligand of the downloaded protein pdb ID; a 5GWK (in blue sticks)-defining receptor pocket and **7c** (in purple sticks) interacting with the pocket residues where hydrogen bonds are represented by a green dashed line and hydrophobic interactions are represented by a pink dashed line.

3. Materials and Methods

3.1. Chemistry

Compounds from **4**, **5a**, and **5b** were the starting materials for preparing compounds from 6a to 8d. The derivatives of acridine **6a–8d** were synthesized as reported in the literature, Scheme 1 [25]. (Compounds **6a–8d** are illustrated individually in Figure S3 and their data analysis results are illustrated in Figures S4–S16).

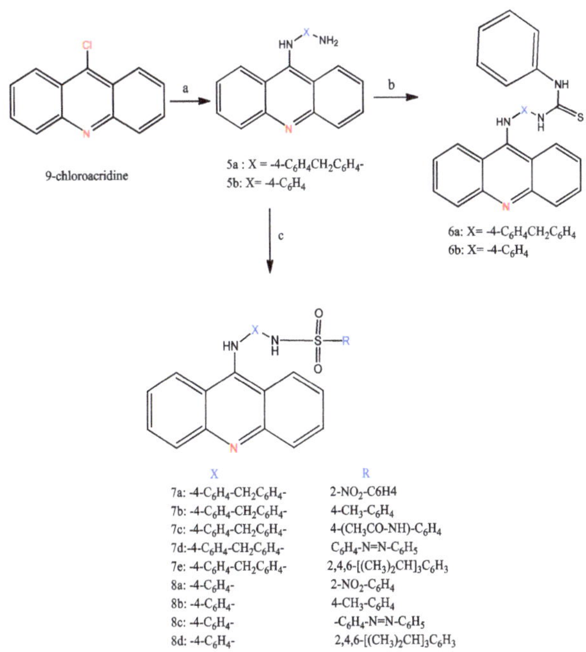

Scheme 1. Synthesis of acridine amalogues: (a) Appropriate amine, H_2N-X-NH_2, EtOH, Er3N, Reflux (below 80 °C), 2.5–7 h. (b) Chiloroform, Ph-N=C=S, stirring at room temperature, (2–8 day). (c) $ArSO_2Cl$, DMF, Et3N, (rt. or reflux).

3.2. Antiproliferative Activity Assays

The antiproliferative activity of the following compounds, **6a**, **6b**, **7a–7e**, **8a–8d**, was assessed using an MTT assay. This study evaluated the cytotoxic effects of these compounds in human hepatic (HepG2), colon (HCT-116), and breast (MCF-7) carcinoma cell lines, as previously reported [25].

Moreover, the most promising compounds (**8b** and **7c**) were evaluated against normal transformed Human Liver Epithelial-2 Cell Line THLE-2, which was procured from the American Type Culture Collection. These cells were cultured in Dulbecco's Modified Eagle's Medium (DMEM) from Invitrogen/Life Technologies, supplemented with 10% Hyclone fetal bovine serum, 10% insulin (10 g/mL Sigma), and 1% penicillin and streptomycin. For experimental purposes, the cells were seeded in 96-well tissue culture plates at an approximate density of 5×10^3 cells per well and were allowed to incubate overnight. The following day, the culture medium was replaced with fresh medium containing varying concentrations of the synthesized compounds to be tested for their toxicity. After 48 h of exposure, the medium was removed, and the cells were washed twice with phosphate-buffered saline (PBS). A new medium containing MTT solution was added, and the cells were further incubated for 4 h. The cellular viability and toxicity in response to each compound were determined using ELISA microplate reader- Bioline by measuring the absorbance at a wavelength of 450 nm.

3.3. Enzyme Inhibitory Assays

The compounds displaying cytotoxic activities underwent further examination for their potential as topoisomerase I and topoisomerase II inhibitors. To evaluate the catalytic activity of Topo-I, supercoiled DNA served as the substrate. Commercial samples of Topo I and supercoiled DNA were provided by TG1018-1A, Inc., TopoGEN, Inc. (Columbus, OH, USA). For the Topo-II assay, a Topo II drug kit, TopoGEN, TG1009, was obtained from the same source, with doxorubicin used as the reference drug.

3.3.1. Topoisomerase I Inhibitory Assay

Camptothecin and the tested compounds were mixed with Topo-I DNA complex and incubated at 37 °C for 30 min. A subsequent increase in temperature to 65 °C for 30 min rendered the enzyme inactive. To visualize the cleavage products after incubation, we used 1% agarose gel electrophoresis to separate the samples.

3.3.2. Topoisomerase II Inhibitory Assay

To assess the Topo-II activity, a typical enzyme reaction was set up using human Topo-II, supercoiled pHot1 DNA as the substrate, the test drug, and the assay buffer. After incubation at 37 °C for 30 min, the mixture underwent a chemical reaction. The reaction was halted by introducing a solution containing 10% sodium dodecyl sulfate and proteinase K and incubating it at 37 °C for 15 min. Following that, the DNA was stained and ran on a 1% agarose gel for one to two hours using a Bio-Rad gel electrophoresis apparatus. Both supercoiled and linear DNA strands were included in the gel to serve as indicators for DNA-Topo-II intercalation.

3.4. Cell Cycle Analysis

Approximately 1×10^5 HepG2 cells were seeded into each well of a six-well plate and incubated for 24 h. Following incubation, the cells were treated with compounds **8b** and **7c** as specified for an additional 24 h. Subsequently, the cells were harvested, fixed in 70% ethanol, and incubated for 12 h at 4 °C. After fixation, the cells were centrifuged, and the cell pellets were washed and resuspended in 0.5 mL of cold phosphate-buffered saline (PBS). The cells were then stained with 5 μg/mL of propidium iodide (PI; Sigma, Panama City, Panama), and the DNA content was determined using a BD FACSCalibur flow cytometer (Becton Dickinson, Franklin Lakes, NJ, USA). Data analysis was performed using CellQuest software, and 10,000 events were analyzed for each sample.

3.5. Apoptosis Assay

An Annexin V-FITC assay was employed to detect apoptosis in the HepG2 cells. In a six-well plate, approximately 2×10^5 cells were incubated with compounds **8b** and **7c** for 24 h. Following incubation, the cells were harvested, washed twice with cold PBS, and resuspended in 500 μL of Annexin V-FITC binding buffer (10 μM HEPES, 140 μM NaCl, and 2.5 μM $CaCl_2$ at pH 7.4) and incubated for 15 min at room temperature. Subsequently, the cells were stained with 5 μL of Annexin V-FITC and incubated in the dark at 4 °C for 30 min, followed by the addition of 5 μL of propidium iodide (PI). Cell apoptosis was detected using a FACSCalibur flow cytometer (Becton-Dickinson, Franklin Lakes, NJ, USA). Data analysis was carried out using CellQuest software, and 10,000 events were analyzed for each sample. The Annexin V assay classified the cells as follows: viable cells were both Annexin V-FITC and PI negative. The cells in early apoptosis were Annexin V-FITC positive and PI negative, whereas the cells in late apoptosis were positive for both Annexin V and PI. The necrotic and dead cells were negative for Annexin V-FITC and positive for PI.

3.6. *Selectivity Index Analysis*

The selectivity index (SI) was obtained from the IC_{50} value of the compound against normal cells divided by the IC_{50} value of the cancer cells [28,29]. Accordingly, each SI value was calculated using the formula given below:

$$SI = IC_{50} \text{ for normal cells}/IC_{50} \text{ for cancer cells.}$$

Compounds were classified as high selectivity if the SI value was >3 and less selective if the SI value was <3 [30].

3.7. *Molecular Docking*

Molecular modeling studies were performed where two proteins were retrieved for docking from the protein data bank; one was the crystal structure topoisomerase-I along with its co-crystallized ligand (PDB ID:1K4T), while the other was topoisomerase2α protein with its co-crystallized ligand, which was downloaded with (PDB ID: 5GWK). Every compound structure was sketched as a 2D structure in a chem-sketch then converted to 3D using open babel and saved in sdf formate, which can be opened in UCSF-chimera 1.17.3 and Discovery Studio. Ligand and protein were then prepared using UCSF-chimera by adding hydrogen and assigning Gasteiger charges, then energy minimization. Grid box was generated using Auto-grid allocated at the macromolecule center. The grid box was then resized to cover the whole protein. Docking was performed using AutoDock vina. The results were saved as pdbqt and then visualized using Biovia Discovery Studio v21.1.0.20298 [31].

4. Conclusions

This work describes the synthesis of three novel series of acridine–sulfonamide hybrids, **6a–b**, **7a–e** and **8a–8d**. An MTT assay was performed against three cancer cell lines where compound **8b** recorded the best results with $IC_{50}s$ values 14.51 ± 1.4, 9.39 ± 0.9, and 8.83 ± 0.9 μM against HepG2, HCT-116, and MCF-7 cell lines, respectively. The compounds' capacity to inhibit topoisomerase activity was established using supercoiled DNA as a substrate where **8b** exhibited the best IC_{50} score of 3.41 μM against Topo-I, and **7c** displayed remarkable Topo-II inhibitory activity with an IC_{50} of 7.33 μM. Upon investigating their impact on apoptosis and cell cycle arrest, compound **8b** dropped the proportion of cells in the S phase from 46.39% to 39.15%. Furthermore, compound **7c** caused a minor decrease in the cell population in the S phase (from 46.39% to 29.27%), and in the G0–G1 phase (from 47.02% to 42.61%). Compound **7c** induced apoptosis to the extent of 55% and necrosis to 88% of the levels observed with SAHA as +ve control, while compound **8b** induced apoptosis to the extent of 79% and necrosis to 46% of that of SAHA. Furthermore, compounds **8b** and **7c** were evaluated on a THLE2 normal cell line using SAHA as a reference, where they exhibited cellular cytotoxicity with IC_{50} of 104 and 55.5 μM, respectively, in comparison to the control drug SAHA IC_{50} = 33.6 μM. In an attempt to explore the selectivity of the tested derivatives to cancer cells, the selectivity index was calculated for both compounds **7c** and **8b** which revealed that both the compounds demonstrated SI > 3 except for **7c** and HCT-116 cells; however, compound **8b** was more selective than **7c** for HepG2, HCT-116, and MCF-7. These findings show that the synthetic compounds have the ability to specifically target cancer cells while causing less damage to healthy cells. Molecular modeling studies revealed that some amino acids residues involved in the interaction of the tested compounds **8b** and **7c** with the receptor pocket were the main residues required for interaction with the co-crystallized ligand as well. Overall, compounds **7c** and **8b** can be considered as promising candidates for further anticancer drug development.

Supplementary Materials: The following supporting information can be downloaded at: https://www.mdpi.com/article/10.3390/ph17111487/s1, Figure S1: dose responsive curve of anticancer activity for acridine; Figure S2a: inhibition power of the synthesized compounds and the control on topo I-DNA cleavage complex formation; Figure S2b: inhibition power of the synthesized compounds and the control on topo II -DNA cleavage complex formation. Figure S3: chemical structures of the synthesized compounds; Figures S4–S16: data analysis results for the synthesized compounds.

Author Contributions: M.B., H.A., A.K., F.A. and D.E. carried out and designed the experiments, evaluated and analyzed the data, and drafted the text. E.I.E. conducted the molecular docking analysis, edited and revised the final draft. Y.S.M. drafted and revised the paper. R.B. funded the research, created the English edition, and revised the final draft. I.E.T.E.S. helped create the study's protocol and was engaged in the conception of the investigation. All authors have read and agreed to the published version of the manuscript.

Funding: Financial support from the Princess Nourah bint Abdulrahman University Researchers Supporting Project number PNURSP2024R304, Princess Nourah bint Abdulrahman University, Riyadh, Saudi Arabia.

Institutional Review Board Statement: Not applicable.

Informed Consent Statement: Not applicable.

Data Availability Statement: The original contributions presented in the study are included in the article/Supplementary Material, further inquiries can be directed to the corresponding authors.

Acknowledgments: The authors gratefully acknowledge the financial support from the Princess Nourah bint Abdulrahman University Researchers Supporting Project number PNURSP2024R304, Princess Nourah bint Abdulrahman University, Riyadh, Saudi Arabia.

Conflicts of Interest: The authors declare no conflict of interest.

References

1. Delgado, J.L.; Hsieh, C.M.; Chan, N.L.; Hiasa, H. Topoisomerases as Anticancer Targets. *Biochem. J.* **2018**, *475*, 373–398. [CrossRef]
2. Buzun, K.; Bielawska, A.; Bielawski, K.; Gornowicz, A. DNA Topoisomerases as Molecular Targets for Anticancer Drugs. *J. Enzym. Inhib. Med. Chem.* **2020**, *35*, 1781–1799. [CrossRef] [PubMed]
3. Sakasai, R.; Iwabuchi, K. The Distinctive Cellular Responses to DNA Strand Breaks Caused by a DNA Topoisomerase I Poison in Conjunction with DNA Replication and RNA Transcription. *Genes Genet. Syst.* **2015**, *90*, 187–194. [CrossRef] [PubMed]
4. Lee, J.H.; Berger, J.M. Cell Cycle-Dependent Control and Roles of DNA Topoisomerase II. *Genes* **2019**, *10*, 859. [CrossRef]
5. Li, M.; Liu, Y. Topoisomerase I in Human Disease Pathogenesis and Treatments. *Genom. Proteom. Bioinform.* **2016**, *14*, 166–171. [CrossRef]
6. Mutschler, E.; Geisslinger, G.; Kroemer, H.K.; Menzel, S.; Ruth, P.; Drożdzik, M.; Kocić, I.; Pawlak, D.; Malinowska, B.; Grotthus, B. Mutschler-Farmakologia i Toksykologia. In *Podręcznik*; medPharm Polska: Wrocław, Poland, 2016.
7. de Almeida, S.M.V.; Lafayette, E.A.; Silva, W.L.; de Lima Serafim, V.; Menezes, T.M.; Neves, J.L.; Ruiz, A.L.T.G.; de Carvalho, J.E.; de Moura, R.O.; Beltrão, E.I.C.; et al. New Spiro-Acridines: DNA Interaction, Antiproliferative Activity and Inhibition of Human DNA Topoisomerases. *Int. J. Biol. Macromol.* **2016**, *92*, 467–475. [CrossRef]
8. de Almeida, S.M.V.; Ribeiro, A.G.; de Lima Silva, G.C.; Alves, J.E.F.; Beltrão, E.I.C.; de Oliveira, J.F.; de Carvalho Junior, L.B.; de Lima, M.d.C.A. DNA Binding and Topoisomerase Inhibition: How Can These Mechanisms Be Explored to Design More Specific Anticancer Agents? *Biomed. Pharmacother.* **2017**, *96*, 1538–1556. [CrossRef]
9. Kathiravan, M.K.; Kale, A.N.; Nilewar, S. Discovery and Development of Topoisomerase Inhibitors as Anticancer Agents. *Mini Rev. Med. Chem.* **2016**, *16*, 1219–1229. [CrossRef] [PubMed]
10. Kathiravan, M.K.; Khilare, M.M.; Nikoomanesh, K.; Chothe, A.S.; Jain, K.S. Topoisomerase as Target for Antibacterial and Anticancer Drug Discovery. *J. Enzym. Inhib. Med. Chem.* **2013**, *28*, 419–435. [CrossRef]
11. Kožurková, M.; Sabolová, D.; Kristian, P. A Review on Acridinylthioureas and Its Derivatives: Biological and Cytotoxic Activity. *J. Appl. Toxicol.* **2017**, *37*, 1132–1139. [CrossRef]
12. Liang, X.; Wu, Q.; Luan, S.; Yin, Z.; He, C.; Yin, L.; Zou, Y.; Yuan, Z.; Li, L.; Song, X. A Comprehensive Review of Topoisomerase Inhibitors as Anticancer Agents in the Past Decade. *Eur. J. Med. Chem.* **2019**, *171*, 129–168. [CrossRef] [PubMed]
13. Panchal, N.B.; Patel, P.H.; Chhipa, N.M.; Parmar, R.S. Acridine a Versatile Heterocyclic Moiety as Anticancer Agent. *Int. J. Pharm. Sci. Res.* **2020**, *10*, 4739–4748.
14. Rupar, J.; Dobričić, V.; Aleksić, M.; Brborić, J.; Čudina, O. A Review of Published Data on Acridine Derivatives with Different Biological Activities. *Kragujev. J. Sci.* **2018**, *40*, 83–101. [CrossRef]
15. Kozurkova, M. Acridine Derivatives as Inhibitors/Poisons of Topoisomerase II. *J. Appl. Toxicol.* **2022**, *42*, 544–552. [CrossRef]

16. Li, F.; Jiang, T.; Li, Q.; Ling, X. Camptothecin (CPT) and its derivatives are known to target topoisomerase I (Top1) as their mechanism of action: Did we miss something in CPT analogue molecular targets for treating human disease such as cancer? *Am. J. Cancer Res.* **2017**, *7*, 2350.
17. Abbas, S.Y.; El-Sharief, M.A.M.S.; Basyouni, W.M.; Fakhr, I.M.I.; El-Gammal, E.W. Thiourea Derivatives Incorporating a Hippuric Acid Moiety: Synthesis and Evaluation of Antibacterial and Antifungal Activities. *Eur. J. Med. Chem.* **2013**, *64*, 111–120. [CrossRef]
18. Saeed, S.; Rashid, N.; Jones, P.G.; Ali, M.; Hussain, R. Synthesis, Characterization and Biological Evaluation of Some Thiourea Derivatives Bearing Benzothiazole Moiety as Potential Antimicrobial and Anticancer Agents. *Eur. J. Med. Chem.* **2010**, *45*, 1323–1331. [CrossRef]
19. Elmongy, E.I.; Alanazi, W.S.; Aldawsari, A.I.; Alfaouri, A.A.; Binsuwaidan, R. Antimicrobial Evaluation of Sulfonamides after Coupling with Thienopyrimidine Coplanar Structure. *Pharmaceuticals* **2024**, *17*, 188. [CrossRef] [PubMed]
20. Elmongy, E.I.; Binjubair, F.A.; Alshehri, O.Y.; Baeshen, K.A.; Almukhalfi, Z.A.; Henidi, H.A. In Silico Screening and Anticancer-Apoptotic Evaluation of Newly Synthesized Thienopyrimidine/Sulfonamide Hybrids. *Int. J. Mol. Sci.* **2023**, *24*, 10827. [CrossRef]
21. Burgeson, J.R.; Moore, A.L.; Boutilier, J.K.; Cerruti, N.R.; Gharaibeh, D.N.; Lovejoy, C.E.; Amberg, S.M.; Hruby, D.E.; Tyavanagimatt, S.R.; Allen III, R.D. SAR Analysis of a Series of Acylthiourea Derivatives Possessing Broad-Spectrum Antiviral Activity. *Bioorg. Med. Chem. Lett.* **2012**, *22*, 4263–4272. [CrossRef]
22. Limban, C.; Vasile, A.; Chirita, I.C.; Caproiu, M. Preparation of New Thiourea Derivatives with Potential Anti-Parasitic and Antimicrobial Activity. *Rev. Chim.* **2010**, *61*, 946–950.
23. Burgess, S.J.; Selzer, A.; Kelly, J.X.; Smilkstein, M.J.; Riscoe, M.K.; Peyton, D.H. A Chloroquine-like Molecule Designed to Reverse Resistance in Plasmodium Falciparum. *J. Med. Chem.* **2006**, *49*, 5623–5625. [CrossRef] [PubMed]
24. Solinas, A.; Faure, H.; Roudaut, H.; Traiffort, E.; Schoenfelder, A.; Mann, A.; Manetti, F.; Taddei, M.; Ruat, M. Acylthiourea, Acylurea, and Acylguanidine Derivatives with Potent Hedgehog Inhibiting Activity. *J. Med. Chem.* **2012**, *55*, 1559–1571. [CrossRef]
25. Sayed, F.A.-A.I.I.E.T. El Synthesis and Antiproliferative Activity of Acridine-Sulfonamide Conjugates. *Eur. J. Pharm. Med. Res.* **2018**, *5*, 59–65.
26. Al-Jewari, H.; AL-Faisal, A.H.M.; Nader, M. InVitro Cytotoxic Activity of the L-Asparaginase Extracted and Purified from Pathogenic Escherichia Coli against Four Leukemic Cell Lines. In Proceedings of the 2nd Annual International Conference of Northest Pharmacy Research 2010, Maha Sarakham, Thailand, 13–14 February 2010; pp. 21–23.
27. Staker, B.L.; Hjerrild, K.; Feese, M.D.; Behnke, C.A.; Burgin, A.B.; Stewart, L. The Mechanism of Topoisomerase I Poisoning by a Camptothecin Analog. *Proc. Natl. Acad. Sci. USA* **2002**, *99*, 15387–15392. [CrossRef]
28. Badisa, R.B.; Ayuk-Takem, L.T.; Ikediobi, C.O.; Walker, E.H. Selective Anticancer Activity of Pure Licamichauxiioic-B Acid in Cultured Cell Lines. *Pharm. Biol.* **2006**, *44*, 141–145. [CrossRef]
29. Sutejo, I.R.; Putri, H.; Meiyanto, E. The Selectivity of Ethanolic Extract of Buah Makassar (*Brucea javanica*) on Metastatic Breast Cancer Cells. *J. Agromed. Med. Sci.* **2016**, *2*, 1–6. [CrossRef]
30. Segun, P.A.; Ogbole, O.O.; Ismail, F.M.D.; Nahar, L.; Evans, A.R.; Ajaiyeoba, E.O.; Sarker, S.D. Resveratrol Derivatives from Commiphora Africana (A. Rich.) Endl. Display Cytotoxicity and Selectivity against Several Human Cancer Cell Lines. *Phytother. Res.* **2019**, *33*, 159–166. [CrossRef]
31. Rafiq, M. Molecular docking using chimera and autodock vina software for nonbioinformaticians. *JMIR Bioinform. Biotechnol.* **2020**, *1*, e14232.

pharmaceuticals

Article

Synthesis, Characterization, and Biological Evaluation of Meldrum's Acid Derivatives: Dual Activity and Molecular Docking Study

Syed Nasir Abbas Bukhari [1,*], Mohamed Abdelwahab Abdelgawad [1], Naveed Ahmed [2], Muhammad Wahab Amjad [3], Muhammad Ajaz Hussain [4], Mervat A. Elsherif [5], Hasan Ejaz [6], Nasser H. Alotaibi [7], Ignjat Filipović [8] and Nenad Janković [9,*]

1 Department of Pharmaceutical Chemistry, College of Pharmacy, Jouf University, Sakaka 72388, Al Jouf, Saudi Arabia
2 Department of Pharmaceutics, College of Pharmacy, Jouf University, Sakaka 72388, Al Jouf, Saudi Arabia
3 Center for Ultrasound Molecular Imaging and Therapeutics, Pittsburgh Heart, Lung, Blood and Vascular Medicine Institute, University of Pittsburgh, Pittsburgh, PA 15260, USA
4 Centre for Organic Chemistry, School of Chemistry, University of the Punjab, Lahore 54590, Pakistan
5 Chemistry Department, College of Science, Jouf University, Sakaka 72388, Al Jouf, Saudi Arabia
6 Department of Clinical Laboratory Sciences, College of Applied Medical Sciences, Jouf University, Sakaka 72388, Al Jouf, Saudi Arabia
7 Department of Clinical Pharmacy, College of Pharmacy, Jouf University, Sakaka 72388, Al Jouf, Saudi Arabia
8 University of Kragujevac, Faculty of Science, Radoja Domanovića 12, 34000 Kragujevac, Serbia
9 University of Kragujevac, Department of Science, Institute for Information Technologies Kragujevac, Jovana Cvijića bb, 34000 Kragujevac, Serbia
* Correspondence: sbukhari@ju.edu.sa (S.N.A.B.); nenad.jankovic@kg.ac.rs (N.J.)

Abstract: In the presented study, eight novel Meldrum's acid derivatives containing various vanillic groups were synthesized. Vanillidene Meldrum's acid compounds were tested against different cancer cell lines and microbes. Out of nine, three showed very good biological activity against *E. coli*, and HeLa and A549 cell lines. It is shown that the *O*-alkyl substituted derivatives possessed better antimicrobial and anticancer activities in comparison with the *O*-acyl ones. The decyl substituted molecule (**3i**) has the highest activity against *E. coli* (MIC = 12.4 μM) and cancer cell lines (HeLa, A549, and LS174 = 15.7, 21.8, and 30.5 μM, respectively). The selectivity index of **3i** is 4.8 (HeLa). The molecular docking study indicates that compound **3i** showed good binding affinity to DNA, *E. coli* Gyrase B, and topoisomerase II beta. The covalent docking showed that **3i** was a Michael acceptor for the nucleophiles Lys and Ser. The best E_b was noted for the topoisomerase II beta-LYS482-**3i** cluster.

Keywords: Meldrum's acid; vanillin; antimicrobial activity; anticancer activity; gyrase B; topoisomerase II

1. Introduction

Meldrum's acid presents a compound that has an active methylene group, which was discovered by Meldrum [1]. Therefore, Meldrum proposed the wrong chemical structure. Four decades later, the structure of Meldrum's acid was solved by Davidson [2]. Its carbonyl function easily can be attacked predominately by oxygen or nitrogen nucleophiles. The C5 position is involved in electrophilic substitution. At high-temperature regimes, Meldrum's acid and its derivatives could suffer from a fragmentation process that leads to reactive ketene formation [3]. Although it was discovered more than a century ago, its use continues to attract the attention of synthetic chemists. Electrophilicity, dienophilicity, and its relatively low acidity (pKa = 4.9) make Meldrum's acid attractive and quite useful for a wide number of synthetic manipulations such as heterocycles synthesis, Friedel-Crafts acylation, total synthesis, alkylidene and asymmetric synthesis, domino reactions, and catalytic additions [4,5]. Furthermore, Meldrum's acid represents an essential molecule in the synthesis of natural products and analogues [6].

Meldrum's acid derivatives have a wide spectrum of biological activities such as antibacterial, antifungal, anticancer properties that make them attractive from the pharmaceutical point of view [7]. The good antimalarial activity of vanillidene Meldrum's acid that was comparable with chloroquine was noted against *P. falciparum* [8]. Some vanillidene Meldrum's acid showed a high antimicrobial effect against *B. subtilis* and *B. cereus* [9]. Bisarylidene Meldrum's acid possesses significant activity against *S. aureus*, *B. cereus*, *E. coli*, and *P. aeruginosa* species [10]. Promising antibiotic potential together with fluoroquinolones for the selected Meldrum's acid arylamino methylene group was achieved. In this communication, authors proved this group's good applicability to fight with reverse resistance in the future [11]. 5-Arilidene derivatives have been published as a promising generation of novel platelet aggregation inhibitors [12]. The importance of aryilidene Meldrum's acid derivatives were applied as key intermediates in synthesis of various molecules such as epoxide [13], carboxylic acid [14], oxopyridines [15], aldehydes [16], and monoalkyl derivatives [17]. In addition, the presence of an α,β-unsaturated carbonyl function is important in the treatment of various diseases [18,19].

Dual-active drugs are a concept noted as an imperative in future drug design. Namely, a good perspective is found in the development of novel drugs that can have double biological behavior (anticancer–antiviral, –antimicrobial, *etc.*) that have the possibility to cure two different diseases [20–24]. To date, there is no evidence about the potential dual activity of Meldrum's acid derivatives. Our continual attention on the development of dual-active drugs prompted us to make a novel set of vanillidene Meldrum's acid derivatives. Going forward, antimicrobial and cytotoxicity evaluation were investigated.

2. Results

Our initial idea was to synthesize a set of benzylidene Meldrum's acid derivatives starting from different vanillic aldehydes that have long-chain fragments. The first reaction was tested with 5′-nitrovanilline (**2a**) and Meldrum's acid (**1**). Applying similar catalytic conditions such as those described in our published paper [9], we isolated product **3a** (79%). In the presence of the complex salt $(PhNH_3)_2CuCl_4$ as the catalyst in a concentration of 1 mg/mL (0.85 mol%), the reaction was very fast (5 min). Therefore, PTSA (*p*-toluene sulfonic acid) as catalyst (5 and 10 mol%) produced **3a** in lower yield (62%) during 12 h. The same experimental conditions were tested in reactions **2b** and **2f** with **1**. In both cases PTSA gave lower yield in comparison with $(PhNH_3)_2CuCl_4$. In some cases, yields were significantly lower. For example, employing PTSA the yield of **3f** was 51%, and 75% when $(PhNH_3)_2CuCl_4$ was loaded. The results prompted us to apply $(PhNH_3)_2CuCl_4$ (1 mg/mL) as the catalyst in reaction **1** and various *O*-acyl or -alkyl vanillic aldehydes.

Applying the catalytic system, good to excellent yields (54–92%) of vanillidene Meldrum's acid derivatives were obtained (Scheme 1). Almost all products were isolated very easily by filtration. However, there were some problems during the workup of compound **3e**. The increased number of oxygen atoms in the side chain made **3e** very soluble in ethanol. The reaction mixture was evaporated, and upon the addition of the DCM product **3e** was precipitated, filtered, and washed with the DCM. Acyl groups (propionyl, cyclopropanecarbonyl, and *p*-metoxybenzoyl) as structural motifs in **3b–d** (61–69%) negatively affected the yield in comparison with long-chain alkyl groups in **3f–i** (75–92%).

Proton and carbon NMR spectra of **3a–i** samples (cca. 40 mg/mL) were carried out in $CDCl_3$ or DMSO-d_6 (Figures S1–S18). In every sample, the methyliden proton was the best measure of condensation efficiency. It can be found at approximately 8.35 ppm in proton NMR. In the ^1H NMR spectrum of **3a** (Figure S1), 1.74 (s), 3.92 (s), 8.14 (s), 8.33 (s), and 8.54 (s) ppm are assigned to 2,2-dimethyl, methoxy, aromatic, and methyliden protons, respectively. The phenolic proton in **3a** is invisible in its NMR, which is presumably caused by the strong electron-withdrawing effect of the nitro group attached to the ortho position. The specific carbon peak located at 155 ppm (Figure S2) was assigned to phenolic carbon, fragment C-OH. Moreover, in all samples, the high-intensity peak at 1.7–1.8 ppm is assigned to the 2,2-dimethyl backbone of Meldrum's acid fragment. As compared to all carbon NMR, the ^{13}C NMR spectrum of **3b–d** (Figures S4, S6 and S8) showed new peaks at around 171 ppm that originated from the acyl carbonyl function.

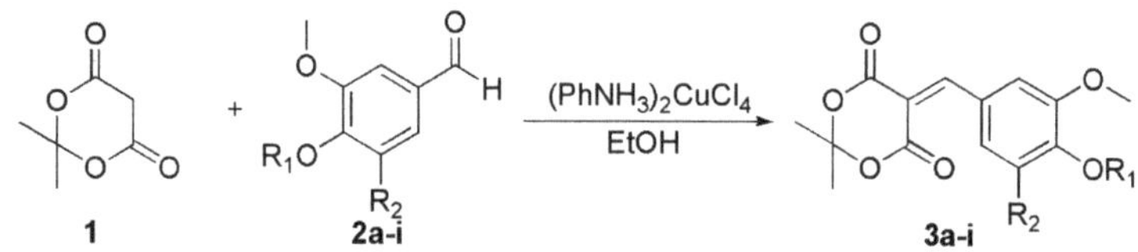

a R_1 = H, R_2 = NO_2 (79%)
b R_1 = propionyl, R_2 = H (63%)
c R_1 = cyclopropanecarbonyl, R_2 = H (61%)
d R_1 = *p*-metoxybenzoyl, R_2 = H (69%)
e R_1 = ~O~O~OH, R_2 = H (54%)
f R_1 = hexyl, R_2 = H (75%)
g R_1 = heptyl, R_2 = H (92%)
h R_1 = octyl, R_2 = H (88%)
i R_1 = decyl, R_2 = H (82%)

Scheme 1. General outline of the synthesis of benzylidene Meldrum's acid **3a–i**.

The strong and broad FT-IR peak at around 3204 cm^{-1} is assigned as an overlapped peak of phenolic OH and methyliden function. The band at 2949 cm^{-1} is related to C-H stretching of a chemically non-equivalent double bond functional within **3a**. Two strong bands at 1751 and 1703 cm^{-1} denote the vibration of a carbonyl group from Meldrum's acid skeleton.

Vanillidene Meldrum's acid derivatives **3a–i** were subjected to biological evaluation. For these experiments, we treated various microbes (*Escherichia coli, Bacillus subtilis, Staphylococcus aureus,* and *Bacillus cereus*) and cancer cell lines (HeLa, K562, A549, LS174, and PaCa-2). Selectivity was checked towards normal human fibroblast (MRC-5). Streptomycin and cis-platinum (cis-DDP) were chosen as reference standards. The data presented in Tables 1 and 2 are average antimicrobial and anticancer activities of **3a–i** delivered after three independent experiments.

Table 1. Antimicrobial activity of **3a–i**. MIC values are presented in μM.

	E. coli	*B. subtilis*	*S. aureus*	*B. cereus*
3a	115.3	-	-	-
3b	89.5	175.3	210.5	174.2
3c	85.1	180.5	205.8	171.5
3d	74.8	169.1	199.6	168.4
3e	38.1	133.1	145.1	80.5
3f	37.5	48.3	73.9	42.3
3g	36.1	45.2	60.4	37.5
3h	14.7	31.5	54.2	64.8
3i	12.4	29.5	50.9	41.5
streptomycin	0.010	0.003	0.005	0.003

Table 2. Cytotoxicity of **3a–i** (MTT in μM).

	HeLa	K562	A549	LS174	PaCa-2	MRC-5
3a	126.1 ± 1.45	151.2 ± 3.47	>200	>200	>200	>200
3b	101.7 ± 2.14	74.5 ± 1.73	41.2 ± 1.74	104.7 ± 4.62	152.7 ± 4.29	>200
3c	84.6 ± 1.84	70.2 ± 1.68	45.7 ± 1.65	99.6 ± 1.44	172.5 ± 5.25	>200
3d	94.5 ± 3.15	105.1 ± 0.92	36.2 ± 0.39	118.9 ± 0.95	145.1 ± 2.92	>200
3e	37.3 ± 1.20	80.5 ± 2.36	25.9 ± 0.46	38.1 ± 0.41	181.5 ± 3.94	110.9 ± 2.19
3f	63.5 ± 0.45	71.9 ± 1.65	39.6 ± 0.74	45.3 ± 0.58	110.5 ± 1.75	144.2 ± 2.84
3g	62.6 ± 0. 71	69.5 ± 1.37	35.1 ± 0.27	47.1 ± 1.69	94.2 ± 1.09	139.1 ± 3.07
3h	18.2 ± 0.11	73.5 ± 1.43	27.4 ± 0.78	36.5 ± 0.24	90.5 ± 0.83	72.8 ± 0.77
3i	15.7 ± 0.28	63.7 ± 2.49	21.8 ± 0.91	30.5 ± 0.35	58.2 ± 0.74	74.6 ± 1.46
cis-DDP	2.36 ± 0.28	5.56±0.23	17.93 ± 0.44	20.8 ± 0.44	25.8 ± 0.65	4.26 ± 0.46

Going deeper inside interactions that should be responsible for antimicrobial or anticancer activity, molecular docking was applied to the most active compounds (**3e**, **3h**, and **3i**). The mentioned compounds were docked to DNA, *Escherichia coli* Gyrase B, and topoisomerase II beta. The energies of binding (E_b) of **3e**, **3h**, and **3i** are presented in Table 3. The E_b of reference compounds ellipticine (for DNA), P3C (for Gyrase B), and etoposide (for topoisomerase II) are −8.98, −7.9, and −14.04 kcal mol^{-1}, respectively. Considering these results, the most promising molecule that showed good binding affinity towards reference compounds is **3i**.

Table 3. Energies of binding (E_b) derived from docking of **3e**, **3h**, and **3i** to DNA (I), *Escherichia coli* Gyrase B (II), and topoisomerase II beta (III).

	E_b (kcal mol^{-1})		
	I	**II**	**III**
3e	−5.61	−5.45	−8.46
3h	−6.94	−7.20	−9.37
3i	−7.74	−7.52	−9.51

3. Discussion

The results of antimicrobial testing of **3a–i** against selected microbes showed significant activity (Table 1) against Gram-positive (*B. subtilis*, *S. aureus*, and *B. cereus*) and Gram-negative bacteria (*E. coli*). All the tested compounds showed significantly better activity against Gram-negative in comparison with Gram-positive strains. From a structural point of view, vanillidene with the *O*-alkyl fragment (**3e–i**) showed to be much more active in contrast to ones with *O*-acyl compounds (**3b–d**) against all treated bacteria. The principal reason could be found in the cell wall structural difference. The Gram-positive bacteria cell wall contains peptidoglycans and teichoic acids, while Gram-negative bacteria have lipopolysaccharides and lipopoliproteins as building units of the cell wall [25]. Hence, the presence of a long alkyl chain in structures **3e–i** make them more lipophilic, and consequently enables easier passage of these molecules through the cell wall. Out of all nine, the compound with the decyl fragment (**3i**) has the most promising activity against *E. coli* (12.4 μM). Compound **3a** has the lowest activity against all bacterial strains.

The cytotoxic potential of **3a–i** was studied against various cancer and normal cell lines. The results are presented as IC_{50} values in Table 2. Several compounds have good activity, while others possessed a moderate to weak cytotoxic effect. Generally, all nine molecules demonstrated the best activity against A549 cell lines in the range of 21.8–45.7 μM. The best activity was shown by **3i**. In particular, this molecule stands out against HeLa, A549, and LS174 cell lines, with values of 15.7, 21.8, and 30.5, respectively. The achieved IC_{50} values are similar to the ones obtained after treatment of HeLa, A549, and LS174 with cis-DDP. The investigation of MRC-5 cell lines suggests no significant toxicity of **3a–g**, while **3h** and **3i** showed moderate toxicity. The three most active compounds possess good selectivity indexes (SI) for HeLa cells (**3h**, **3e**, and **3i**, with values of 2.9, 4.0, and 4.8, respectively). Significant SIs were obtained for A549 cells at 4.3, 2.6, and 3.6 for **3h**, **3e**, and **3i**, respectively. Acyl fragments in structures of **3b–d** influenced the obtained IC_{50} value being at least double compared to ones delivered to alkyl substituted vanillidene Meldrum's acid (**3e–i**). The compounds of interest are **3e**, **3h**, and especially **3i** (Figure 1), which has the highest cytotoxicity and antimicrobial activity against all treated cell lines and bacteria. Considering all results noted during antimicrobial and cytotoxicity testing, we employed molecular docking to investigate the affinity of **3e**, **3h**, and **3i** to DNA, *E. coli* Gyrase B, and topoisomerase II beta at the molecular level.

The selected compounds (**3e**, **3h**, and **3i**,) were computationally evaluated for intercalating ability by docking to a six base pair DNA structure (PDB: 1z3f) against ellipticine, a co-crystalized substrate from the used file [26]. The results are presented in Table 3. Testing of each compound produced results of higher energy when compared against the reference substrate, indicating weaker binding. In each case, the pseudo planar part of the resulting lowest energy structure was packed into an intercalating cavity with slightly different positioning, indicating that each compound may bind in this way. The compound with the longest aliphatic chain (**3i**) showed the most promising results and is displayed in Figure 2.

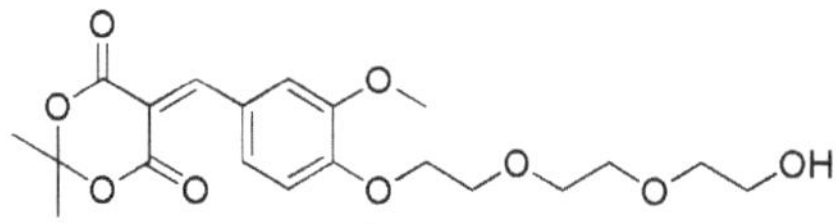

3e

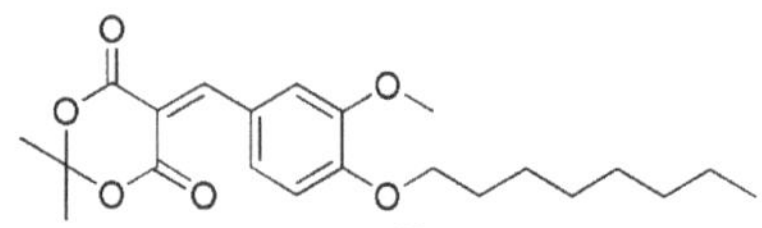

3h

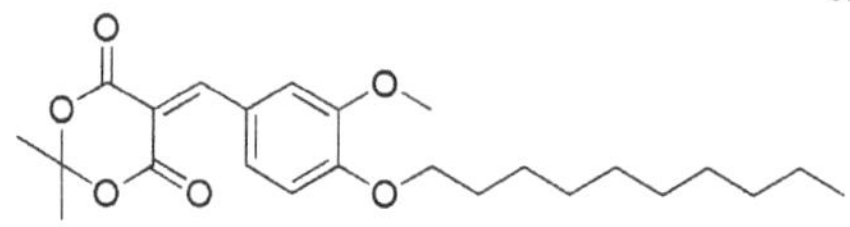

3i

Figure 1. Structures of compounds **3e**, **3h**, and **3i**.

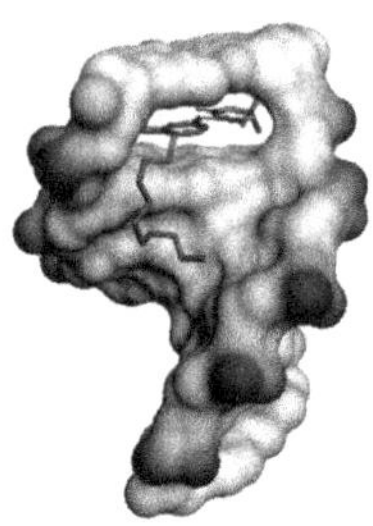

Figure 2. Compound **3i** bound to DNA in intercalative mode.

It was decided based on our biological tests to computationally examine new compounds for their binding affinity towards enzyme Gyrase B of *E. Coli*. Molecular docking of our compounds was performed on the crystal structure of Gyrase B (PDB:6f86) where the target cavity was the binding place of 4-(4-bromo-1*H*-pyrazol-1-yl)-6-[(ethylcarbamoyl)amino]-*N*-(pyridine-3-yl)pyridine-3-carboxamide (P3C). Since the inhibitory activity of P3C is known (IC_{50} = 0.037 μM [27]), it serves as a good reference for comparison of obtained docking results for new inhibitors. Binding energies are presented in Table 3. Based on the obtained binding energies, we can conclude that most of our compounds would generally show weaker inhibitory activity when compared to P3C, but the result obtained from docking **3i** is comparable to the result displayed by the reference compound. The position of the docked structure for **3i** is similar to the position of PC3, and this similarity is evident from Figure 3 and Figure S19. The difference is, for example, that **3i** does not possess a side group bound to a central aromatic ring like the bromopyrazolyl moiety of P3C, which allows interactions with and in the vicinity of ILE94, while **3i** has a long aliphatic chain that extends much deeper into the cavity of the enzyme than the ethylcarbamoyl group of P3C.

To further examine the potential anticancer application of examined compounds, molecular docking experiments were performed on the crystal structure of type II topoisomerase beta (TOP2β) in a complex with DNA and the anticancer drug etoposide (PDB: 3qx3) [28]. Benzylidene Meldrum's acid was noted as the key precursor to novel topoisomerase II inhibitors [29]. One of them is α-lapachone, which is approved as a topoisomerase II inhibitor [30,31]. Redocking of etoposide was used to obtain the reference value of binding energy. The main reason for choosing the selected binding site was comparison with the selected substrate (etoposide) that binds in that exact site. Topoisomerase II is symmetric, and there are two identical sides. Docking experiments were conducted on both sides and identical results (geometry and energy of binding) were obtained. For better clarity we present only one of the identical binding modes.

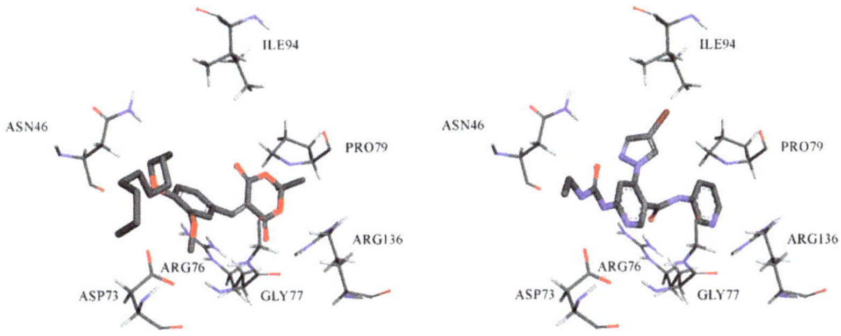

Figure 3. The lowest energy structures that resulted from docking of **3i** to active site of Gyrase B of *E. coli*. (**Left**): **3i**, (**right**): P3C.

The results derived from docking of the examined compounds are compared against the reference value in Table 3. Significantly higher binding energies of tested compounds when compared to etoposide suggest much weaker binding in the same pocket. Upon investigation of structures obtained from the best docking results, it was found that contrary to etoposide, which intercalates with DNA and interacts with the amino acid residues of the enzyme that can be found on both sides of the DNA chain, our compounds position themselves in such a way that only allows interactions with residues from a single side of the chain (Figure 4 and Figure S20).

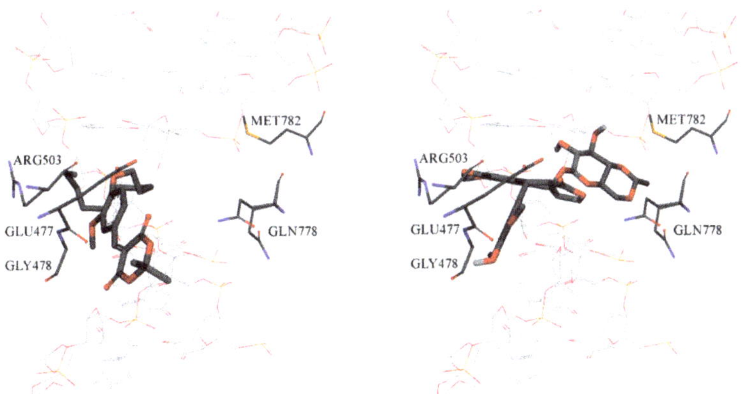

Figure 4. Comparison of binding modes for lowest energy **3i** structure (**left**) and lowest energy etoposide structure (**right**).

Docking of covalently bound ligands is important for gaining insight into enzymatic processes and designing superior covalent ligands. AutoDock 4 was used to implement two methods for docking covalently bound ligands: a grid-based method and a modification of the flexible side chain approach [32]. Covalent docking has also been applied to oligopeptidase [33] and proteasome [34] inhibitors and glycoside hydrolyses ligands [35]. Benzylidene Meldrum's acid was used as a valuable Michael acceptor [36,37]. Consequently, in this paper, covalent docking was employed to simulate the formation of the covalent adduct using the flexible side chain technique proposed by Bianco et al. [32]. This protocol needs to adapt the residue taking part in the covalent bond by attaching the ligand to its side chain; this modified residue is then considered flexible during the docking calculation. Docking was attained for the ligand, keeping the remodeled cysteine and serine residues flexible. This permitted us to sample the torsional flexibility of the ligand within the Gyrase B and topoisomerase II beta. It was shown that the best conformational compatibility between the subjected ligand Michael acceptor (**3i**) and substrate Gyrase B or topoisomerase II beta had residues LYS139 (Gyrase B), LYS482, and SER480 (topoisomerase II beta). The molecular interactions between **3i** and the amino acids lysine and serine in Gyrase B and topoisomerase II beta were estimated by using covalent docking calculations. The obtained results for E_b are presented in Table 4. The more negative E_b values indicate that the investigated ligand **3i** inhibits the receptor better.

Table 4. Binding energies for best docking conformation of **3i** with Gyrase B and topoisomerase II beta.

	E_b (kcal mol^{-1})		
	Gyrase B-LYS139	**Topoisomerase-LYS482**	**Topoisomerase-SER480**
3i	−5.64	−7.40	−5.24

As shown in Table 4, the investigated ligand strongly binds to the target receptors. The docking analyses revealed that covalent interactions existed between the investigated molecule and target receptors. These interactions occur between **3i** and the amino acids LYS and SER in positions 482 and 480 in the primary structure of the topoisomerase II beta-receptor. On the other hand, **3i** forms the covalent bond with the amino acid LYS139 from Gyrase B (Figure 5). Additionally, the docking results show that several non-covalent interactions occurred between the investigated molecule and target receptors. The important interactions are hydrogen bonds, carbon–hydrogen bonds, alkyl–alkyl, and alkyl–π interactions (Figure 5). The obtained results indicate that the **3i** ligand could act as a potential covalent inhibitor.

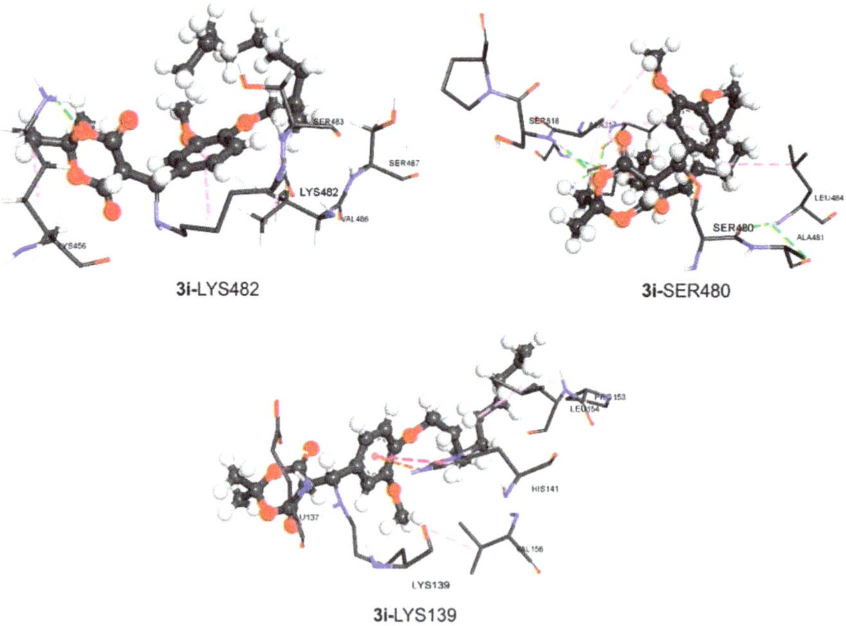

Figure 5. The docking interactions of the most stable conformation of **3i** and investigated receptors.

The α- and β-lapachones were described as covalent inhibitors of topoisomerase II [38]. The mentioned compounds have also been investigated as irreversible catalytic inhibitors of topoisomerase II [30]. The high affinity of **3i** to form a covalent bond with topoisomerase II beta makes it similar to the aforementioned Meldrum's acid analogs.

4. Materials and Methods

The IR spectra were recorded by a Perkin–Elmer Spectrum One FT-IR spectrometer on a KBr pellet. The NMR spectra of compounds **3a–i** were performed in DMSO-d_6 and $CDCl_3$ with TMS as the internal standard on a Varian Gemini 200 MHz NMR spectrometer (^{1}H at 200 and ^{13}C at 50 MHz). Abbreviations for the NMR signals that were used are as follows: s = singlet, d = doublet, dd = double of double, t = triplet, and m = multiplet. The ^{1}H and ^{13}C spectra are given in Supplementary Information (Figures S1–S18).

4.1. Synthesis of Benzylidene Meldrum's Acid Derivatives (3)

In the 10 mL round-bottomed flask, Meldrum's acid (1.5 mmol) was dissolved in 5 mL of absolute ethanol. Selected aldehyde was added at room temperature. Immediately after, $(PhNH_3)_2CuCl_4$ (1 mg/mL, 0.85 mol%) was loaded as catalyst. Precipitation of products occurred in the interval of 5 to 10 min. One hour later, amorphous powder was filtrated and washed out with 96% ethanol and water. Knoevenagel adducts were isolated in good purity grade for NMR measurements. NMR data for **3a–i** are given below.

5-(4′-hydroxy-3′-methoxy-5′-nitrobenzylidene)-2,2-dimethyl-1,3-dioxane-4,6-dione **3a**

Green solid; yield: 79% (383 mg); IR ν 3204, 2949, 1751, 1703, 1577, 1541 cm^{-1}; ^{1}H NMR (200 MHz, DMSO-d$_6$) δ 8.54 (s, 1H), 8.33 (s, 1H), 8.14 (s, 1H), 3.92 (s, 3H), 1.74 (s, 6H) ppm; ^{13}C NMR (50 MHz, DMSO-d$_6$) δ 162.8, 160.0, 155.1, 147.3, 124.2, 121.8, 120.2, 113.5, 104.5, 56.9, 27.2 ppm; ESI-MS (*m*/*z*): [M + Na]$^+$ = 346.

2′-methoxy-4′-((2,2-dimethyl-4,6-dioxo-1,3-dioxan-5-ylidene)methyl)phenyl propionate **3b**

Light green solid; yield: 63% (316 mg); IR ν 3465, 2991, 2944, 1762, 1733, 1598, 1582, 1512 cm^{-1}; ^{1}H NMR (200 MHz, $CDCl_3$) δ 8.36 (s, 1H), 8.22 (d, *J* = 2.0 Hz, 1H), 7.63–7.51 (m, 1H), 7.15 (d, *J* = 8.2 Hz, 1H), 3.90 (s, 3H), 2.65 (q, *J* = 7.5 Hz, 2H), 1.80 (s, 6H), 1.29 (t, *J* = 7.5 Hz, 3H) ppm; ^{13}C NMR (50 MHz, $CDCl_3$) δ 171.7, 163.3, 159.8, 157.1, 151.1, 144.6, 130.2, 129.1, 123.0, 116.9, 114.1, 104.5, 56.1, 27.6, 27.4, 9.1 ppm; ESI-MS (*m*/*z*): [M + H]$^+$ = 335.

2′-methoxy-4′-((2,2-dimethyl-4,6-dioxo-1,3-dioxan-5-ylidene)methyl)phenyl cyclopropanecarboxylate **3c**

Light green solid; yield: 61% (317 mg); IR ν 3469, 2985, 1753, 1733, 1513, 1379 cm^{-1}; ^{1}H NMR (200 MHz, $CDCl_3$) δ 8.36 (s, 1H), 8.22 (d, *J* = 2.0 Hz, 1H), 7.57 (dd, *J* = 8.4, 2.0 Hz, 1H), 7.17 (d, *J* = 8.3 Hz, 1H), 3.91 (s, 3H), 1.96–1.80 (m, 7H), 1.23–1.01 (m, 4H) ppm; ^{13}C NMR (50 MHz, $CDCl_3$) δ 172.1, 163.3, 159.8, 157.1, 151.2, 144.6, 130.2, 129.1, 123.1, 116.9, 114.1, 104.46, 56.1, 27.6, 12.9, 9.5 ppm; ESI-MS (*m*/*z*): [M + H]$^+$ = 347.

2′-methoxy-4′-((2,2-dimethyl-4,6-dioxo-1,3-dioxan-5-ylidene)methyl)phenyl 4-methoxybenzoate **3d**

Green solid; yield: 69% (426 mg); IR ν 3446, 2993, 1732, 1606, 1577 cm^{-1}; ^{1}H NMR (200 MHz, $CDCl_3$) δ 8.40 (s, 1H), 8.26 (d, *J* = 1.8 Hz, 1H), 8.15 (t, *J* = 5.7 Hz, 2H), 7.62 (dd, *J* = 8.3, 1.9 Hz, 1H), 7.30–7.26 (m, 1H), 6.98 (t, *J* = 5.7 Hz, 2H), 3.89 (d, *J* = 1.9 Hz, 6H), 1.81 (s, 6H) ppm; ^{13}C NMR (50 MHz, $CDCl_3$) δ 164.0, 163.7, 163.3, 159.8, 157.2, 151.4, 144.9, 132.5, 130.2, 129.2, 123.2, 121.1, 117.0, 114.1, 113.9, 104.5, 56.1, 55.5, 27.6 ppm; ESI-MS (*m*/*z*): [M + H]$^+$ = 413.

5-(4′-(2″-(2‴-(2⁗-hydroxyethoxy)ethoxy)ethoxy)-3′-methoxybenzylidene)-2,2-dimethyl-1,3-dioxane-4,6-dione **3e**

Green solid; yield: 54% (332 mg); IR ν 3583, 3439, 2939, 2913, 1745, 1708, 1578, 1559 cm^{-1}; ^{1}H NMR (200 MHz, $CDCl_3$) δ 8.35 (s, 1H), 8.28 (d, *J* = 2.0 Hz, 1H), 7.63 (dd, *J* = 8.6, 2.0 Hz, 1H), 6.97 (d, *J* = 8.5 Hz, 1H), 4.33–4.25 (m, 2H), 3.95–3.85 (m, 5H), 3.76–3.59 (m, 8H), 2.58 (s, 1H), 1.80 (s, 6H) ppm; ^{13}C NMR (50 MHz, $CDCl_3$) δ 163.9, 160.4, 158.0, 153.9, 148.9, 132.2, 125.2, 116.2, 111.9, 110.7, 104.1, 72.5, 70.9, 70.3, 69.2, 68.5, 61.7, 55.9, 27.5 ppm; ESI-MS (*m*/*z*): [M + H]$^+$ = 411.

5-(4′-hexyloxy-3′-methoxybenzylidene)-2,2-dimethyl-1,3-dioxane-4,6-dione **3f**

Yellow solid; yield: 75% (407 mg); IR ν 3444, 2956, 2937, 1748, 1713, 1558, 1523 cm^{-1}; ^{1}H NMR (200 MHz, $CDCl_3$) δ 8.36 (s, 1H), 8.29 (d, *J* = 2.1 Hz, 1H), 7.64 (dd, *J* = 8.6, 2.1 Hz, 1H), 6.94 (d, *J* = 8.6 Hz, 1H), 4.13 (t, *J* = 6.9 Hz, 2H), 3.94 (s, 3H), 1.93–1.82 (m, 8H), 1.56–1.26

(m, 6H), 0.94–0.87 (m, 3H) ppm; ^{13}C NMR (50 MHz, $CDCl_3$) δ 164.1, 160.5, 158.2, 154.5, 148.9, 132.6, 124.7, 116.0, 111.4, 110.2, 104.0, 69.2, 56.0, 31.5, 28.8, 27.5, 25.5, 22.5, 14.0 ppm; ESI-MS (m/z): $[M + H]^+$ = 363.

5-(4′-heptyloxy-3′-methoxybenzylidene)-2,2-dimethyl-1,3-dioxane-4,6-dione **3g**

Yellow solid; yield: 92% (519 mg); IR ν 3436, 2949, 2923, 1746, 1708, 1548, 1522 cm^{-1}; ^{1}H NMR (200 MHz, $CDCl_3$) δ 8.36 (s, 1H), 8.29 (d, J = 2.1 Hz, 1H), 7.64 (dd, J = 8.6, 2.1 Hz, 1H), 6.94 (d, J = 8.5 Hz, 1H), 4.13 (t, J = 6.9 Hz, 2H), 3.94 (s, 3H), 2.06–1.77 (m, 8H), 1.77–1.03 (m, 8H), 1.03–0.79 (m, 3H) ppm; ^{13}C NMR (50 MHz, $CDCl_3$) δ 164.1, 160.5, 158.2, 154.5, 148.9, 132.6, 124.7, 116.1, 111.4, 110.3, 104.0, 69.2, 56.0, 31.7, 29.0, 28.9, 27.5, 25.8, 22.6, 14.1 ppm; ESI-MS (m/z): $[M + H]^+$ = 377.

5-(3′-methoxy-4′-octyloxybenzylidene)-2,2-dimethyl-1,3-dioxane-4,6-dione **3h**

Yellow solid; yield: 88% (515 mg); IR ν 3439, 2950, 2930, 1745, 1708, 1549, 1523 cm^{-1}; ^{1}H NMR (200 MHz, $CDCl_3$) δ 8.36 (s, 1H), 8.30 (d, J = 2.1 Hz, 1H), 7.63 (d, J = 8.5 Hz, 1H), 6.94 (d, J = 8.6 Hz, 1H), 4.13 (t, J = 6.8 Hz, 2H), 3.94 (s, 3H), 2.05–1.62 (m, 8H), 1.60–1.16 (m, 10H), 0.89 (m, 3H) ppm; ^{13}C NMR (50 MHz, $CDCl_3$) δ 164.0, 160.5, 158.1, 154.5, 148.9, 132.5, 124.7, 116.1, 111.4, 110.3, 103.9, 69.2, 56.0, 31.7, 29.2, 29.1, 28.8, 27.5, 25.8, 22.6, 14.0 ppm; ESI-MS (m/z): $[M + H]^+$ = 391.

5-(4′-decyloxy-3′-methoxybenzylidene)-2,2-dimethyl-1,3-dioxane-4,6-dione **3i**

Yellow solid; yield: 82% (514 mg); IR ν 3439, 2952, 2928, 1746, 1709, 1545, 1521 cm^{-1}; ^{1}H NMR (200 MHz, $CDCl_3$) δ 8.36 (s, 1H), 8.30 (d, J = 2.1 Hz, 1H), 7.64 (dd, J = 8.6, 2.1 Hz, 1H), 6.94 (d, J = 8.6 Hz, 1H), 4.13 (t, J = 6.8 Hz, 2H), 3.94 (s, 3H), 2.00–1.68 (m, 8H), 1.62–1.16 (m, 14H), 0.88 (t, J = 6.4 Hz, 3H) ppm; ^{13}C NMR (50 MHz, $CDCl_3$) δ 164.0, 160.5, 158.1, 154.5, 148.9, 132.5, 124.7, 116.0, 111.4, 110.3, 104.0, 69.2, 56.0, 31.9, 29.5, 29.3, 29.3, 28.8, 27.5, 25.8, 22.6, 14.1 ppm; ESI-MS (m/z): $[M + H]^+$ = 419.

4.2. Determination of Antimicrobial Activity

The following strains of bacteria were used as test organisms in this study: *Staphylococcus aureus* (ATCC 25923), *Bacillus subtilis* (ATCC 6633), *Bacillus cereus* (ATCC 10987), and *Escherichia coli* (ATCC 25922). All the bacteria used were obtained from the American Type Culture Collection (ATCC). Cultures were stored at 4 °C and subcultured every 15 days. The minimal inhibitory concentration (MIC) was determined by the broth microdilution method using 96-well micro-titer plates [39]. A series of dilutions of the tested compounds was used in the experiment against every microorganism. The starting solutions of tested compounds were obtained by measuring off a certain quantity of the compounds and dissolving it in 5% DMSO. The MIC was determined with resazurin. The inoculated plates were incubated at 37 °C for 24 h. Resazurin is an oxidation–reduction indicator used for the evaluation of microbial growth. It is a blue non-fluorescent dye that becomes pink and fluorescent when reduced to resorufin by oxidoreductases within viable cells. The boundary dilution without any changing in color of resazurin was defined as the MIC for the tested microorganism at a given concentration. As a positive control of growth inhibition, streptomycin was used. A 5% DMSO solution was used as a negative control for the influence of the solvents.

4.3. Evaluation of Cytotoxicity

Cell line cells were obtained from the American Type Culture Collection. Human cervical adenocarcinoma HeLa, human chronic myelogenous leukemia K562, non-small cell lung carcinoma A549, human colon carcinoma LS174, pancreatic carcinoma PaCa-2, and normal human lung fibroblast MRC-5 cell lines were grown in RPMI-1640 medium with 10% fetal bovine serum, L-glutamine, and penicillin–streptomycin solution. Cells were plated into the 96-well cell culture plates. Cells were incubated at 37 °C in CO_2 incubator. Adherent cell lines were incubated for 20 h before addition of tested compounds. Two hours before addition of compounds, K562 cells were seeded into 96-well plates. The tested compounds were applied at five different concentrations ranging from 10 to 300 μM for 72 h. Thereafter, the survival of cells was determined by MTT assay in accordance with the protocol established by Mosmann [40] and Ohno and Abe [41]. The solution of MTT was added to blank and cell samples. The plates were incubated for 5 h at 37 °C, and then 10% solution of sodium dodecyl sulfate was added to the wells. The absorbance was measured at 570 nm using Thermo Scientific Multiskan EX plate reader. Three independent experiments were performed.

4.4. Molecular Docking

Molecular docking experiments were prepared using AutoDockTools and were performed using AutoDock 4 [42]. Structures of target molecules were obtained from rscb.org, and were prepared by first removing co-crystalized substrates, ions, and water molecules, followed by calculating Gasteiger charges and removing non-polar hydrogens. Structures of investigated molecules were optimized using PM7 method of MOPAC2016 [43], but Gasteiger charges were calculated for use in docking experiments. Each co-crystalized substrate was re-docked to its original parent molecule to obtain reference value of binding energy and to validate the applied method. Grid maps were calculated using cube-shaped grid boxes 60 pts wide for protein targets and 40 pts wide for DNA target (1 pt = 0.375 A). Grid boxes were centered using coordinates of co-crystalized substrates. Each experiment consisted of 10 hybrid genetic algorithm–local search runs, with 2.5×10^7 energy evaluations per run.

The two-point attractor and flexible side chain approaches [32] are two types of covalent docking methods that can be used with AutoDock 4. Both methods use precalculated grid maps and atom probes to speed up the scoring process. However, they simulate ligand conformations in different ways during the scoring process. In this paper, we used the flexible side chain method. This approach uses tethered docking to mimic the way covalent ligands are bound in the pocket. For this plan to work, the electrophilic center of the ligand must bond to the two nucleophilic atoms at the ends of the protein. By figuring out the right SMARTS pattern, these two atoms are put on top of the right residue atoms in the protein to make the structure that is wanted. The bound ligand is then treated as a flexible residue, and the standard AutoDock 4 method for flexible residues is used to look at its different shapes in the pocket. With the aid of the scripts provided by the AutoDock 4 website [44], the ei ligand and cysteine and serine residues were overlapped. Subsequently, the receptor grid maps were calculated with the AutoGrid4 software, mapping the receptor interaction energies using the ligand atom types as probes. The grid of 60 Å × 60 Å × 60 Å with 0.375 Å spacing was centered on the coordinates of the ligand originally present in the Gyrase B and topoisomerase II beta. Finally, the actual docking was attained for the ei ligand, keeping the cysteine and serine residues as flexible. This permitted the sampling of the torsional flexibility within the receptors.

5. Conclusions

In this paper, eight novel Meldrum's acid derivatives containing vanillic fragments were synthesized under soft reaction conditions in good to excellent yields (54–92%). All compounds showed good to moderate anticancer and antimicrobial activities. Vanillidene Meldrum's acid derivatives with an *O*-acyl group attached on the vanillic motif have lower activity against cancer cell lines and bacteria. However, the presence of the *O*-alkyl fragment significantly influenced the biological activity, both anticancer and antimicrobial. The most active compound that showed dual activity was **3i**. This molecule contains a decyl chain that is probably responsible for the good activity against *E. coli* (12.4 μM) and cancer cell lines (IC_{50} for HeLa, LS174, and A549: 15.7, 20.8, and 21.8 μM, respectively). Significant dual activity was displayed by compounds **3e** and **3h**. Very good selectivity indices were accomplished with **3e**, **3h** and **3i**. The highest selectivity index was for **3i** against HeLa (4.8), while the next highest was **3h** against A549 cell lines (4.3). In further investigation, the most active compounds (**3e**, **3h**, and **3i**) were subjected to molecular docking. The binding affinities of mentioned compounds to DNA, *E. coli* Gyrase B, and topoisomerase II beta were tested. Compounds **3e**, **3h**, and **3i** showed good non-covalent interactions and consequently good energy of binding. Therefore, **3i** possessed the highest affinity and was subjected to a covalent docking study. Under this investigation, **3i** was played as the Michael acceptor in the reaction with the amino or hydroxy group from Lys and Ser residues, respectively. The lowest binding energy was realized for the topoisomerase II beta-LYS482-**3i** cluster ($E_b = -7.40$ kcal mol^{-1}). From molecular docking and experimental outputs, it was shown that **3i** is a molecule of interest. Going forward, we believe that the presented concept and dual-active compounds described herein have bright futures.

Supplementary Materials: The following supporting information can be downloaded at: https://www.mdpi.com/article/10.3390/ph16020281/s1, Figures S1–S18: NMR spectra of **3a–i**.

Author Contributions: Conceptualization, writing—original draft preparation, and supervision, S.N.A.B., M.W.A. and N.J.; methodology and visualization, M.A.A., N.A., M.A.H., M.A.E., H.E., N.H.A. and I.F. All authors have read and agreed to the published version of the manuscript.

Funding: This work was funded by the Deanship of Scientific Research at Jouf University under Grant Number (DSR2022-RG-0148).

Institutional Review Board Statement: Not applicable.

Informed Consent Statement: Not applicable.

Data Availability Statement: Data is contained within the article and supplementary material.

Acknowledgments: The authors are indebted to the Deanship of Scientific Research at Jouf University, Saudi Arabia, and Dejan Milenković (University of Kragujevac, IITKG, Serbia) for their extraordinary support.

Conflicts of Interest: The authors declare no conflict of interest.

References

1. Meldrum, A.N. A β-lactonic acid from acetone and malonic acid. *J. Chem. Soc. Trans.* **1908**, *93*, 598–601. [CrossRef]
2. Davidson, D.; Bernhard, S.A. The Structure of Meldrum's supposed β-Lactonic Acid. *JACS* **1948**, *70*, 3426–3428. [CrossRef] [PubMed]
3. Xu, F.; Armstrong, J.D.; Zhou, G.X.; Simmons, B.; Hughes, D.; Ge, Z.; Grabowski, E.J.J. Mechanistic evidence for an α-Oxoketene pathway in the formation of β-Ketoamides/Esters via Meldrum's acid adducts. *JACS* **2004**, *126*, 13002–13009. [CrossRef] [PubMed]
4. Brosge, F.; Singh, P.; Bolm, C. Selected applications of Meldrum's acid—A tutorial. *Org. Biomol. Chem.* **2021**, *19*, 5014–5027. [CrossRef] [PubMed]
5. Dumas, A.M.; Fillion, E. Meldrum's acids and 5-Alkylidene Meldrum's acids in catalytic carbon−carbon bond-forming processes. *Acc. Chem. Res.* **2010**, *43*, 440–454. [CrossRef]
6. Ivanov, A.S. Meldrum's acid and related compounds in the synthesis of natural products and analogs. *Chem. Soc. Rev.* **2008**, *37*, 789–811. [CrossRef]
7. Tokala, R.; Bora, D.; Shankaraiah, N. Contribution of Knoevenagel condensation products toward the development of anticancer agents: An updated review. *ChemMedChem* **2022**, *17*, e202100736. [CrossRef]
8. Sandhu, H.S.; Sapra, S.; Gupta, M.; Nepali, K.; Gautam, R.; Yadav, S.; Kumar, R.; Jachak, S.M.; Chugh, M.; Gupta, M.K.; et al. Synthesis and biological evaluation of arylidene analogues of Meldrum's acid as a new class of antimalarial and antioxidant agents. *Bioorg. Med. Chem.* **2010**, *18*, 5626–5633. [CrossRef]
9. Janković, N.; Muškinja, J.; Ratković, Z.; Bugarčić, Z.; Ranković, B.; Kosanić, M.; Stefanović, S. Solvent-free synthesis of novel vanillidene derivatives of Meldrum's acid: Biological evaluation, DNA and BSA binding study. *RSC Adv.* **2016**, *6*, 39452–39459. [CrossRef]
10. Elham, M.; Enayatollah, S.; Dadkhoda, G.; Shahla, S. Uncatalyzed synthesis of new antibacterial bisarylidene Meldrum's acid derivatives functionalized with ether groups. *Lett. Org. Chem.* **2019**, *16*, 818–824.
11. Da Silva, M.M.C.; de Araújo-Neto, J.B.; de Araújo, A.C.J.; Freitas, P.R.; de M Oliveira-Tintino, C.D.; Begnini, I.M.; Rebelo, R.A.; da Silva, L.E.; Mireski, S.L.; Nasato, M.C.; et al. Potentiation of antibiotic activity by a Meldrum's acid arylamino methylene derivative against multidrug-resistant bacterial strains. *Indian J. Microbiol.* **2021**, *61*, 100–103. [CrossRef]
12. Abdelaziz, M.; Azuaje, J.; Alberto, C.; Ernesto, C.; Matilde, Y.; Carmen, L.; Vicente, Y.; Carlos, C.; Eddy, S. Discovery and preliminary SAR of 5-Arylidene-2,2-dimethyl-1,3-dioxane- 4,6-diones as platelet aggregation inhibitors. *Comb. Chem.* **2012**, *15*, 551–554.
13. Takashi, T.; Kunio, S.; Hideo, A. A facile epoxidation of 5-Methylene-1,3-dioxane-4,6-diones with hydrogen peroxide without catalyst. *Heterocycles* **1994**, *38*, 2631.

14. Kadam, A.J.; Desai, U.V.; Mane, R.B. Microwave assisted hydrolysis of Meldrum's acid derivatives and decarboxylation of derived malonic acids. *J. Labbeled. Comp. Radiopharm.* **1999**, *42*, 835. [CrossRef]
15. Rodriguez, H.; Martin, O.; Ochoa, E.; Suarez, M.; Reyes, O.; Garay, H.; Albericio, F.; Martin, N. High-throughput preparation of alkyl 4-aryl substituted-2-methyl-6-thioxo-1,4,5,6-tetrahydropyridine-3-carboxylates under microwave irradiation. *ARKIVOC* **2011**, *ix*, 125–141. [CrossRef]
16. Frost, C.G.; Hartley, B.C. Tandem Molybdenum Catalyzed Hydrosilylations: An Expedient Synthesis of β-Aryl Aldehydes. *Org. Lett.* **2007**, *9*, 4259–4261. [CrossRef]
17. Huang, X.; Xie, L. One Pot Synthesis of Monosubstituted Isopropylidene Malonates. *Synth. Commun.* **1986**, *16*, 1701. [CrossRef]
18. Madasu, C.; Xu, Y.-M.; Wijeratne, E.M.K.; Liu, M.X.; Molnár, I.; Gunatilaka, A.A.L. Semi-synthesis and cytotoxicity evaluation of pyrimidine, thiazole, and indole analogues of argentatins A–C from guayule (*Parthenium argentatum*) resin. *Med. Chem. Res.* **2022**, *31*, 1088–1098. [CrossRef]
19. Arshad, L.; Jantan, I.; Bukhari, S.N.A.; Haque, M.A. Immunosuppressive Effects of Natural α,β-Unsaturated Carbonyl-Based Compounds, and Their Analogs and Derivatives, on Immune Cells: A Review. *Front. Pharmacol.* **2017**, *8*, 22. [CrossRef]
20. Janković, N.; Milović, E.; Đorović Jovanović, J.; Marković, Z.; Vraneš, M.; Stanojković, T.; Matić, I.; Đorđić Crnogorac, M.; Klisurić, O.; Cvetinov, M.; et al. A new class of half-sandwich ruthenium complexes containing Biginelli hybrids: Anticancer and anti-SARS-CoV-2 activities. *Chem. Biol. Interact.* **2022**, *363*, 110025. [CrossRef]
21. Xu, Y.; Wang, F.; Guo, H.; Wang, S.; Ni, S.; Zhou, Y.; Wang, Z.; Bao, H.; Wang, Y. Antitussive and anti-inflammatory dual-active agents developed from natural product lead compound 1-Methylhydantoin. *Molecules* **2019**, *24*, 2355. [CrossRef] [PubMed]
22. Milović, E.; Janković, N.; Petronijević, J.; Joksimović, N.; Kosanić, M.; Stanojković, T.; Matić, I.; Grozdanić, N.; Klisurić, O.; Stefanović, S. Synthesis, characterization, and biological evaluation of tetrahydropyrimidines: Dual-Activity and mechanism of action. *Pharmaceutics* **2022**, *14*, 2254. [CrossRef] [PubMed]
23. Hegazy, G.E.; Abu-Serie, M.M.; Abo-Elela, G.M.; Ghozlan, H.; Sabry, S.A.; Soliman, N.A.; Abdel-Fattah, Y.R. In vitro dual (anticancer and antiviral) activity of the carotenoids produced by haloalkaliphilic archaeon Natrialba sp. M6. *Sci. Rep.* **2020**, *10*, 5986. [CrossRef] [PubMed]
24. Mihaela Aldea, M.; Michot, J.-M.; Danlos, F.-X.; Ribas, A.; Soria, J.-C. Repurposing of Anticancer Drugs Expands Possibilities for Antiviral and Anti-Inflammatory Discovery in COVID-19. *Cancer Discov.* **2021**, *11*, 1336–1344. [CrossRef] [PubMed]
25. Beveridge, T.J. Structures of Gram-Negative Cell Walls and Their Derived Membrane Vesicles. *J. Bacterio.* **1999**, *181*, 4725–4733. [CrossRef]
26. Canals, A.; Purciolas, M.; Aymamí, J.; Coll, M. The anticancer agent ellipticine unwinds DNA by intercalative binding in an orientation parallel to base pairs. *Acta Cryst. Des.* **2005**, *61*, 1009–1012. [CrossRef]
27. Narramore, S.; Stevenson, C.E.M.; Maxwell, A.; Lawson, D.M.; Fishwick, C.W.G. New insights into the binding mode of pyridine-3-carboxamide inhibitors of *E. coli* DNA gyrase. *Bioorg. Med. Chem.* **2019**, *27*, 3546–3550. [CrossRef]
28. Wu, C.-C.; Li, T.-K.; Farh, L.; Lin, L.-Y.; Lin, T.-S.; Yu, Y.-J.; Yen, T.J.; Chiang, C.-W.; Chan, N.-L. Structural Basis of Type II Topoisomerase Inhibition by the Anticancer Drug Etoposide. *Science* **2011**, *333*, 459–462. [CrossRef]
29. Liu, K.; Ren, Z.-L.; Wang, W.; Gong, J.-X.; Chu, M.-J.; Ma, Q.-W.; Wang, J.-C.; Lv, X.H. Novel coumarin-pyrazole carboxamide derivatives as potential topoisomerase II inhibitors: Design, synthesis and antibacterial activity. *Eur. J. Med. Chem.* **2018**, *157*, 81–87. [CrossRef]
30. Krishnan, P.; Bastow, K.F. Novel mechanisms of DNA topoisomerase II inhibition by pyranonaphthoquinone derivatives-eleutherin, α-lapachone, and β-lapachone. *Biochem. Pharmacol.* **2000**, *60*, 1367–1379. [CrossRef]
31. Krishnan, P.; Bastow, K.F. Novel mechanism of cellular DNA topoisomerase II inhibition by the pyranonaphthoquinone derivatives alpha-lapachone and beta-lapachone. *Cancer Chemother. Pharmacol.* **2001**, *47*, 187–198. [CrossRef]
32. Bianco, G.; Forli, S.; Goodsell, D.S.; Olson, A.J. Covalent Docking Using Autodock: Two-Point Attractor and Flexible Side Chain Methods. *Protein Sci.* **2016**, *25*, 295–301. [CrossRef]
33. Lawandi, J.; Toumieux, S.; Seyer, V.; Campbell, P.; Thielges, S.; Juillerat Jeanneret, L.; Moitessier, N. Constrained peptidomimetics reveal detailed geometric requirements of covalent prolyl oligopeptidase inhibitors. *J. Med. Chem.* **2009**, *52*, 6672–6684. [CrossRef]
34. Zhang, S.; Shi, Y.; Jin, H.; Liu, Z.; Zhang, L. Covalent complexes of proteasome model with peptide aldehyde inhibitors MG132 and MG101: Docking and molecular dynamics study. *J. Mol. Model.* **2009**, *15*, 1481–1490. [CrossRef]
35. Moura-Tamames, S.A.; Ramos, M.J.; Fernandes, P.A. Modelling beta-1,3-exoglucanase-saccharide interactions: Structure of the enzyme-substrate complex and enzyme binding to the cell wall. *J. Mol. Graph. Model.* **2009**, *27*, 908–920. [CrossRef]
36. Pair, E.; Cadart, T.; Levacher, V.; Brière, J.-F. Meldrum's Acid: A Useful Platform in Asymmetric Organocatalysis. *ChemCatChem* **2016**, *8*, 1882–1890. [CrossRef]
37. Bernasconi, C.F.; Murray, C.J. Nucleophilic addition to olefins. 18. Kinetics of the addition of primary amines and.alpha.-effect nucleophiles to benzylidene Meldrum's acid. *J. Am. Chem. Soc.* **1986**, *108*, 5251–5257. [CrossRef]
38. Frydman, B.; Marton, L.J.; Sun, J.S.; Neder, K.; Witiak, D.T.; Liu, A.A.; Wang, H.-M.; Mao, Y.; Wu, H.-Y.; Sanders, M.M.; et al. Induction of DNA Topoisomerase II-mediated DNA Cleavage by β-Lapachone and Related Naphthoquinones. *Cancer Res.* **1997**, *57*, 620–627.
39. Sarker, S.D.; Nahar, L.; Kumarasamy, Y. Microtitre plate-based antibacterial assay incorporating resazurin as an indicator of cell growth, and its application in the in vitro antibacterial screening of phytochemicals. *Methods* **2007**, *42*, 321–324. [CrossRef]

40. Mosmann, T. Rapid colorimetric assay for cellular growth and survival: Application to proliferation and cytotoxicity assays. *J. Immunol. Methods.* **1983**, *65*, 55–63. [CrossRef]
41. Ohno, M.; Abe, T. Rapid colorimetric assay for the quantification of leukemia inhibitory factor (LIF) and interleukin-6 (IL-6). *J. Immunol. Methods* **1991**, *145*, 199–203. [CrossRef] [PubMed]
42. Morris, G.M.; Goodsell, D.S.; Huey, R.; Olson, A.J. Distributed automated docking of flexible ligands to proteins: Parallel applications of AutoDock 2.4. *J. Comput.-Aided Mol. Des.* **1996**, *10*, 293–304. [CrossRef] [PubMed]
43. Stewart, J.J.P. MOPAC: A semiempirical molecular orbital program. *J. Comput.-Aided Mol. Des.* **1990**, *4*, 1–103. [CrossRef] [PubMed]
44. AutoDock. CCSB AutoDock Suite. Available online: http://autodock.scripps.edu/ (accessed on 10 January 2023).

pharmaceuticals

Article

Gel-Free Tools for Quick and Simple Screening of Anti-Topoisomerase 1 Compounds

Josephine Geertsen Keller [1,†], Kamilla Vandsø Petersen [2], Karol Mizielinski [1], Celine Thiesen [2], Lotte Bjergbæk [2], Rosa M. Reguera [3], Yolanda Pérez-Pertejo [3], Rafael Balaña-Fouce [3], Angela Trejo [4], Carme Masdeu [4], Concepcion Alonso [4], Birgitta R. Knudsen [1,2] and Cinzia Tesauro [1,*,†]

1 VPCIR Biosciences ApS, 8000 Aarhus C, Denmark
2 Department of Molecular Biology and Genetics, Aarhus University, 8000 Aarhus C, Denmark
3 Department of Biomedical Sciences, Faculty of Veterinary Medicine, University of León, 24071 León, Spain
4 Department of Organic Chemistry, Faculty of Pharmacy, University of Basque Country (UPV/EHU), 01006 Vitoria-Gasteiz, Spain
* Correspondence: ct@vpcir.com
† These authors contributed equally to this work.

Abstract: With the increasing need for effective compounds against cancer or pathogen-borne diseases, the development of new tools to investigate the enzymatic activity of biomarkers is necessary. Among these biomarkers are DNA topoisomerases, which are key enzymes that modify DNA and regulate DNA topology during cellular processes. Over the years, libraries of natural and synthetic small-molecule compounds have been extensively investigated as potential anti-cancer, anti-bacterial, or anti-parasitic drugs targeting topoisomerases. However, the current tools for measuring the potential inhibition of topoisomerase activity are time consuming and not easily adaptable outside specialized laboratories. Here, we present rolling circle amplification-based methods that provide fast and easy readouts for screening of compounds against type 1 topoisomerases. Specific assays for the investigation of the potential inhibition of eukaryotic, viral, or bacterial type 1 topoisomerase activity were developed, using human topoisomerase 1, *Leishmania donovani* topoisomerase 1, monkeypox virus topoisomerase 1, and *Mycobacterium smegmatis* topoisomerase 1 as model enzymes. The presented tools proved to be sensitive and directly quantitative, paving the way for new diagnostic and drug screening protocols in research and clinical settings.

Keywords: human topoisomerase 1; *Mycobacterium smegmatis* topoisomerase 1; *Leishmania donovani* topoisomerase 1; monkeypox virus topoisomerase 1; enzyme activity; rolling circle amplification; drug screening

1. Introduction

DNA modifying enzymes, such as DNA topoisomerases, play an essential role in DNA replication, transcription, chromosome segregation, and recombination [1–3], as they are responsible for the removal of superhelical tension in genomic DNA [1,4]. Both eukaryotic and prokaryotic DNA topoisomerases are targets of small-molecule compounds with potential effects against human cancers or pathogens, such as eukaryotic parasites or bacteria causing diseases [5–11]. Hence, the need for new tools to investigate the clinical relevance of potential drugs against topoisomerases is highly relevant. Notably, the eukaryotic type 1B topoisomerases (TOP1B) are of high interest due to their potential as targets of compounds that are effective against cancer or pathogen-borne diseases [9,10,12,13]. Human topoisomerase 1 (hTOP1) helps to maintain genomic DNA integrity during cellular processes [2] and its activity is increased in several cancer types [14–16]. Other TOP1B include the monkeypox virus TOP1 (mpxvTOP1), which is a conserved enzyme playing an essential role in the replication and spreading of the monkeypox virus [17,18], and the *Leishmania donovani* TOP1 (LdTOP1), that is an essential enzyme involved in the replication of the *L. donovani* parasite, the causative agent of the neglected tropical disease visceral leishmaniasis [19]. Moreover, the *Mycobacterium tuberculosis* TOP1 (mtTOP1), a type 1A topoisomerase (TOP1A) present in the bacteria responsible for tuberculosis, is a relevant target for small-molecule compounds with anti-pathogen potential [20].

Common to these topoisomerases, is that they remove the topological tension in genomic DNA during cellular processes such as replication and transcription. This is achieved by the enzyme introducing a transient single-stranded break in DNA, thereby generating the formation of a cleavage intermediate with the enzyme linked to the 3′-end (for the TOP1B enzymes) or to the 5′-end (for the TOP1A enzymes) of the nicked DNA. Subsequently, the nicked DNA is religated and the enzyme leaves the DNA intact [1,4]. This catalytic cycle can be targeted by small-molecule compounds acting either as TOP1 inhibitors or TOP1 poisons [21]. TOP1 inhibitors act by preventing the catalytic activity by inhibiting the DNA binding and/or cleavage, whereas TOP1 poisons act by prolonging the half-life of the enzyme–DNA cleavage complex. This can result in accumulation of TOP1-bound nicks in the genome, which ultimately can lead to cell death by collision with the DNA replication and transcription machinery. Reduced TOP1 activity levels are detrimental for single-cell eukaryotes such as *Trypanosoma brucei* and *L. major* [22–24], while TOP1 appears only to be essential during early developmental stages in higher eukaryotes and mammals [25,26]. Hence, TOP1 inhibitors are of particular interest as potential anti-parasitic drugs [10], while TOP1 poisons are of high interest as potential anti-cancer agents, since they convert TOP1 activity into a cell killer [27]. Examples of TOP1-targeting anti-cancer drugs include the water-soluble camptothecin (CPT) derivatives, topotecan, and irinotecan [28].

In order to screen libraries of natural and synthetic small-molecule compounds for potential TOP1-targeting properties, it is necessary to have access to assays that allow easy and fast investigation of the effect of these compounds on TOP1 activity. Several assays, including the relaxation assay [29], electrophoretic mobility shift assay (EMSA) [30,31], the DNA suicide cleavage-ligation assay [32,33], and the in vivo complex of enzymes (ICE) assay [34], have been developed. These assays can be used to assess the inhibitory effect of new compounds. However, these assays all rely on gel electrophoresis, which requires DNA intercalating agents, or they require special equipment and expertise. Moreover, most of these assays only perform optimally when using relatively large amounts of purified TOP1 enzyme. To enable specific detection of hTOP1 activity in small and even crude samples, rolling-circle-enhanced enzyme activity detection (REEAD), was previously developed [35]. This assay relies on the hTOP1-mediated circularization of a specific DNA substrate. The closed circles are then amplified by isothermal rolling circle amplification (RCA), generating 10^3 tandem repeats, which subsequently can be visualized either by a fluorescent-based readout or a chemiluminescent-based readout.

TB is currently the second-leading infectious disease, with 1.6 million deaths in 2022 [36]. The number of new cases may potentially rise in the near future due to growing worldwide mobility. This emphasizes how critical it is to establish tools for rapid TB diagnosis and for the screening of candidate molecules for the treatment of TB, counteracting the emergent antibiotic resistance.

Taking the challenges in measuring the activities of TOP1A and TOP1B into account, this study presents the easy and fast REEAD assay as a drug screening tool for compounds against hTOP1 and LdTOP1. It also shows how this assay can be adapted to investigate the activity of mpxvTOP1, as a model for poxviruses, even in crude samples. Moreover, an alternative use of the previously developed assay for the detection of mtTOP1 activity [37] is presented, using the non-pathogen *M. smegmatis* TOP1 (MsTOP1) as a model enzyme. The assay is termed TB enzyme activity detection (TB-EAD), and this study demonstrates the drug screening potential of this assay.

2. Results

2.1. Detection of hTOP1, LdTOP1, and mpxvTOP1 Activities Using the REEAD Assay

TOP1 inhibitors are significant compounds for their anti-cancer, anti-bacterial, and anti-parasitic capacity [6,9–11,13,38]. Efficient and sensitive methods for the detection and quantification of their inhibitory capacity are therefore highly desirable, as traditional control systems are expensive and require skilled personnel. Here, an assay capable of detecting the activity and the drug response of TOP1B is presented. The principle of this assay is illustrated in Figure 1 and relies on the TOP1-mediated circularization of an open substrate (Figure 1A) with or without the presence of TOP1-targeting drugs. The hTOP1- and LdTOP1-specific substrates fold into a dumbbell shape containing two single-stranded loops and a double-stranded stem. The double-stranded stem contains an hTOP1/LdTOP1-preferred cleavage site and one of the single-stranded loops contains a sequence complementary to a surface-anchored primer suitable for RCA. The mpxvTOP1-specific substrate folds into a half-dumbbell shape, with one single-stranded loop containing a sequence complementary to the surface-anchored primer and a double-stranded region containing a mpxvTOP1-preferred cleavage site. Upon TOP1-mediated cleavage and ligation, the circularized substrates are hybridized to the surface-anchored primers, in well-defined rectangular areas (termed wells) of a glass slide. The hybridized circles act as a template for isothermal RCA mediated by Phi29 polymerase (Figure 1B(i)). During RCA, biotinylated nucleotides are incorporated into the rolling circle products (RCPs) (Figure 1B(ii)). This allows binding of horseradish peroxidase (HRP)-conjugated anti-biotin antibody and visualization of the generated RCPs using an enhanced chemiluminescence (ECL)-based readout (Figure 1B(iii)). The presented REEAD assay provides a sensitive detection of TOP1 activity, as one TOP1-mediated cleavage-ligation reaction generates a single RCP, thereby making the assay directly quantitative.

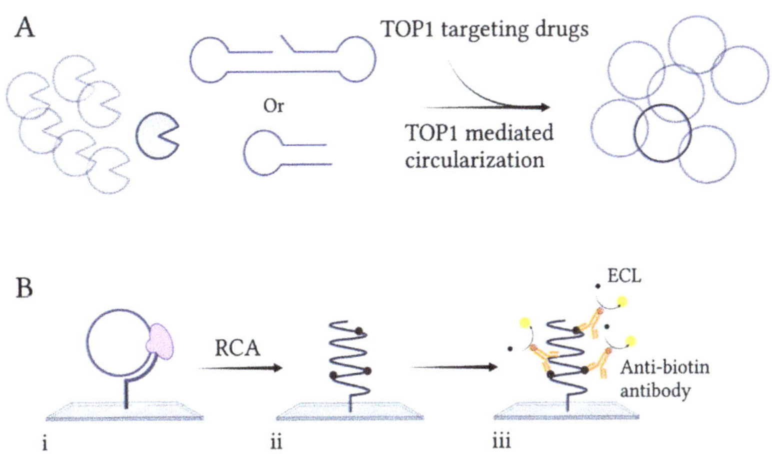

Figure 1. Schematic illustration of the REEAD assay. (**A**) TOP1 will cleave and ligate a specific substrate, thereby generating closed circular substrates. (**B**) (**i**) The closed circles are hybridized to a primer attached to a glass surface and can be amplified by RCA using the Phi29 polymerase (purple). (**ii**) In the presence of biotin-coupled nucleotides, the generated RCPs will be biotinylated, allowing the binding of HRP-conjugated anti-biotin antibodies. (**iii**) Visualization of the RCPs is carried out by addition of the ECL components, luminol and H_2O_2, that are converted into detectable light by HRP. Created with BioRender.com.

2.2. Using REEAD as a Drug Screening Tool for Drugs against hTOP1 and LdTOP1

To demonstrate the functionality of the REEAD assay as a drug screening tool, the activity of hTOP1 and LdTOP1 in the presence of TOP1-targeting compounds was measured. The two indenoisoquinoline derivatives, LMP400 and LMP744 (see the structure of the compounds in Supplementary Figure S1), which are known hTOP1 [39] and LdTOP1 inhibitors [40] were used, and the hTOP1-targeting anti-cancer drug CPT was included as a control (as indicated in Figure 2). Twelve nanograms of recombinant purified hTOP1 or LdTOP1 (see Supplementary Figure S2A for protein purifications [41,42]) was incubated with the specific substrate in the presence of 80 μM of the compounds or 5% of the solvent DMSO. The generated circles were amplified by RCA in the presence of biotin-labeled nucleotides, and the HRP-conjugated anti-biotin antibody was subsequently bound to the biotinylated RCPs. Visualization was performed using the ECL-based readout, and Figure 2A,B show the results of these analyses. Figure 2A,B upper panels show representative images of the intensities of the biotinylated RCPs when visualized using ECL. As evident from the graphical depictions, and as expected from the literature [5,43], CPT inhibits both the hTOP1 (Figure 2A lower panel) and LdTOP1 (Figure 2B lower panel) activities significantly. LMP400 has a stronger inhibitory effect on LdTOP1 compared to hTOP1, while LPM744 seemed to be more potent against hTOP1 rather than LdTOP1 when compared to CPT. However, the scope of these experiments was not to directly compare the inhibitory effects of the two compounds, but merely to demonstrate the ability of the REEAD assay to be used as a screening tool for compounds that act against human, parasite, or other eukaryotic TOP1.

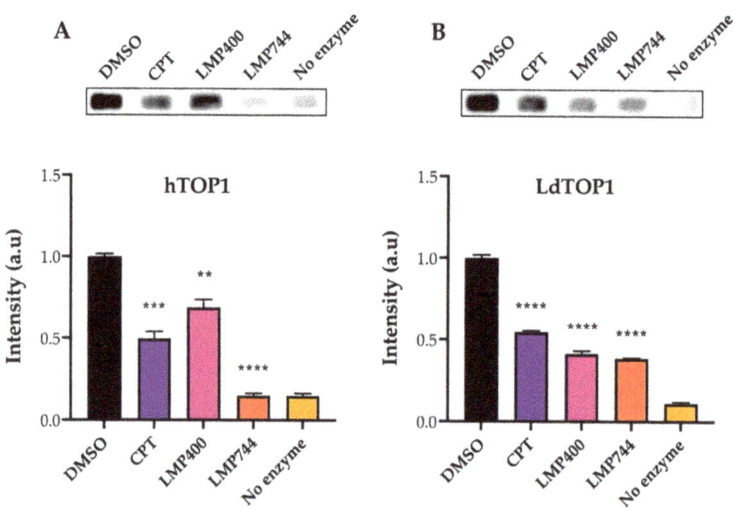

Figure 2. REEAD as a drug screening tool. (**A**) Top panel: image obtained after analyzing activity of purified hTOP1 in the presence of 5% of the solvent DMSO, or 80 μM of either of the compounds CPT, LPM400, or LMP744, using REEAD with the ECL-based readout. Lower panel: graphical depiction of the quantification of the results obtained when analyzing hTOP1 activity in the presence of 80 μM of the compounds indicated on the figure. A negative control, without enzyme, was included. Plotted data are normalized to the intensity obtained when measuring hTOP1 activity in the presence of DMSO and represent average +/− standard error of the mean (SEM) from six independent experiments. One-way ANOVA with Brown–Forsythe and Welch correction. Asterisks indicate significant difference compared to DMSO, ** $p < 0.005$; *** $p < 0.0005$; **** $p < 0.0001$. a.u: arbitrary units. (**B**) same as (**A**), except that LdTOP1 was used instead of hTOP1. Plotted data represent average +/− SEM from three independent experiments. One-way ANOVA with Brown–Forsythe and Welch correction, **** $p < 0.0001$. a.u: arbitrary units.

The observed inhibitory effect of the compounds was moreover demonstrated to be dose dependent (see Supplementary Figure S3A,B). The results presented are in line with the literature [5,43] and clearly demonstrate the drug screening ability of the REEAD assay.

2.3. The REEAD Assay Can Be Used to Measure Poxvirus TOP1B Activity

Poxviruses encapsidate a type 1B topoisomerase essential for viral growth. As a model for a poxvirus TOP1B, mpxvTOP1 was used. To screen for the activity of mpxvTOP1, a substrate that folds into a half-dumbbell shape, containing the specific cleavage sequence recognized by the poxviruses TOP1, as first identified in the vaccinia virus TOP1 [44], was designed (see Figure 1A). To validate the specificity of the mpxvTOP1 REEAD, the ability of hTOP1 and mpxvTOP1 to cleave/ligate the half-dumbbell-shaped substrate was compared. Recombinant purified mpxvTOP1 (0–50 ng) (see Supplementary Figure S2B for protein purification) or hTOP1 was incubated with the half-dumbbell substrate to generate closed circles that subsequently were amplified by RCA in the presence of biotin-labeled nucleotides. The biotinylated RCPs were again visualized using the ECL-based readout. The upper panel of Figure 3A shows representative images of the intensities of the biotinylated RCPs when measuring 0–50 ng of purified mpxvTOP1 or hTOP1, and the lower panel of Figure 3A shows a graphical depiction of the resulting quantifications.

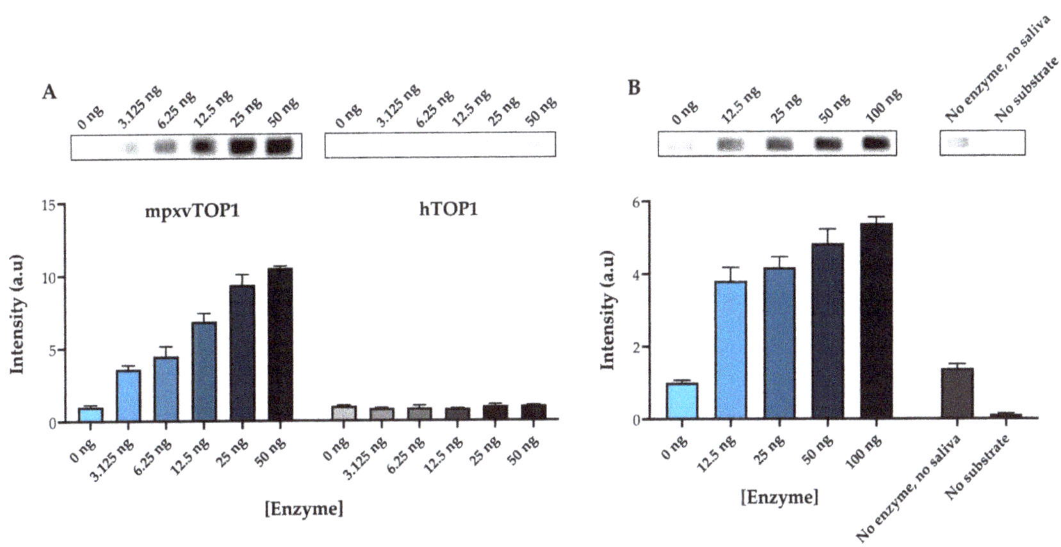

Figure 3. mpxvTOP1 REEAD assay. (**A**) Top panels: image obtained after analyzing the activity of 0–50 ng of purified mpxvTOP1 and hTOP1 using mpxvREEAD with ECL-based readout. Lower panel: graphical quantification of the results obtained when analyzing the activity of 0–50 ng of purified mpxvTOP1 or hTOP1 using the mpxvREEAD. Plotted data are normalized to the intensity obtained when analyzing 0 ng of purified enzyme and represent average +/− SEM from three independent experiments. a.u: arbitrary units. (**B**) Top panels: image obtained after analyzing the activity of 0–50 ng purified mpxvTOP1 containing a fixed amount of saliva spike using mpxvREEAD with ECL-based readout. Lower panel: graphical quantification of the results obtained when analyzing the activity of 0–50 ng of purified mpxvTOP1 containing a fixed amount of saliva spike using mpxvREEAD. Plotted data are normalized to the intensity obtained when analyzing 0 ng of purified enzyme and represent average +/− SEM from three independent experiments. a.u: arbitrary units.

Figure 3A shows that the signal intensity rises when analyzing a higher amount of purified mpxvTOP1, but not when examining increasing amounts of purified hTOP1. This result clearly demonstrates the specificity of the mpxvTOP1 REEAD assay. However, it should be noted that this assay is not specific for mpxvTOP1 compared to other poxvirus TOP1B, but can be used to detect all poxvirus TOP1 in purified form.

The REEAD assay has previously proven to be a powerful tool for the simple and fast detection of TOP1 activity in crude biological samples [15,45–50]. Next, it was investigated if the mpxvREEAD assay could be used to detect mpxvTOP1 in a crude extract. For this purpose, we performed a titration experiment, spiking in purified mpxvTOP1 in saliva, as a crude biological specimen. Using a fixed amount of saliva and increasing amount of purified mpxvTOP1, it was possible to detect mpxvTOP1 activity above background level, as indicated in Figure 3B. Two negative controls were included, one containing only the DNA substrate ("No enzyme, no saliva" in Figure 3B), as a measure of the possible nonspecific incorporation of biotin-labeled nucleotides during RCA of a non-circularized substrate, and one containing only saliva and purified mpxvTOP1 enzyme but no DNA substrate ("No substrate" in Figure 3B), as a measure of the potential nonspecific binding of the anti-biotin antibody to the wells. Both these controls gave intensities around or below the intensity observed when the reactions were performed in the absence of purified mpxvTOP1 (0 ng) with a fixed amount of saliva. The result presented in Figure 3B shows that the mpxvTOP1 REEAD can be used to detect mpxvTOP1 in a crude extract. Again, it is important to highlight that, due to the designed substrate characteristic, this assay cannot be used for diagnostic purposes as it does not allow for discrimination between TOP1B from different poxviruses.

2.4. Detection of MsTOP1 Activity Using the TB-EAD Assay

As mentioned, TOP1 inhibitors are relevant compounds for their anti-bacterial or anti-parasitic capacity. Here, the ability of the previously designed TB-EAD assay [37] to be used as a drug screening tool for new, more effective anti-TB drugs was investigated. TB-EAD is schematically depicted in Figure 4. The assay is based on the use of a single-stranded DNA substrate containing a strong TOP1 site (STS) [51], an RCA primer annealing sequence, and a probe annealing sequence. The substrate is hybridized to a surface-anchored RCA primer, and upon MsTOP1-mediated cleavage and ligation, the substrate is circularized (Figure 4A). This step occurs with or without the presence of the compound to be tested for inhibition ability. Starting from the surface-anchored primer, the closed circle is amplified by RCA mediated by Phi29 polymerase (Figure 4B(i)), generating a long tandem repeat product. By hybridization with fluorescently labeled probes (Figure 4B(ii)) to the generated RCPs, they can be visualized in a fluorescence microscope. The TB-EAD assay is highly sensitive and can detect MsTOP1 activity at a single molecule level, since each detected fluorescent spot in the microscope corresponds to a single catalytic cycle reaction of MsTOP1. Hence, the TB-EAD assay is also directly quantitative.

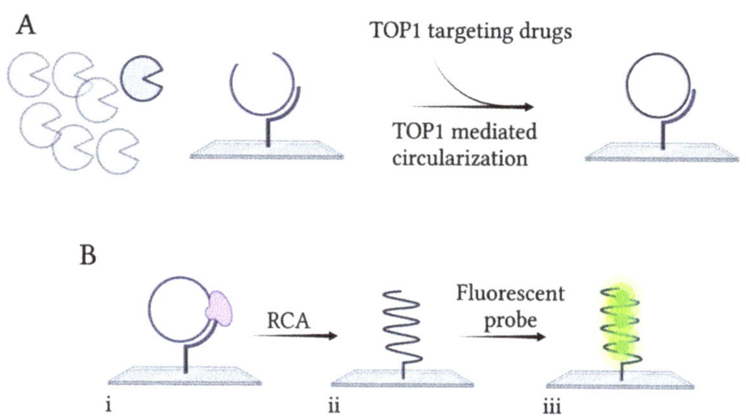

Figure 4. Schematic illustration of the TB-EAD assay. (**A**) The substrate is hybridized to a glass surface containing a complementary primer. MsTOP1 will then cleave and ligate the substrate, thereby making a closed circle. (**B**) (**i**) Following circularization, the circle is amplified by RCA generating tandem repeats (**ii**). (**iii**) Fluorescent probes are then hybridized to the RCPs, enabling visualization of the products in a fluorescence microscope. Created with BioRender.com.

2.5. The TB-EAD Assay Can Be Used as a Drug Screening Tool for Drugs against MsTOP1

To demonstrate that the TB-EAD assay can be used as a drug screening tool for MsTOP1-targeting compounds, the activity of purified MsTOP1 was tested (see Supplementary Figure S2C for protein purification) in the presence of increasing concentrations of novel synthesized compounds, as indicated in Figure 5A. In the design of methods for the detection and quantification of this enzymatic inhibition, some heterocyclic compounds are key species due to their behavior against this target [52], as previously published [53–56]. Incidentally, it is important to mention that the recently optimized Povarov reaction [57,58] is a very suitable strategy, whose multicomponent version gives access to different families of heterocycles in a simple and fast manner. For this study, six compounds targeting MsTOP1 were synthesized (see Supplementary Figure S4A,B for the synthesis of the compounds and Supplementary Figure S5 and Supplementary Information S6.3 for structural elucidation) to be used to test the drug screening ability of our REEAD-based assay. In total, 100 μM or 150 μM of the compounds, or 5% of the solvent DMSO, was added to the reaction mixtures during the circularization step of the substrate. Subsequently, the generated circles were amplified by RCA and visualized by hybridization to fluorescent probes. The number of signals was quantified, normalized to DMSO, and is graphically depicted in Figure 5A. Incubation with three compounds (**SF2**, **SF3** and **SF5**) and the starting reagent (**sulfadoxine 1**) showed a decrease in MsTOP1 activity already at the lowest dose, of 100 μM, whereas incubation with the compound **SF10** only showed a significant inhibition at the highest dose, of 150 μM. Incubation with the compound **HINSF1** did not result in any significant changes in the MsTOP1 activity.

One of the compounds showing the highest inhibitory effect on MsTOP1 activity was also tested against hTOP1 using the REEAD assay, shown in Figure 1. This was performed to show the specificity of the compounds against MsTOP1 alone. Indeed, when screening for new compounds against a specific TOP1 target, it is necessary to consider the specificity and potential off-target effects. In this case, having the possibility of testing both the human and bacterial enzyme, it was possible to assess if the tested compound could function as an anti-bacterial drug. Increasing concentrations of **SF5** were added to the circularization reaction of hTOP1, as indicated in Figure 5B. The circles were amplified by RCA and the resulting RCPs were visualized using the ECL-based readout. As evident from the graphical depiction, the two highest concentrations of **SF5** used show some inhibitory effect on hTOP1 activity, but to a lower extent compared to the effect on MsTOP1 activity. This demonstrated that **SF5** shows a small off-target effect. The results presented clearly demonstrate the ability of the TB-EAD assay to be used as a drug screening tool for compounds targeting MsTOP1.

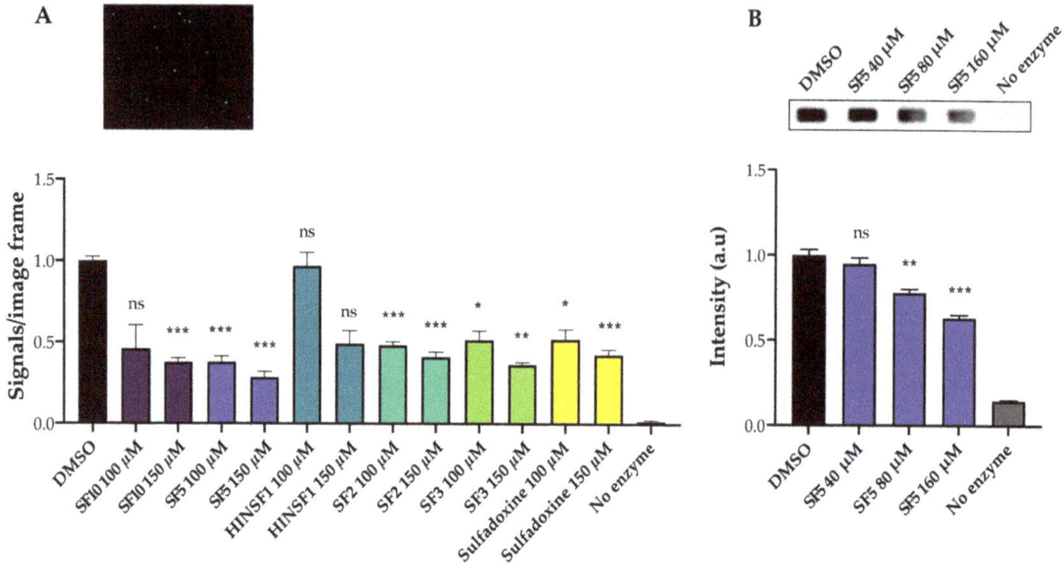

Figure 5. TB-EAD as a drug screening tool. (**A**) Top panel: representative microscopic image when analyzing purified mtTOP1 using the TB-EAD assay. Lower panel: graphical quantification of the results obtained when analyzing purified MsTOP1 activity using the TB-EAD assay in the presence of 100–150 µM of the compounds **SF10**, **SF5**, **HINSF1**, **SF2**, **SF3**, **sulfadoxine 1**, or 5% of the solvent DMSO. A negative control without enzyme was included. Plotted data are normalized to the number of signals obtained when measuring purified MsTOP1 activity in the presence of DMSO, and represent average +/− SEM from three independent experiments. One-way ANOVA with Brown–Forsythe and Welch correction. Asterisks indicate significant difference compared to DMSO, ns: non-significant; * $p < 0.05$; ** $p < 0.005$; *** $p < 0.0005$. (**B**) Top panel: image obtained after analyzing the activity of purified hTOP1 in the presence of 40–160 µM of the compound **SF5**, or 5% of the solvent DMSO, as indicated using the REEAD assay with ECL-based readout. Lower panel: graphical quantification of the results obtained when analyzing hTOP1 activity in the presence of 40–160 µM of the compound **SF5**. A negative control without enzyme was included. Plotted data are normalized to the intensity obtained when measuring hTOP1 activity in the presence of DMSO and represent average +/− SEM from three independent experiments. One-way ANOVA with Brown–Forsythe and Welch correction. * indicates significant difference compared to DMSO, ns: non-significant; ** $p < 0.005$; *** $p < 0.0005$.

3. Discussion and Conclusions

The ongoing demand for the discovery of novel small-molecule compounds with anti-cancer or anti-pathogen effects requires the development of easily accessible tools for measuring the interaction between a vast number of natural or synthesized new drugs and their potential cellular targets. Due to their key role in the maintenance of the genome integrity, topoisomerases have been targets of investigations of inhibitor compounds for the treatment of several types of cancers [12,13] since their discovery. In addition, topoisomerase inhibitors have also been investigated for their potential as anti-bacterial or anti-parasite infections agents [8,10]. These studies typically involve in vitro experiments using cell cultures along with different gel-based assays, which are labor and time consuming and require specialized equipment [29–34]. The presented study shows the advantages of RCA-based assays for the quick and simple screening of libraries of small-molecule compounds with potential effects as TOP1 inhibitors. As a proof of principle, we presented the REEAD-based assay for the measurement of the inhibitory effect of small-molecule compounds against eukaryotic and poxvirus TOP1B, using commercially available drugs with well-known effects, and using human, leishmanial, and monkeypox virus TOP1 as model enzymes. The presented tool will help with pre-screening of a large number of compounds to measure their potential inhibitory ability. More assays can be combined to test for target specificity, while other tools can be applied to investigate the mechanisms of the compounds in detail [59] and identify TOP1 inhibitors versus TOP1 poisons.

We demonstrated that the REEAD assay, with a novel developed ECL-based readout, is a valid alternative to the relaxation and radiolabeled oligonucleotide-based assays when rapid screening of a panel of a compounds is required. Moreover, the assay is easy and adaptable to every laboratory setting and is directly quantitative.

One of the most challenging diseases to fight, and with a high and growing spread around the globe, is TB. Most cases of TB can be treated with a combination of antibiotics [60], but some strains of TB have developed resistance and more resistant strains are emerging [61]. This highlights the importance of developing more TB drug discovery pipelines. We have previously developed an assay that can be used as a diagnostic tool for the identification of TB infection directly from saliva from patients [37]. Here, we demonstrated that the same tool is particularly useful also for the identification of novel anti-TB drugs, using mtTOP1 as the target.

We believe that the presented results provide solid tools that, in principle, might significantly shorten the time required to move from the identification of a possible inhibitor into a clinical trial.

4. Materials and Methods

4.1. *Reagents*

All chemicals were purchased from Sigma Aldrich, Søborg, Denmark.

4.2. *DNA Oligonucleotides*

DNA oligonucleotides were synthesized by LGC Biosearch Technologies, Lystrup Denmark. The sequences were as follow:

- 5′amine REEAD primer: 5′-/5AmMC6/CCAACCAACCAACCAAGGAGCCAAA CATGTGCATTGAGG
- hTOP1/LdTOP1 dumbbell substrate: 5′-AGAAAAATTTTTAAAAAAACTGTGAA GATCGCTTATTTTTTTAAAAATTTTTCTAAGTCTTTTAGATCCCTCAATGCACATG TTTGGCTCCGATCTAAAAGACTTAGA
- mpxvTOP1 half-dumbbell substrate: 5′-ATTGTATCGGAATAAGGGCGACAGACT CACTGTGAAGATCGCTTATCCTCAATGCACATGTTTGGCTCCGAGTCTGTCGCC CTTATT
- MsTOP1 substrate: 5′-CAGTGAGCGAGCTTCCGCTTGACATCCCATATCTCTACT GTGAAGATCGCTTATTCTCTCCTCAATGCACATGTTTGGCTCCTCTCTGAGCTTC CGCT
- Fluorescent probe: 5′-FAM-CCTCAATGCACATGTTTGGCTCC

4.3. *REEAD*

4.3.1. Preparation of Slides

A custom-designed silicone grid (Grace-bio lab, Bend, Oregon, USA), termed the Well-maker, was attached to a CodeLink Activated HD slide (Surmodics, Saint Paul, Minnesota) and the slides were coupled with 10 μM of the 5′amine REEAD primer in 300 mM Na_3PO_4, pH 8. The slide was incubated overnight in a humidity chamber with saturated NaCl. The slide was subsequently blocked in 50 mM Tris, 50 mM Tris-HCl, and 50 mM ethanolamine pH 9, for 30 min at 50 °C and washed in 4× SSC, 0.1% SDS for 30 min at 50 °C.

4.3.2. Circularization for Drug Screening

Circularization of the hTOP1- or LdTOP1-specific substrates was carried out by incubating purified hTOP1 or LdTOP1 with 0.1 μM of the specific substrate in the presence of 80 μM of the drugs to be tested (as indicated in the figure legend of Figure 2A,B) in a buffer containing 10 mM Tris-HCl pH 7.5, 5 mM EDTA, and 50 mM NaCl for 30 s for hTOP1 and 1 min for LdTOP1, at 37 °C. The reaction was stopped by the addition of 0.5% SDS. Subsequently, the circles were hybridized on the REEAD primer coupled slides for 1 h at 37 °C. The slide was then washed in wash buffer 1 (100 mM Tris-HCl pH 7.5, 150 mM NaCl, 0.3% SDS) for 1 min at room temperature, wash buffer 2 (100 mM Tris-HCl pH 7.5, 150 mM NaCl, 0.05% Tween20) for 1 min at room temperature, and finally dehydrated for 1 min in 70% EtOH.

4.3.3. Circularization for mpxvTOP1-Specificity Test

Specific circularization of the mpxvTOP1 substrate was carried out by incubating either purified mpxvTOP1 or hTOP1 with 0.1 μM of the mpxvTOP1-specific substrate in a buffer containing 10 mM Tris-HCl pH 7.5, 5 mM EDTA, and 50 mM NaCl, for 1 h at 37 °C. The reaction was stopped by increasing the NaCl concentration to 250 mM. The hybridization of the circles to the slides and washing of the slides were performed as described above.

Alternatively, mpxvTOP1-mediated circularization of the specific substrate was carried out in the presence of a fixed amount of saliva. Here, increasing amounts of purified mpxvTOP1 (as indicated in the figure legend of Figure 3) were incubated with 0.5 μM of the specific substrate in the presence of 2 μL of saliva. The circularization was performed as described above.

4.3.4. RCA and Detection of RCPs

RCA was performed in 1× Phi29 buffer (50 mM Tris-HCl pH 7.5, 10 mM $MgCl_2$, 10 mM $(NH_4)_2SO_4$, 4 mM DTT) supplemented with 0.2 μg BSA, 100 μM dATP, 100 μM dTTP, 100 μM dGTP, 90 μM dCTP, 10 μM biotin-dCTP, and 1 Unit Phi29 polymerase. The reaction was carried out for 2 h at 37 °C in a humidity chamber. The slide was then washed with wash buffers 1 and 2 and 70% EtOH, as previously. Subsequently, the slide was blocked in 1× TBST (20 mM Tris-HCl pH 9, 150 mM NaCl, 0.05% Tween20 pH 9) supplemented with 5% skimmed dry milk and 5% BSA for 1 h at room temperature, followed by incubation with 1:300 HRP-conjugated anti-biotin antibody in 1× TBST supplemented with 5% skimmed dry milk and 5% BSA, for 1 h at room temperature. The slide was washed 3 × 3 min in 1× TBST before addition of 2 μL of 1:1 ECL mixture to allow chemiluminescence readout using a CCD camera.

4.4. *TB-EAD*

4.4.1. Preparation of slides

Slides were prepared as in REEAD (see step Section 4.3.1).

4.4.2. Circularization

Ten picomoles of the MsTOP1 substrate, in 10 mM Tris-HCl pH 7.5, 1 mM EDTA, and 200 mM NaCl, was hybridized to the REEAD primer on the slides, for 1 h at 37 °C in a humidity chamber. The slides were washed for 1 min in wash buffer 1, 1 min in wash buffer 2, and 1 min in 70% EtOH, as described previously. Then, 100 ng of purified MsTOP1 was added to the slides in a buffer containing 10 mM Tris-HCl pH 7.5, 10 mM $MgCl_2$, 10 mM $MnCl_2$, 200 mM NaCl, 1 mM DTT, 0.1% Tween20, 100 μg/mL BSA in the presence of DMSO, or 100 μM or 150 μM of each of the drugs, as indicated in the figure legend of Figure 5. The reactions were incubated for 90 minutest 37 °C in a humidity chamber before the slides were washed in wash buffers 1 and 2 and 70% EtOH, as previously.

4.4.3. RCA and Detection of RCPs

RCA was performed in 1× Phi29 buffer (50 mM Tris-HCl pH 7.5, 10 mM $MgCl_2$, 10 mM $(NH_4)_2SO_4$, 4 mM DTT) supplemented with 0.2 μg BSA, 1 mM dNTP, and 1 unit of Phi29 polymerase. The reaction was carried out for 1 h at 37 °C in a humidity chamber, followed by wash in wash buffers 1 and 2 and 70% EtOH, as previously. Two picomoles of the fluorescent probe were added to the slides in 2× SSC, 20% formamide, and 5% glycerol for 30 min in a humidity chamber at 37 °C. Subsequently, the slides were washed in wash buffer 1 for 10 min, wash buffer 2 for 5 min, and 1 min in 70% EtOH. The slides were mounted with Vectashield (Vector laboratories, Burlington, ON, Canada) and a cover glass, and analyzed in a fluorescence microscope with a 60× magnification objective. Twelve images were acquired for each sample and quantified using Image J Fiji.

4.5. Statistical Analysis

Data were analyzed using the GraphPad Prism software and expressed as mean +/− standard error of the mean (SEM). Statistical significance was assessed using one-way ANOVA test applying Brown–Forsythe and Welch correction.

Supplementary Materials: The following supporting information can be downloaded at: https://www.mdpi.com/article/10.3390/ph16050657/s1. Figure S1: Chemical structure of LMP400 and LMP744. Figure S2: Coomassie stains of purified enzymes. Figure S3: Dose-dependent inhibition of hTOP1 and LdTOP1 activities. Figure S4: Synthesis of compounds targeting mtTOP1. Figure S5: Structural elucidation of new compounds.

Author Contributions: Conceptualization, C.T. (Cinzia Tesauro), R.M.R., Y.P.-P., R.B.-F., B.R.K. and C.A.; methodology, C.T. (Cinzia Tesauro), J.G.K., K.V.P., C.M. and A.T.; data analysis, J.G.K.; K.V.P., C.T. (Cinzia Tesauro) and K.M.; writing—original draft preparation, J.G.K.; writing—review and editing, J.G.K., C.T. (Celine Thiesen), R.B.-F., L.B. and C.T. (Cinzia Tesauro). All authors have read and agreed to the published version of the manuscript.

Funding: This research was funded in part by the Ministerio de Ciencia e Innovación, Spain (PID2021-122558OB-I00, UE) and by Gobierno Vasco, Universidad del País Vasco (GV, IT1701-22; UPV).

Institutional Review Board Statement: Not applicable.

Informed Consent Statement: Not applicable.

Data Availability Statement: Data is contained within the article; raw data are available upon request from the corresponding author.

Acknowledgments: We acknowledge laboratory technician Noriko Y. Hansen for the assistance in purification of MsTOP1.

Conflicts of Interest: The authors C.T., J.G.K, and K.M are employees of VPCIR Biosciences ApS. C.T. and B.R.K. are shareholders and/or share option holders. The other authors declare that they have no competing interests.

References

1. Champoux, J.J. DNA Topoisomerases: Structure, Function, and Mechanism. *Annu. Rev. Biochem.* **2001**, *70*, 369–413. [CrossRef]
2. Leppard, J.B.; Champoux, J.J. Human DNA Topoisomerase I: Relaxation, Roles, and Damage Control. *Chromosoma* **2005**, *114*, 75–85. [CrossRef] [PubMed]
3. Wang, J.C. DNA Topoisomerases. *Annu. Rev. Biochem.* **1996**, *65*, 635–692. [CrossRef] [PubMed]
4. Wang, J.C. Cellular Roles of DNA Topoisomerases: A Molecular Perspective. *Nat. Rev. Mol. Cell Biol.* **2002**, *3*, 430–440. [CrossRef] [PubMed]
5. Pommier, Y. Topoisomerase I Inhibitors: Camptothecins and Beyond. *Nat. Rev. Cancer* **2006**, *6*, 789–802. [CrossRef]
6. Ho, J.S.Y.; Mok, B.W.Y.; Campisi, L.; Jordan, T.; Yildiz, S.; Parameswaran, S.; Wayman, J.A.; Gaudreault, N.N.; Meekins, D.A.; Indran, S.V.; et al. TOP1 Inhibition Therapy Protects against SARS-CoV-2-Induced Lethal Inflammation. *Cell* **2021**, *184*, 2618–2632. [CrossRef] [PubMed]
7. Collin, F.; Karkare, S.; Maxwell, A. Exploiting Bacterial DNA Gyrase as a Drug Target: Current State and Perspectives. *Appl. Microbiol. Biotechnol.* **2011**, *92*, 479–497. [CrossRef]
8. Balaña-Fouce, R.; Álvarez-Velilla, R.; Fernández-Prada, C.; García-Estrada, C.; Reguera, R.M. Trypanosomatids Topoisomerase Re-Visited. New Structural Findings and Role in Drug Discovery. *Int. J. Parasitol. Drugs Drug Resist.* **2014**, *4*, 326–337. [CrossRef]
9. Balaña-Fouce, R.; Redondo, C.M.; Pérez-Pertejo, Y.; Díaz-González, R.; Reguera, R.M. Targeting Atypical Trypanosomatid DNA Topoisomerase I. *Drug Discov. Today* **2006**, *11*, 733–740. [CrossRef]
10. García-Estrada, C.; Prada, C.F.; Fernández-Rubio, C.; Rojo-Vázquez, F.; Balaña-Fouce, R. DNA Topoisomerases in Apicomplexan Parasites: Promising Targets for Drug Discovery. *Proc. Biol. Sci.* **2010**, *277*, 1777–1787. [CrossRef]

11. Álvarez-Bardón, M.; Pérez-Pertejo, Y.; Ordóñez, C.; Sepúlveda-Crespo, D.; Carballeira, N.M.; Tekwani, B.L.; Murugesan, S.; Martinez-Valladares, M.; García-Estrada, C.; Reguera, R.M.; et al. Screening Marine Natural Products for New Drug Leads against Trypanosomatids and Malaria. *Mar. Drugs* **2020**, *18*, 187. [CrossRef] [PubMed]
12. Cinelli, M.A. Topoisomerase 1B Poisons: Over a Half-Century of Drug Leads, Clinical Candidates, and Serendipitous Discoveries. *Med. Res. Rev.* **2019**, *39*, 1294–1337. [CrossRef] [PubMed]
13. Pommier, Y.; Leo, E.; Zhang, H.; Marchand, C. DNA Topoisomerases and Their Poisoning by Anticancer and Antibacterial Drugs. *Chem. Biol.* **2010**, *17*, 421–433. [CrossRef]
14. Proszek, J.; Roy, A.; Jakobsen, A.K.; Frøhlich, R.; Knudsen, B.R.; Stougaard, M. Topoisomerase I as a Biomarker: Detection of Activity at the Single Molecule Level. *Sensors* **2014**, *14*, 1195–1207. [CrossRef]
15. Jakobsen, A.K.; Lauridsen, K.L.; Samuel, E.B.; Proszek, J.; Knudsen, B.R.; Hager, H.; Stougaard, M. Correlation between Topoisomerase I and Tyrosyl-DNA Phosphodiesterase 1 Activities in Non-Small Cell Lung Cancer Tissue. *Exp. Mol. Pathol.* **2015**, *99*, 56–64. [CrossRef]
16. Roy, A.; Tesauro, C.; Frøhlich, R.; Hede, M.S.; Nielsen, M.J.; Kjeldsen, E.; Bonven, B.; Stougaard, M.; Gromova, I.; Knudsen, B.R. Decreased Camptothecin Sensitivity of the Stem-Cell-like Fraction of Caco2 Cells Correlates with an Altered Phosphorylation Pattern of Topoisomerase I. *PLoS ONE* **2014**, *9*, e99628. [CrossRef]
17. Perry, K.; Hwang, Y.; Bushman, F.D.; van Duyne, G.D. Insights from the Structure of a Smallpox Virus Topoisomerase-DNA Transition State Mimic. *Structure* **2010**, *18*, 127. [CrossRef]
18. Estep, R.D.; Messaoudi, I.; O'Connor, M.A.; Li, H.; Sprague, J.; Barron, A.; Engelmann, F.; Yen, B.; Powers, M.F.; Jones, J.M.; et al. Deletion of the Monkeypox Virus Inhibitor of Complement Enzymes Locus Impacts the Adaptive Immune Response to Monkeypox Virus in a Nonhuman Primate Model of Infection. *J. Virol.* **2011**, *85*, 9527–9542. [CrossRef]
19. Villa, H.; Marcos, A.R.O.; Reguera, R.M.; Balaña-Fouce, R.; García-Estrada, C.; Pérez-Pertejo, Y.; Tekwani, B.L.; Myler, P.J.; Stuart, K.D.; Bjornsti, M.A.; et al. A Novel Active DNA Topoisomerase I in Leishmania Donovani. *J. Biol. Chem.* **2003**, *278*, 3521–3526. [CrossRef]
20. Ahmed, W.; Menon, S.; Godbole, A.A.; Karthik, P.V.D.N.B.; Nagaraja, V. Conditional Silencing of Topoisomerase I Gene of Mycobacterium Tuberculosis Validates Its Essentiality for Cell Survival. *FEMS Microbiol. Lett.* **2014**, *353*, 116–123. [CrossRef]
21. Pommier, Y. Diversity of DNA Topoisomerases I and Inhibitors. *Biochimie* **1998**, *80*, 255–270. [CrossRef]
22. Gutiérrez-Corbo, C.; Álvarez-Velilla, R.; Reguera, R.M.; García-Estrada, C.; Cushman, M.; Balaña-Fouce, R.; Pérez-Pertejo, Y. Topoisomerase IB Poisons Induce Histone H2A Phosphorylation as a Response to DNA Damage in Leishmania Infantum. *Int. J. Parasitol. Drugs Drug Resist.* **2019**, *11*, 39–48. [CrossRef] [PubMed]
23. Balaña-Fouce, R.; García-Estrada, C.; Pérez-Pertejo, Y.; Reguera, R.M. Gene Disruption of the DNA Topoisomerase IB Small Subunit Induces a Non-Viable Phenotype in the Hemoflagellate Leishmania Major. *BMC Microbiol.* **2008**, *8*, 113. [CrossRef] [PubMed]
24. Bakshi, R.P.; Shapiro, T.A. RNA Interference of Trypanosoma Brucei Topoisomerase IB: Both Subunits Are Essential. *Mol. Biochem. Parasitol.* **2004**, *136*, 249–255. [CrossRef]
25. Zhang, C.X.; Chen, A.D.; Gettel, N.J.; Hsieh, T.S. Essential Functions of DNA Topoisomerase I in Drosophila Melanogaster. *Dev. Biol.* **2000**, *222*, 27–40. [CrossRef] [PubMed]
26. Morham, S.G.; Kluckman, K.D.; Voulomanos, N.; Smithies, O. Targeted Disruption of the Mouse Topoisomerase I Gene by Camptothecin Selection. *Mol. Cell Biol.* **1996**, *16*, 6804–6809. [CrossRef]
27. Burgess, D.J.; Doles, J.; Zender, L.; Xue, W.; Ma, B.; McCombie, W.R.; Hannon, G.J.; Lowe, S.W.; Hemann, M.T. Topoisomerase Levels Determine Chemotherapy Response in Vitro and in Vivo. *Biol. Sci.* **2008**, *105*, 9053–9058. [CrossRef]
28. Pommier, Y. Camptothecins and Topoisomerase I: A Foot in the Door. Targeting the Genome beyond Topoisomerase I with Camptothecins and Novel Anticancer Drugs: Importance of DNA Replication, Repair and Cell Cycle Checkpoints. *Curr. Med. Chem. Anticancer. Agents* **2004**, *4*, 429–434. [CrossRef]
29. Nitiss, J.L.; Soans, E.; Rogojina, A.; Seth, A.; Mishina, M. Topoisomerase Assays. *Curr. Protoc. Pharmacol.* **2012**, *57*, 3. [CrossRef]
30. Tesauro, C.; Fiorani, P.; D'annessa, I.; Chillemi, G.; Turchi, G.; Desideri, A. Erybraedin C, a Natural Compound from the Plant Bituminaria Bituminosa, Inhibits Both the Cleavage and Religation Activities of Human Topoisomerase I. *Biochem. J.* **2010**, *425*, 531–539. [CrossRef]
31. Keller, J.G.; Hymøller, K.M.; Thorsager, M.E.; Hansen, N.Y.; Erlandsen, J.U.; Tesauro, C.; Simonsen, A.K.W.; Andersen, A.B.; VandsøPetersen, K.; Holm, L.L.; et al. Topoisomerase 1 Inhibits MYC Promoter Activity by Inducing G-Quadruplex Formation. *Nucleic Acids Res.* **2022**, *11*, 6332–6342. [CrossRef]
32. Christiansen, K.; Westergaard, O. Characterization of Intra- and Intermolecular DNA Ligation Mediated by Eukaryotic Topoisomerase I. Role of Bipartite DNA Interaction in the Ligation Process. *J. Biol. Chem.* **1994**, *269*, 721–729. [CrossRef]
33. Svejstrup, J.Q.; Christiansen, K.; Andersen, A.H.; Lund, K.; Westergaard, O. Minimal DNA Duplex Requirements for Topoisomerase I-Mediated Cleavage in Vitro. *J. Biol. Chem.* **1990**, *265*, 12529–12535. [CrossRef]
34. Anand, J.; Sun, Y.; Zhao, Y.; Nitiss, K.C.; Nitiss, J.L. Detection of Topoisomerase Covalent Complexes in Eukaryotic Cells. *Methods Mol. Biol.* **2018**, *1703*, 283–299. [PubMed]
35. Stougaard, M.; Lohmann, J.S.; Mancino, A.; Celik, S.; Andersen, F.F.; Koch, J.; Knudsen, B.R. Single-Molecule Detection of Human Topoisomerase I Cleavage-Ligation Activity. *ACS Nano* **2009**, *3*, 223–233. [CrossRef]

36. World Health Organization. Global Tuberculosis Report. Available online: https://www.who.int/publications/i/item/9789240061729 (accessed on 23 January 2023).
37. Franch, O.; Han, X.; Marcussen, L.B.; Givskov, A.; Andersen, M.B.; Godbole, A.A.; Harmsen, C.; Nørskov-Lauritsen, N.; Thomsen, J.; Pedersen, F.S.; et al. A New DNA Sensor System for Specific and Quantitative Detection of Mycobacteria. *Nanoscale* **2019**, *11*, 587–597. [CrossRef]
38. Prada, C.F.; Álvarez-Velilla, R.; Balaña-Fouce, R.; Prieto, C.; Calvo-Álvarez, E.; Escudero-Martínez, J.M.; Requena, J.M.; Ordóñez, C.; Desideri, A.; Pérez-Pertejo, Y.; et al. Gimatecan and Other Camptothecin Derivatives Poison Leishmania DNA-Topoisomerase IB Leading to a Strong Leishmanicidal Effect. *Biochem. Pharmacol.* **2013**, *85*, 1433–1440. [CrossRef] [PubMed]
39. Cushman, M. Design and Synthesis of Indenoisoquinolines Targeting Topoisomerase I and Other Biological Macromolecules for Cancer Chemotherapy. *J. Med. Chem.* **2021**, *64*, 17572–17600. [CrossRef]
40. Balanã-Fouce, R.; Prada, C.F.; Requena, J.M.; Cushman, M.; Pommier, Y.; Álvarez-Velilla, R.; Escudero-Martínez, J.M.; Calvo-Álvarez, E.; Pérez-Pertejo, Y.; Reguera, R.M. Indotecan (LMP400) and AM13-55: Two Novel Indenoisoquinolines Show Potential for Treating Visceral Leishmaniasis. *Antimicrob. Agents Chemother.* **2012**, *56*, 5264–5270. [CrossRef]
41. Lisby, M.; Krogh, B.O.; Boege, F.; Westergaard, O.; Knudsen, B.R. Camptothecins Inhibit the Utilization of Hydrogen Peroxide in the Ligation Step of Topoisomerase I Catalysis. *Biochemistry* **1998**, *37*, 10815–10827. [CrossRef] [PubMed]
42. Knudsen, B.R.; Straub, T.; Boege, F. Separation and Functional Analysis of Eukaryotic DNA Topoisomerases by Chromatography and Electrophoresis. *J. Chromatogr. B Biomed. Appl.* **1996**, *684*, 307–321. [CrossRef]
43. Staker, B.L.; Hjerrild, K.; Feese, M.D.; Behnke, C.A.; Burgin, A.B.; Stewart, L. The Mechanism of Topoisomerase I Poisoning by a Camptothecin Analog. *Proc. Natl. Acad. Sci. USA* **2002**, *99*, 15387–15392. [CrossRef]
44. Shuman, S. Vaccinia Virus DNA Topoisomerase: A Model Eukaryotic Type IB Enzyme. *Biochim. Biophys. Acta* **1998**, *1400*, 321–337. [CrossRef]
45. Keller, J.G.; Petersen, K.V.; Knudsen, B.R.; Tesauro, C. Simple and Fast DNA-Based Tool to Investigate Topoisomerase 1 Activity, a Biomarker for Drug Susceptibility in Colorectal Cancer. In *Recent Understanding of Colorectal Cancer Treatment*; Intech Open: London, UK, 2022.
46. Keller, J.G.; Mizielinski, K.; Petersen, K.V.; Stougaard, M.; Knudsen, B.R.; Tesauro, C. Simple and Fast Rolling Circle Amplification-Based Detection of Topoisomerase 1 Activity in Crude Biological Samples. *J. Vis. Exp.* **2022**, *190*, e64484.
47. Keller, J.G.; Stougaard, M.; Knudsen, B.R. Enzymatic Activity in Single Cells. In *Methods in Enzymology*; Elsevier: Amsterdam, The Netherlands, 2019; Volume 628, pp. 43–57. ISBN 9780128170908.
48. Keller, J.G.; Tesauro, C.; Coletta, A.; Graversen, A.D.; Ho, Y.P.; Kristensen, P.; Stougaard, M.; Knudsen, B.R. On-Slide Detection of Enzymatic Activities in Selected Single Cells. *Nanoscale* **2017**, *9*, 13546–13553. [CrossRef] [PubMed]
49. Tesauro, C.; Keller, J.G.; Gromova, I.; Gromov, P.; Frøhlich, R.; Erlandsen, J.U.; Andersen, A.H.; Stougaard, M.; Knudsen, B.R. Different Camptothecin Sensitivities in Subpopulations of Colon Cancer Cells Correlate with Expression of Different Phospho-Isoforms of Topoisomerase I with Different Activities. *Cancers* **2020**, *12*, 1240. [CrossRef] [PubMed]
50. Jakobsen, A.K.; Yuusufi, S.; Madsen, L.B.; Meldgaard, P.; Knudsen, B.R.; Stougaard, M. TDP1 and TOP1 as Targets in Anticancer Treatment of NSCLC: Activity and Protein Level in Normal and Tumor Tissue from 150 NSCLC Patients Correlated to Clinical Data. *Lung Cancer* **2022**, *164*, 23–32. [CrossRef]
51. Sikder, D.; Nagaraja, V. Determination of the Recognition Sequence of Mycobacterium Smegmatis Topoisomerase I on Mycobacterial Genomic Sequences. *Nucleic Acids Res.* **2000**, *28*, 1830–1837. [CrossRef]
52. Martín-Encinas, E.; Selas, A.; Palacios, F.; Alonso, C. The Design and Discovery of Topoisomerase I Inhibitors as Anticancer Therapies. *Expert. Opin. Drug Discov.* **2022**, *17*, 581–601. [CrossRef]
53. Alonso, C.; Fuertes, M.; Martín-Encinas, E.; Selas, A.; Rubiales, G.; Tesauro, C.; Knudssen, B.K.; Palacios, F. Novel Topoisomerase I Inhibitors. Syntheses and Biological Evaluation of Phosphorus Substituted Quinoline Derivates with Antiproliferative Activity. *Eur. J. Med. Chem.* **2018**, *149*, 225–237. [CrossRef]
54. Selas, A.; Fuertes, M.; Melcón-Fernández, E.; Pérez-Pertejo, Y.; Reguera, R.M.; Balaña-Fouce, R.; Knudsen, B.R.; Palacios, F.; Alonso, C. Hybrid Quinolinyl Phosphonates as Heterocyclic Carboxylate Isosteres: Synthesis and Biological Evaluation against Topoisomerase 1b (Top1b). *Pharmaceuticals* **2021**, *14*, 784. [CrossRef]
55. Selas, A.; Ramírez, G.; Palacios, F.; Alonso, C. Design, Synthesis and Cytotoxic Evaluation of Diphenyl(Quinolin-8-Yl)Phosphine Oxides. *Tetrahedron Lett.* **2021**, *70*, 153019. [CrossRef]
56. Fuertes, M.; Selas, A.; Trejo, A.; Knudsen, B.R.; Palacios, F.; Alonso, C. Synthesis of Hybrid Phosphorated Indenoquinolines and Biological Evaluation as Topoisomerase I Inhibitors and Antiproliferative Agents. *Bioorg Med. Chem. Lett.* **2022**, *57*, 128517. [CrossRef] [PubMed]
57. Ghashghaei, O.; Masdeu, C.; Alonso, C.; Palacios, F.; Lavilla, R. Recent Advances of the Povarov Reaction in Medicinal Chemistry. *Drug Discov. Today Technol.* **2018**, *29*, 71–79. [CrossRef]
58. Trejo, A.; Masdeu, C.; Serrano-Pérez, I.; Pendrola, M.; Juanola, N.; Ghashghaei, O.; Jiménez-Galisteo, G.; Lavilla, R.; Palacios, F.; Alonso, C.; et al. Efficient AntiMycolata Agents by Increasing the Lipophilicity of Known Antibiotics through Multicomponent Reactions. *Antibiotics* **2023**, *12*, 83. [CrossRef] [PubMed]
59. Petersen, K.V.; Selas, A.; Hymøller, K.M.; Mizielinski, K.; Thorsager, M.; Stougaard, M.; Alonso, C.; Palacios, F.; Pérez-Pertejo, Y.; Reguera, R.M.; et al. Simple and Fast Dna Based Sensor System for Screening of Small-Molecule Compounds Targeting Eukaryotic Topoisomerase 1. *Pharmaceutics* **2021**, *13*, 1255. [CrossRef]

60. World Health Organization. WHO Consolidated Guidelines on Tuberculosis: Module 4: Treatment: Drug-Resistant Tuberculosis Treatment. Available online: https://www.who.int/publications/i/item/9789240007048 (accessed on 23 January 2023).
61. Koch, A.; Cox, H.; Mizrahi, V. Drug-Resistant Tuberculosis: Challenges and Opportunities for Diagnosis and Treatment. *Curr. Opin. Pharmacol.* **2018**, *42*, 7–15. [CrossRef]

pharmaceuticals

Article

Hepatoprotection by Methylene Blue Against Doxorubicin Toxicity Through Coordinated Modulation of Oxidative Stress, ER Stress, and Apoptotic Pathways

Enas S. Gad [1,2], Ahmed M. Ashour [3,*], Amany M. Gad [1,4], Ali Khames [5], Shaimaa G. Ibrahim [6], Mohamed H. A. Gadelmawla [7,*] and Mona Mansour [8]

1 Department of Pharmacology and Toxicology, Faculty of Pharmacy, Sinai University, Kantara Branch, Ismailia 41636, Egypt; enasgad1988@gmail.com (E.S.G.); amany.gad@su.edu.eg (A.M.G.)
2 Department of Pharmaceutical Sciences, College of Clinical Pharmacy, King Faisal University, Al-Ahsa 13889, Saudi Arabia
3 Department of Pharmacology and Toxicology, College of Pharmacy, Umm Al-Qura University, P.O. Box 13578, Makkah 21955, Saudi Arabia
4 Department of Pharmacology, Egyptian Drug Authority (EDA)-Formerly NODCAR, Giza 12654, Egypt
5 Department of Pharmacology and Toxicology, Faculty of Pharmacy, Sohag University, Sohag 82511, Egypt; ali.khames@pharm.sohag.edu.eg
6 Department of Pharmacology and Toxicology, Faculty of Pharmacy, October 6 University, Giza 12585, Egypt; drshaimaa.gomaa@yahoo.com
7 Department of Life Sciences, Faculty of Biotechnology, Sinai University, Kantara Branch, Ismailia 41636, Egypt
8 Department of Pharmacology and Toxicology, Faculty of Pharmacy (Girls), Al-Azhar University, Cairo 11884, Egypt; mona.mansour@azhar.edu.eg
* Correspondence: amashour@uqu.edu.sa (A.M.A.); mohamed.hassany@su.edu.eg or mhassany_dna@yahoo.com (M.H.A.G.)

Abstract: Background and Objectives: Doxorubicin (DOX) is a potential chemotherapeutic whose clinical application is limited by hepatotoxicity mediated through apoptosis, endoplasmic reticulum (ER) stress, and oxidative stress (OS). This study aimed to assess the hepatoprotective impact of methylene blue (MB) against DOX-induced liver injury. **Methods:** Forty rats were arbitrarily divided equally into four groups: control, DOX (15 mg/kg, i.p., single dose), MB (4 mg/kg, i.p., daily for 7 days), and DOX + MB (same regimen, MB initiated 1 h post DOX). Serum ALT, AST, and γ-GT were measured, along with hepatic TAC and HO-1. ELISA quantified PERK, GRP78, and CHOP. Immunohistochemistry assessed Caspase-3, p53, NF-κB, and Nrf2. Histopathological evaluation was performed using H&E staining. **Results:** DOX administration significantly elevated ALT, AST, γ-GT, HO-1, PERK, GRP78, and CHOP while reducing TAC and Nrf2 expression. Strong Caspase-3, p53, and NF-κB immunoreactivity and severe histopathological damage were observed. MB treatment markedly reversed these changes, restoring antioxidant status, downregulating ER stress markers, preserving Nrf2 expression, and improving hepatic architecture. **Conclusions:** MB exerts significant hepatoprotection against DOX-induced injury, likely via attenuation of OS, ER stress, apoptosis, and inflammation.

Keywords: methylene blue; hepatotoxicity; doxorubicin; inflammation; endoplasmic reticulum stress; apoptosis

1. Introduction

Doxorubicin (DOX) is a key chemotherapeutic agent with strong effectiveness towards hematological and various solid tumors [1]. Nevertheless, DOX utilization clinically is limited due to adverse impacts on multiple organs, including the bone marrow, intestinal epithelium, heart, liver, and kidneys, as well as the potential for inducing cancer cell resistance [2]. Both powdered and liquid versions of DOX are accessible, and it is usually given intravenously [3]. The administration routes of both doxorubicin (intraperitoneal) and methylene blue (intraperitoneal) were selected based on their established efficacy in preclinical rodent models. While intravenous administration is the standard in human clinical practice, the intraperitoneal route provides a reliable and practical alternative for achieving systemic exposure in rats, ensuring translational relevance for early-stage mechanistic studies. Despite its potent chemotherapeutic activity, doxorubicin's conventional application has been reduced due to serious adverse impacts on different organs, leading to fertility issues, nephro-, cardio-, and hepatotoxicity, and the initiation of diabetic cardiac injury [4–6]. Its biphasic toxicity results in both acute and sub-acute hepatic damage [7]. The generation of reactive oxygen species (ROS) during DOX metabolism in the liver, causing an imbalance in redox potential, is the main molecular mechanism behind DOX-induced hepatotoxicity (DIHT). Oxidative stress (OS), reduced antioxidant enzyme values, apoptosis, inflammation, and mitochondrial malfunction are caused by this imbalance. Metabolism of doxorubicin in the liver generates superoxide radicals and peroxynitrite radicals, initiating lipid peroxidation and causing hepatic damage, evidenced by the production serum AST and ALT (hepatic enzymes). These enzyme levels serve as biomarkers for hepatotoxicity [8,9]. The drug also depletes antioxidant molecules like glutathione peroxidase (GPx), catalase (CAT), superoxide dismutase (SOD), and glutathione (GSH), thereby halting the activity of defensive mechanism [10,11]. Genes that produce apoptotic enzymes like Caspase-3 and antioxidant enzymes like Nrf2 and HO-1 are elevated within DOX therapy. ROS are produced when malondialdehyde values are elevated, leading to a cascade of apoptotic events [8,12]. The lipophilic nature of DOX and its DNA-binding capacity contribute to its capacity to accumulate in hepatocytic nuclei, leading to DNA damage [13]. The hepatotoxicity caused by DOX has been lessened by several drugs, according to a number of studies. It has been demonstrated that a number of natural products and medications have hepatoprotective properties that lessen the negative effects of DOX.

Methylene blue (MB), a thiazine dye, is utilized in tissue staining, histological studies, and medical interventions [14]. It is effective in treating conditions such as encephalopathy [15], methemoglobinemia, and poisonings caused by carbon monoxide, cyanide, and nitrates [16]. Additionally, MB helps prevent septic shock hypotension [17], and protects against renal and hepatic injury [18]. It also acts as a bacteriostatic disinfectant for the genitourinary tract [19]. Among its notable therapeutic properties is its impact on central nervous system (CNS) conditions. Furthermore, its effects on liver diseases, kidney disorders, lung damage, and cardiovascular conditions have been explored [20]. MB is FDA-approved for treating methemoglobinemia of various origins [21]. One of its most intriguing characteristics is its influence on mitochondrial function and regulation and ROS generation [15]. This study aims to assess the mechanisms by which methylene blue can mitigate liver damage induced by DOX, with a focus on the roles of OS and apoptosis.

2. Results

2.1. Effect of MB on DOX-Induced Hepatic Dysfunction

The serum levels of alanine aminotransferase (ALT), aspartate aminotransferase (AST), and gamma-glutamyl transferase (γ-GT) were ultimately upregulated in the DOX group (15 mg/kg, IP, once) by 3.9-, 3.1-, and 4.8-fold, respectively, compared to the control group (Figure 1). MB-treated rats (4 mg/kg/IP/daily for 7 days) exhibited a remarkable decline in liver enzyme values by 53.4%, 50.2%, and 64.4%, respectively, in comparison to the DOX group ($p < 0.05$).

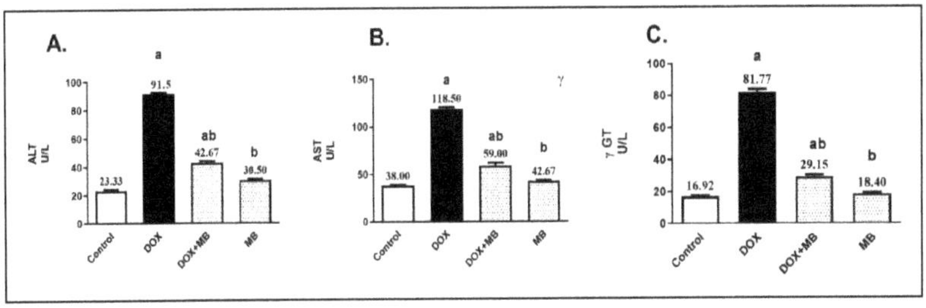

Figure 1. MB effects on serum ALT (**A**), AST (**B**), and γ-GT (**C**) levels. Data are expressed as mean ± S.E.M (n = 10). a: significance towards control, b: significance towards DOX group at $p < 0.05$. Abbreviations: ALT, alanine aminotransferase; AST, aspartate aminotransferase; γ-GT, gamma-glutamyl transferase; DOX, doxorubicin; S.E.M., standard error of the mean.

2.2. MB Effects on Hepatic OS Markers Against DIHT

The oxidative stress (OS) impact of MB against DOX-induced ROS production was evaluated by measuring the contents of total antioxidant capacity (TAC) and heme oxygenase-1 (HO-1). Induction of DOX hepatotoxicity provided a significant decrease in hepatic TAC by about 73%, as well as a significant increase in hepatic HO-1 by 4.8-fold compared to the control group (Figure 2). MB-treated rats revealed a remarkable elevation in the hepatic content of TAC by 154.9% and prevented the elevation in HO-1 content by about 61.1% compared to the DOX group ($p < 0.05$).

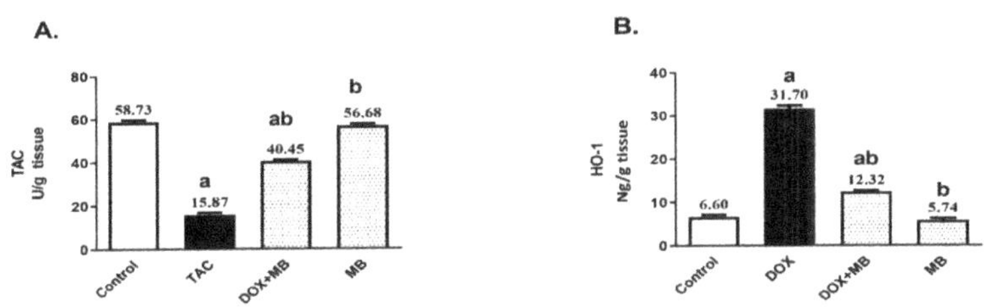

Figure 2. Effects of MB on total antioxidant capacity (TAC) (**A**) and HO-1 (**B**) contents. Data are expressed as mean ± S.E.M (n = 10). a: significance towards control, b: significance towards DOX group at $p < 0.05$. Abbreviations: TAC, total antioxidant capacity; HO-1, heme oxygenase-1; DOX, doxorubicin; S.E.M., standard error of the mean.

2.3. Effects of MB on PERK/GRP78/CHOP Pathway Markers

Doxorubicin (DOX) substantially elevated the hepatic contents of PERK, GRP78, and CHOP by about 5-, 4.8- and 4.4-fold, respectively, compared to the control group, as presented in Figure 3. Treatment with MB revealed a remarkable reduction in the hepatic contents of PERK, GRP78, and CHOP by 56.06%, 58.5%, and 61.23%, respectively, in comparison to the DOX group ($p < 0.05$).

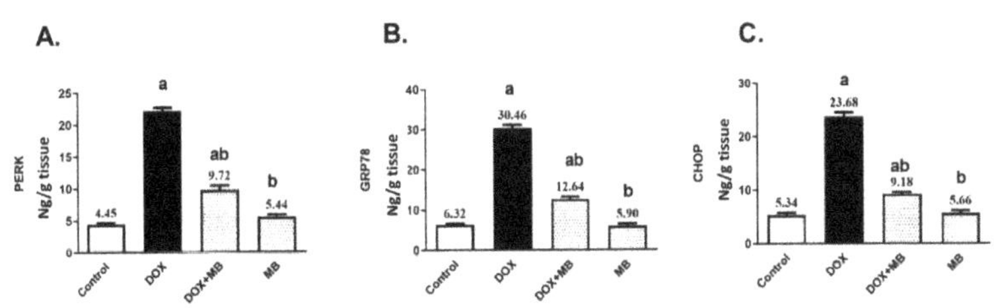

Figure 3. MB effects on PERK (**A**), GRP78 (**B**), and CHOP (**C**) contents. Data are expressed as mean ± S.E.M (n = 10). a: significance towards control, b: significance towards DOX group at $p < 0.05$.

Abbreviations: DOX, doxorubicin; PERK, protein kinase RNA-like endoplasmic reticulum kinase; GRP78, glucose-regulated protein 78; CHOP, C/EBP homologous protein; S.E.M., standard error of the mean.

2.4. Evaluation of Histological Changes

Histopathological examinations from the normal control group (Figure 4) revealed normal histologic structure of hepatic parenchyma in both centrilobular and portal areas. Likewise, apparently normal liver sections were detected in the MB-alone group. The DOX group exhibited marked histopathological alterations, and excessive hepatocellular vacuolation was noticed. Focal areas of necrosis with inflammatory cell infiltration were detected. The portal areas were heavily infiltrated with mononuclear inflammatory cells. Concerning the DOX + MB group marked improvement was detected, as hepatocytes in the centrilobular areas were apparently normal with mild mononuclear inflammatory cell infiltration in the portal areas.

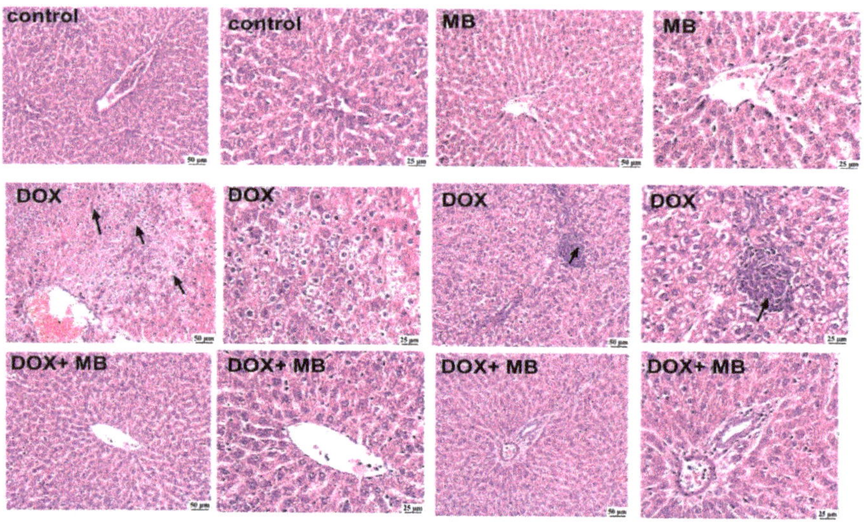

Figure 4. The photomicrographs show the liver tissue of all groups. The control group displayed normal hepatic architecture with well-arranged hepatocyte cords radiating from the central vein, intact cytoplasm, and clearly defined nuclei. The MB group revealed apparently normal hepatocytes in the centrilobular and periportal areas, indicating no morphological alterations. In contrast, the DOX group exhibited marked hepatic injury characterized by diffuse hepatocellular vacuolation, cytoplasmic degeneration, and focal necrosis accompanied by mononuclear inflammatory cell infiltration (arrows). The DOX + MB group demonstrated a remarkable improvement in hepatic architecture, showing nearly normal hepatocytes around the central and portal areas, with restoration of cellular integrity and minimal vacuolar degeneration (100 and 400×, H&E). Abbreviations: DOX, doxorubicin; H&E, hematoxylin and eosin.

2.5. Evaluation of Immunohistochemistry Changes

The IHC analysis demonstrated weak expression of Caspase-3, P53, and NF-κB in hepatic tissues of the control and MB groups. The DOX group showed severe expressions of Caspase-3, P53, and NF-κB. The DOX + MB group revealed a weak Caspase-3, P53, and NF-κB immunoreactivity approximately like the control tissues. However, for NRF-2, severe immunoreactivity was shown in the control and MB groups, moderate expression in the DOX + MB group, and weak expression in the DOX group (Figures 5 and 6).

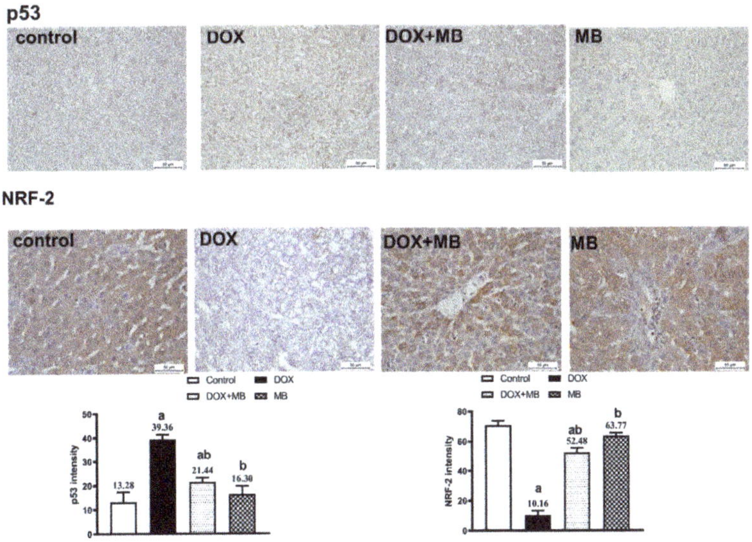

Figure 5. Protective effect of MB on hepatic immunoreactivity of p53 and NRF-2 in DOX-evoked hepatic damage in rats. Photomicrographs of liver tissues from all groups. The immunoreactivity of P53 and NRF-2 were visible in the tissues as a brown color generated by DAB chromogen (DAB, ×400). The control group exhibited normal hepatic histoarchitecture with faint p53 and moderate NRF-2 staining. The MB-alone group showed mild NRF-2 nuclear localization and negligible p53 expression. The DOX-treated group revealed intense p53 immunostaining in hepatocyte nuclei and cytoplasm, accompanied by a marked reduction in NRF-2 expression. In contrast, the DOX + MB group displayed attenuated p53 staining and enhanced NRF-2 immunoreactivity. Quantitative analysis of immunostaining intensity for p53 and NRF-2 is expressed as mean ± S.E.M. (n = 10). a: significant difference compared with the control group; b: significant difference compared with the DOX group at $p < 0.05$. Abbreviations: NRF-2, nuclear factor erythroid 2-related factor 2; DAB, 3,3′-diaminobenzidine; S.E.M., standard error of the mean.

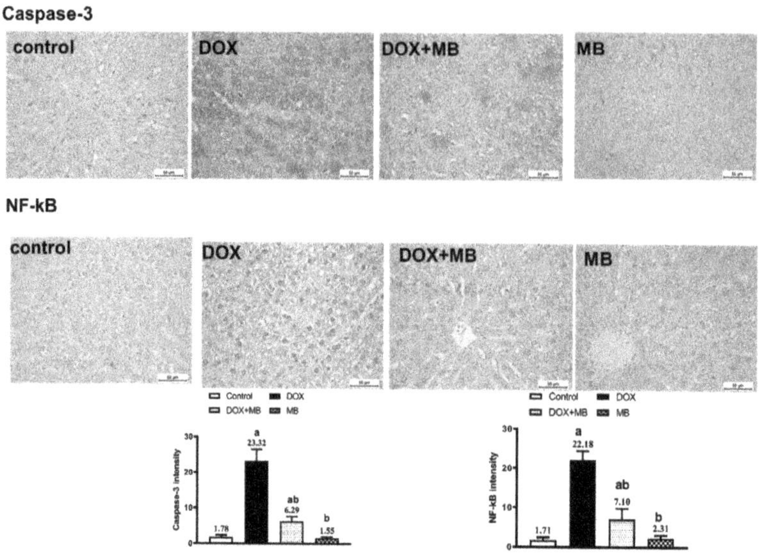

Figure 6. Protective effect of MB on hepatic immunoreactivity of Caspase-3 and NF-κB in DOX-treated rats evoked hepatic damage in rats. Photomicrographs of hepatic tissues of all groups. Caspase-3 and
NF-κB reactivity were detected in tissues using DAB chromogen, resulting in a brown color (DAB, ×400). The control group exhibited weak cytoplasmic Caspase-3 and NF-κB staining. The MB-alone group displayed minimal immunoreactivity similar to the control. In contrast, the DOX group showed strong Caspase-3 and NF-κB staining. The DOX+MB group demonstrated markedly reduced immunostaining intensity for both markers. Quantitative analysis of Caspase-3 and NF-κB immunoreactivity is expressed as mean ± S.E.M. (n = 10). a: significant difference compared with the control group; b: significant difference compared with the DOX-treated group at $p < 0.05$. Abbreviations: NF-κB, nuclear factor kappa B; DAB, 3,3′-diaminobenzidine; S.E.M., standard error of the mean.

3. Discussion

This study has certain limitations. First, the use of the intraperitoneal route for drug administration may limit the translational relevance of the findings, and future studies employing intravenous delivery are recommended. Second, the analysis was restricted to a specific set of biomarkers, while inclusion of additional antioxidant markers such as SOD and catalase could provide a more comprehensive understanding. Finally, reliance on ELISA and immunohistochemistry without complementary molecular techniques, such as Western blotting or qPCR, limits mechanistic validation; thus, future work will aim to incorporate these methods to strengthen the molecular evidence.

Doxorubicin (DOX) is a widely prescribed anticancer agent with proven efficacy against numerous solid as well as hematological malignancies [22]. Its therapeutic usage is markedly limited by its dose-dependent and cumulative hepatotoxicity [23]. In certain cases, the hepatic damage provoked by DOX can become a serious clinical concern, potentially compromising the patient's tolerance to further chemotherapy. DIHT is well documented in preclinical models. Among the principal mechanisms implicated, ER stress has a pivotal role. The substantial amount of unfolded/misfolded proteins inside the ER lumen during DOX exposure disrupts cellular homeostasis, initiating a cascade of signaling events that culminate in hepatocyte dysfunction and cell death [24,25].

The timing of methylene blue (MB) administration is crucial for optimizing its cardioprotective effects. In this study, MB was co-administered with doxorubicin (DOX), but whether pre- or post-treatment would yield similar benefits remains to be explored. Pre-treatment may enhance cellular defenses before DOX injury, while post-treatment better simulates clinical scenarios. Further studies comparing these strategies could improve the translational relevance of MB.

In the present study, DOX administration produced a clear biochemical signature of hepatocellular damage, elucidated by marked elevations in serum ALT, GGT, and AST, and this agreed with Gotama et al. [26] and Wu et al. [27]. These changes indicate loss of hepatocyte membrane integrity, mitochondrial perturbation, and oxidative stress-related impairment of hepatic function. Mechanistically, such elevations are consistent with DOX-induced ROS generation via redox cycling, which triggers lipid peroxidation and disrupts phospholipid bilayers. In addition, ER stress and disturbances in protein folding and calcium homeostasis further compromise cellular stability, facilitating enzyme leakage into the circulation. The observed pattern of increased ALT, AST, and GGT therefore reflects the combined effects of oxidative damage, ER and mitochondrial dysfunction, and subsequent hepatocyte necrosis/apoptosis, findings supported by the histopathological alterations detected in the DOX group [28].

In this study, DOX administration induced a marked oxidative imbalance, as evidenced by a pronounced reduction in hepatic TAC together with a substantial increase in HO-1 content, and this aligned with Barakat et al. [8] and Saleh et al. [29,30]. The decline in TAC reflects depletion of the endogenous antioxidant reserve, indicating that the hepatic defense system was overwhelmed by excessive ROS production. The elevation in HO-1 represents an adaptive cellular stress response, as this inducible enzyme is upregulated under oxidative challenge to degrade pro-oxidant heme and generate cytoprotective molecules. However, persistent or exaggerated HO-1 induction may also signify ongoing oxidative injury rather than successful adaptation. Together, the decrease in TAC and the rise in HO-1 demonstrate that DOX triggers severe oxidative stress in hepatic tissue, disrupting redox [30,31].

In the present study, HO-1 was selected as a representative antioxidant marker due to its dual role as both an inducible antioxidant enzyme and a cytoprotective mediator under oxidative stress conditions. HO-1 is particularly responsive to redox imbalance and has been shown to be upregulated in cardiac tissue following doxorubicin-induced injury, making it a sensitive indicator of oxidative stress modulation. While we acknowledge the established roles of other classical antioxidant enzymes such as superoxide dismutase (SOD), catalase (CAT), glutathione peroxidase (GPx), and peroxiredoxins (Prx), the scope of the current study was limited to evaluating early signaling responses, with HO-1 serving as a key target. Future studies are warranted to provide a more comprehensive profile of the antioxidant defense system by including a panel of enzymatic markers to fully delineate the redox-modulating effects of methylene blue.

In the present work, DOX administration markedly activated ER stress signaling, as demonstrated by significant increases in hepatic PERK, GRP78, and CHOP, and this is in accord with Yarmohammadi et al. [32] and Kopsida et al. [33]. PERK is a key transducer of the unfolded protein response, and its activation reflects ER matrix accumulated misfolded proteins. GRP78, an ER chaperone, is upregulated to stabilize protein folding and mitigate stress, while CHOP is a downstream transcription factor that shifts the ER stress response towards apoptosis when stress is severe or prolonged [34,35]. The concurrent elevation of these markers in DOX-treated animals indicates sustained ER stress that has transitioned from adaptive to pro-apoptotic signaling. Such activation is known to exacerbate hepatocellular injury by promoting inflammatory responses, amplifying oxidative stress, and triggering apoptotic pathways, thereby contributing to the overall hepatotoxic profile of DOX, as shown in Figure 7.

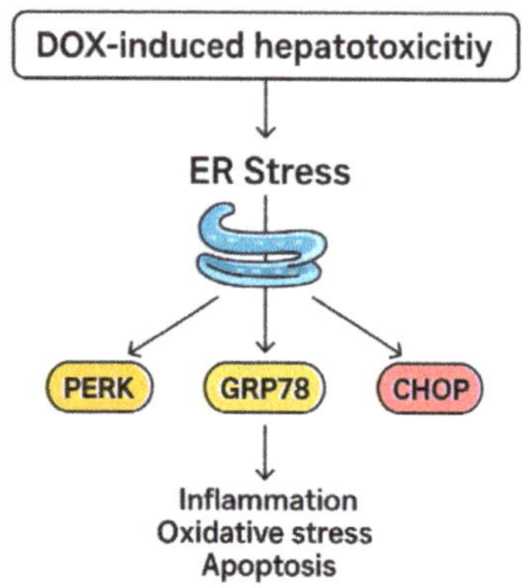

Figure 7. Proposed mechanism of DIHT.

In the present study, immunohistochemical assessment provided further insight into the mechanistic pathway underlying DIHT. Hepatic tissues from DOX-intoxicated animals exhibited strong immunoreactivity for Caspase-3 and p53, indicating activation of the apoptotic pathway, and this is in agreement with Lin et al. [36] and Khafaga and El-Sayed [37]. p53 acts as a central tumor suppressor protein that responds to DNA damage by initiating pro-apoptotic signaling, while Caspase-3 serves as a key executioner enzyme that orchestrates the final stages of programmed cell death [38–40]. The concurrent overexpression of NF-κB in the DOX group reflects an associated inflammatory response, as it regulates the generation of numerous pro-inflammatory mediators and can further amplify tissue injury. In contrast, Nrf2 immunoreactivity was markedly reduced in the DOX group, consistent with impaired stimulation of the intracellular antioxidant defense [41]. The suppression of Nrf2 compromises transcriptional induction of cytoprotective genes, thereby exacerbating oxidative and inflammatory damage. Together, these findings confirm that DOX-induced hepatic injury is mediated by a combination of apoptosis, inflammation, and diminished antioxidant defense capacity [42,43].

The observed pattern of Nrf2 expression—high in the control and MB-treated groups, suppressed in the DOX group, and partially restored with DOX + MB co-treatment—is consistent with the redox-sensitive regulation of this transcription factor. Under basal conditions, Nrf2 contributes to cellular homeostasis by regulating antioxidant gene expression. DOX-induced oxidative stress is known to impair the Nrf2 pathway, leading to reduced nuclear translocation and degradation of the protein. Co-administration of MB likely alleviates this oxidative burden, stabilizing Nrf2 and enabling its partial activation, which contributes to the restoration of downstream antioxidant responses, including TAC. This biphasic response of Nrf2 under oxidative stress is well documented.

Histopathological analysis revealed normal hepatic architecture in both the control and MB groups. DOX-treated livers showed diffuse hepatocellular vacuolation, focal necrosis, and dense mononuclear inflammatory infiltration in portal areas, consistent with severe oxidative and ER stress-mediated injury. These changes reflect combined cytotoxic, metabolic, and inflammatory mechanisms underlying DIHT.

In the present study, MB co-administration markedly reduced the DOX-induced elevations in ALT, AST, and GGT, indicating effective preservation of hepatocyte integrity and function. This improvement suggests that MB was able to limit the extent of cellular membrane disruption and enzyme leakage into the circulation. The normalization of these hepatic enzyme levels reflects attenuation of DOX-mediated hepatocellular damage and supports the hepatoprotective role of MB in maintaining biochemical markers of liver function within near-normal ranges.

MB counteracts DOX-induced hepatic injury by restoring the antioxidant defense system, replenishing TAC, and reducing the oxidative burden signaled by HO-1 overexpression. With a strengthened redox balance, the excessive ROS generation that fuels cellular damage is curtailed, preserving intracellular homeostasis, and this is endorsed by Poteet et al. [44]. This reduction in oxidative stress dampens ER stress signaling, leading to lower activation of PERK, reduced expression of the chaperone GRP78, and suppression of the pro-apoptotic factor CHOP [45].

The absence of a significant antioxidant response following MB treatment alone suggests that its redox-modulating effects are conditionally activated under oxidative stress states. MB likely functions as a redox sensor or stress-responsive modulator, exerting its antioxidant potential primarily in environments of elevated ROS, such as those induced by DOX. In the absence of oxidative insult, endogenous antioxidant systems remain balanced, and MB does not significantly alter basal redox homeostasis. This context-dependent activity highlights MB's potential selectivity in targeting pathological oxidative stress without disrupting normal cellular function, which is a favorable characteristic for therapeutic use.

As the ER stress cascade is restrained, the downstream triggers of hepatocellular apoptosis and inflammation are weakened. The apoptotic signals driven by p53 and Caspase-3 diminish, while inflammatory amplification via NF-κB is blunted. At the same time, Nrf2 activity is partially restored, allowing sustained transcription of antioxidant and cytoprotective genes. Through this sequence of interlinked events beginning with oxidative stress suppression, progressing through ER stress modulation, and culminating in reduced apoptosis and inflammation [46], MB preserves hepatocyte structure and function, protecting the liver from progressive injury, as shown in Figure 8.

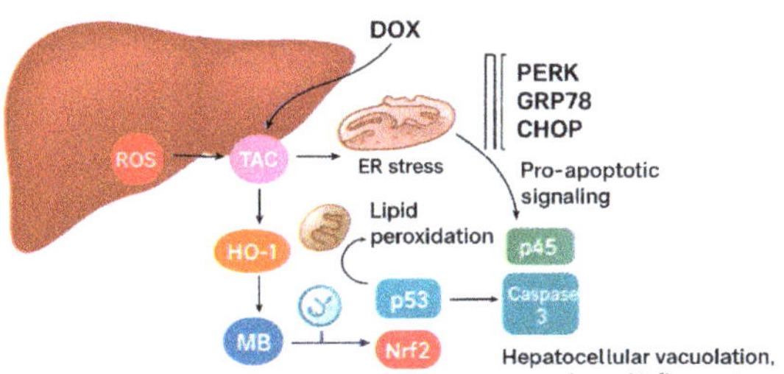

Figure 8. The proposed mechanism of MB's hepatoprotective effect.

4. Materials and Methods

4.1. Experimental Animals and Design

Forty adult male albino rats (160–180 g) were purchased from the laboratory animal unit (NODCAR, Egypt). The rats received a standard control water and diet ad libitum for 7 days to become used to their new environment. This study was performed after the Ethics Committee for Animal Experimentation at Sinai University [SU.REC.2024 (33 A)] approval was granted (13 January 2025). The animals were arbitrarily allocated equally to 4 groups (n = 10); the control group received intraperitoneal (i.p.) saline injection once (15 mL/kg; i.p.). Diseased group: treated with a single i.p. injection of DOX at a dose of 15 mg/kg [47]. MB control group: received a daily dose of MB for 7 consecutive days (4 mg/kg; i.p.) [48]. Treatment group: rats received 4 mg/kg of MB daily for 7 consecutive days; the first dose of MB started after injection with DOX by one hour.

Doxorubicin (DOX) and methylene blue (MB) were administered intraperitoneally, a route commonly used in rodent studies to mimic systemic exposure comparable to intravenous administration in humans. This approach is widely accepted for evaluating cardiotoxicity and therapeutic modulation in preclinical models.

4.2. Sample Collection

Blood samples were drawn from the retro-orbital venous sinus of each animal into plain test tubes. After allowing 10 min for clotting, to separate the serum, the samples were centrifuged at 3000 rpm (4 °C) and then frozen at −80 °C for later examination of biochemical indicators. After the end of the treatment, rats received an intraperitoneal injection (1 mL) of anesthesia (0.3 mL of xylazine and 0.7 mL of ketamine) and were sacrificed using a sterile surgical blade. Prior to homogenization, one gram of hepatic tissue underwent triple washing with cold NaCl solution (0.9%) and was then homogenized in cold phosphate-buffered saline (PBS) (9 mL; pH 7.5) [49]. The hepatic homogenates were subjected to cold centrifugation for approximately 15 min at 3000 rpm, after which the supernatants were extracted to assess antioxidant, oxidative stress, and inflammatory levels. A small piece of liver was preserved in formalin for further histological and immunohistochemical examination.

4.3. Biochemical Analysis

4.3.1. Serum Hepatic Markers

The serum aspartate aminotransferase (AST) and alanine aminotransferase (ALT) values were analyzed calorimetrically utilizing examining kits obtained from (Teco Diagnostics, Anaheim, CA, USA). Serum gamma glutamyl transferase (GGT) was evaluated utilizing a specific rat ELISA kit (My BioSource, San Diego, CA, USA, Cat #: MBS9343646). All techniques were performed according to the manufacturers' protocols.

4.3.2. Oxidative and Antioxidant Stress Markers

The homogenized liver samples were employed to elucidate the OS degree through measurement of the tissue concentrations of OH- radical using the specific ELISA kit (Enzo life science, Inc., 10 Executive Blvd, Farmingdale, NY, USA, Cat #: ADI-EKS-810A). The Rat Total Antioxidant Status (TAC) of the hepatic tissue samples was assessed using a quantification method with the specific ELISA kits (My BioSource, San Diego, CA, USA, Cat #: MBS1600693). The procedure was carried out according to the manufacturer's protocols in the attached brochures.

TAC assay was performed to provide an integrated parameter reflecting the cumulative antioxidant defense status in cardiac tissue. This assay was selected because it evaluates the overall capacity of both enzymatic and non-enzymatic antioxidants, offering a broader picture of oxidative stress than measuring individual antioxidants alone. Given that oxidative damage plays a central role in doxorubicin-induced cardiotoxicity, TAC assessment directly supports this study's aim of evaluating methylene blue's potential cardioprotective and antioxidant effects.

4.3.3. Estimation of Hepatic Tissue Concentrations of Apoptotic Biomarkers

The rat hepatic tissue levels of the apoptotic pathway markers PERK/GRP/CHOP were analyzed using specific rat ELISA kits purchased from My BioSource (San Diego, CA, USA, Cat #: MBS2511166) for PERK and from BT LAB (Shanghai, China) for GRP (Cat #:LS-F8683) and CHOP (Cat #: LS-F11285). The quantification method followed the manufacturers' protocols.

4.4. Histopathological Examination

The hepatic tissues underwent preservation in 10% formalin solution before being subjected to conventional histological procedures. These procedures involved successive immersion in elevating ethanol concentration for dehydration, followed by paraffin wax embedding. Subsequently, the paraffin-embedded tissue blocks were sliced into 4 μm sections and stained using routine work stain, hematoxylin and eosin (H&E) [50].

4.5. Immunohistochemistry (IHC)

We used IHC to examine the presence and location of Caspase-3, NRF-2, P53, and NF-κB proteins in liver tissues. Sections were processed, and antigens were retrieved by boiling in citrate buffer (pH 6.0). Hydrogen peroxide (3%) was administered for 20 min to inhibit tissue endogenous peroxidase activity. Then, to avoid non-specific antibody interaction with tissue, 5% bovine serum albumin was employed as a blocking agent. Next, 4 μm liver sections were immunostained with anti-Caspase-3, NRF-2, P53, and NF-κB primary antibodies for 90 min [51]. Following that, the samples were incubated for 30 min with HRP-labeled 2^{ry} antibodies. The targeted proteins were stained with a diaminobenzidine (DAB) kit (ScyTek Laboratories, Inc., Logan, UT, USA), and hematoxylin was utilized as a counterstain. Tissue segment images were acquired with digital imaging equipment coupled to a light microscope (Leica Flexacam i5, Leica, Wetzlar, Germany). Signal intensity for Caspase-3, NRF-2, P53, and NF-κB immunoreactivity was determined in all groups at 400× magnification [52].

4.6. Assessment of IHC Staining

The quantitative IHC was performed by employing ImageJ Fiji software version 1.2. The intensity for positive Caspase-3, NRF-2, P53, and NF-κB immunoreactions was determined at magnification 400× for all groups [53].

4.7. Statistical Analysis

All data were gathered and analyzed with SPSS, version 22.0 (IBM Corporation, Armonk, NY, USA). Data were expressed as mean ± S.E.M. The distinction between groups was statistically examined using GraphPad Prism 5 (La Jolla, CA, USA) utilizing one-way ANOVA followed by the Tukey–Kramer Multiple Comparison Test. p values < 0.05 were considered significant.

5. Conclusions

MB exhibits remarkable hepatoprotective potency against DIHT. This protection is mediated via restoration of TAC; reduction of HO-1; suppression of ER stress signaling via downregulation of PERK, GRP78, and CHOP; and inhibition of apoptosis and inflammation through decreased p53, Caspase-3, and NF-κB expression. Additionally, partial restoration of Nrf2 activity reinforces antioxidant defenses, collectively preserving hepatocellular structure and function.

Supplementary Materials: The following supporting information can be downloaded at: https://www.mdpi.com/article/10.3390/ph18111625/s1, Table S1. The animal body weight values throughout the experiment.

Author Contributions: A.K.—design of the experiment, execution of experiments, sample collection, data handling, manuscript writing; A.M.G. and A.M.A.—design of the experiment, revision and approval of manuscript; E.S.G.—design of the experiment, revision and approval of manuscript; M.M. and S.G.I.—sample collection, data handling, design of the experiment, revision and approval of manuscript; M.H.A.G. and A.M.A.—histopathological and immunohistochemical examination, revision and approval of manuscript; A.M.A. and M.H.A.G.—design of the experiment, execution of experiments, sample collection, data handling, manuscript writing; A.M.A.—funding. All authors have read and agreed to the published version of the manuscript.

Funding: This research work was funded by Umm Al-Qura University, Saudi Arabia, under grant number 25UQU4310007GSSR01.

Institutional Review Board Statement: The animal study protocol was approved by the Sinai University Research Ethics Committee (SU. REC.2024. 33 A) (13 January 2025).

Data Availability Statement: The original contributions presented in this study are included in the article/Supplementary Material. Further inquiries can be directed to the corresponding author(s).

Acknowledgments: The authors extend their appreciation to Umm Al-Qura University, Saudi Arabia, for funding this research work through grant number 25UQU4310007GSSR01.

Conflicts of Interest: The authors declare that they have no known competing financial interests or personal relationships that could have appeared to influence the work reported in this paper.

References

1. Jacevic, V.; Djordjevic, A.; Srdjenovic, B.; Milic-Tores, V.; Segrt, Z.; Dragojevic-Simic, V.; Kuca, K. Fullerenol Nanoparticles Prevents Doxorubicin-Induced Acute Hepatotoxicity in Rats. *Exp. Mol. Pathol.* **2017**, *102*, 360–369. [CrossRef]
2. Martindale, W.; Sweetman, S.C. (Eds.) *Martindale: The Complete Drug Reference*, 36th ed.; Pharmaceuticale Press, PhP: London, UK; Chicago, IL, USA, 2009; ISBN 978-0-85369-840-1.
3. Clinton, J.W.; Kiparizoska, S.; Aggarwal, S.; Woo, S.; Davis, W.; Lewis, J.H. Drug-Induced Liver Injury: Highlights and Controversies in the Recent Literature. *Drug Saf.* **2021**, *44*, 1125–1149. [CrossRef]
4. Renu, K.; Valsala Gopalakrishnan, A. Deciphering the Molecular Mechanism during Doxorubicin-Mediated Oxidative Stress, Apoptosis through Nrf2 and PGC-1α in a Rat Testicular Milieu. *Reprod. Biol.* **2019**, *19*, 22–37. [CrossRef]
5. Renu, K.; Sruthy, K.B.; Parthiban, S.; Sugunapriyadharshini, S.; George, A.; Pichiah, P.B.T.; Suman, S.; Abilash, V.G.; Arunachalam, S. Elevated Lipolysis in Adipose Tissue by Doxorubicin via PPARα Activation Associated with Hepatic Steatosis and Insulin Resistance. *Eur. J. Pharmacol.* **2019**, *843*, 162–176. [CrossRef]
6. Renu, K.; Abilash, V.G.; Pichiah, P.B.T.; Arunachalam, S. Molecular Mechanism of Doxorubicin-Induced Cardiomyopathy—An Update. *Eur. J. Pharmacol.* **2018**, *818*, 241–253. [CrossRef]
7. Henninger, C.; Huelsenbeck, J.; Huelsenbeck, S.; Grösch, S.; Schad, A.; Lackner, K.J.; Kaina, B.; Fritz, G. The Lipid Lowering Drug Lovastatin Protects against Doxorubicin-Induced Hepatotoxicity. *Toxicol. Appl. Pharmacol.* **2012**, *261*, 66–73. [CrossRef]
8. Barakat, B.M.; Ahmed, H.I.; Bahr, H.I.; Elbahaie, A.M. Protective Effect of Boswellic Acids against Doxorubicin-Induced Hepatotoxicity: Impact on Nrf2/HO-1 Defense Pathway. *Oxidative Med. Cell Longev.* **2018**, *2018*, 8296451. [CrossRef]
9. Zhao, X.; Zhang, J.; Tong, N.; Chen, Y.; Luo, Y. Protective Effects of Berberine on Doxorubicin-Induced Hepatotoxicity in Mice. *Biol. Pharm. Bull.* **2012**, *35*, 796–800. [CrossRef]
10. Mansouri, E.; Jangaran, A.; Ashtari, A. Protective Effect of Pravastatin on Doxorubicin-Induced Hepatotoxicity. *Bratisl. Lek. Listy* **2017**, *118*, 273–277. [CrossRef]
11. Indu, R.; Azhar, T.S.; Nair, A.; Nair, C.K.K. Amelioration of Doxorubicin Induced Cardio-and Hepato-Toxicity by Carotenoids. *J. Cancer Res. Ther.* **2014**, *10*, 62–67. [CrossRef]
12. Enomoto, A.; Itoh, K.; Nagayoshi, E.; Haruta, J.; Kimura, T.; O'Connor, T.; Harada, T.; Yamamoto, M. High Sensitivity of Nrf2 Knockout Mice to Acetaminophen Hepatotoxicity Associated with Decreased Expression of ARE-Regulated Drug Metabolizing Enzymes and Antioxidant Genes. *Toxicol. Sci.* **2001**, *59*, 169–177. [CrossRef]
13. Tacar, O.; Sriamornsak, P.; Dass, C.R. Doxorubicin: An Update on Anticancer Molecular Action, Toxicity and Novel Drug Delivery Systems. *J. Pharm. Pharmacol.* **2013**, *65*, 157–170. [CrossRef]
14. Heidari, R.; Moezi, L.; Asadi, B.; Ommati, M.M.; Azarpira, N. Hepatoprotective Effect of Boldine in a Bile Duct Ligated Rat Model of Cholestasis/Cirrhosis. *PharmaNutrition* **2017**, *5*, 109–117. [CrossRef]
15. Gureev, A.P.; Sadovnikova, I.S.; Popov, V.N. Molecular Mechanisms of the Neuroprotective Effect of Methylene Blue. *Biochemistry* **2022**, *87*, 940–956. [CrossRef]
16. Gores, G.J.; Miyoshi, H.; Botla, R.; Aguilar, H.I.; Bronk, S.F. Induction of the Mitochondrial Permeability Transition as a Mechanism of Liver Injury during Cholestasis: A Potential Role for Mitochondrial Proteases. *Biochim. Biophys. Acta* **1998**, *1366*, 167–175. [CrossRef] [PubMed]
17. Palmeira, C.M.; Rolo, A.P. Mitochondrially-Mediated Toxicity of Bile Acids. *Toxicology* **2004**, *203*, 1–15. [CrossRef] [PubMed]
18. Arduini, A.; Serviddio, G.; Tormos, A.M.; Monsalve, M.; Sastre, J. Mitochondrial Dysfunction in Cholestatic Liver Diseases. *Front. Biosci. (Elite Ed.)* **2012**, *4*, 2233–2252. [CrossRef]

19. Schulz, S.; Schmitt, S.; Wimmer, R.; Aichler, M.; Eisenhofer, S.; Lichtmannegger, J.; Eberhagen, C.; Artmann, R.; Tookos, F.; Walch, A.; et al. Progressive Stages of Mitochondrial Destruction Caused by Cell Toxic Bile Salts. *Biochim. Biophys. Acta* **2013**, *1828*, 2121–2133. [CrossRef] [PubMed]
20. Heidari, R.; Ahmadi, A.; Ommati, M.M.; Niknahad, H. Methylene Blue Improves Mitochondrial Function in The Liver of Cholestatic Rats. *Trends Pharm. Sci. Technol.* **2020**, *6*, 73–86. [CrossRef]
21. Bužga, M.; Machytka, E.; Dvořáčková, E.; Švagera, Z.; Stejskal, D.; Máca, J.; Král, J. Methylene Blue: A Controversial Diagnostic Acid and Medication? *Toxicol. Res.* **2022**, *11*, 711–717. [CrossRef]
22. Saddik, M.S.; Elsayed, M.M.A.; Abdel-Rheem, A.A.; El-Mokhtar, M.A.; Mosa, E.S.; Al-Hakkani, M.F.; Al-Shelkamy, S.A.; Khames, A.; Daha, M.A.; Abdel-Aleem, J.A. A Novel C@Fe@Cu Nanocomposite Loaded with Doxorubicin Tailored for the Treatment of Hepatocellular Carcinoma. *Pharmaceutics* **2022**, *14*, 1845. [CrossRef] [PubMed]
23. Khames, A.; Khalaf, M.M.; Gad, A.M.; Abd El-Raouf, O.M.; Kandeil, M.A. Nicorandil Combats Doxorubicin-Induced Nephrotoxicity via Amendment of TLR4/P38 MAPK/NFκ-B Signaling Pathway. *Chem. Biol. Interact.* **2019**, *311*, 108777. [CrossRef]
24. Zhang, J.; Guo, J.; Yang, N.; Huang, Y.; Hu, T.; Rao, C. Endoplasmic Reticulum Stress-Mediated Cell Death in Liver Injury. *Cell Death Dis.* **2022**, *13*, 1051. [CrossRef]
25. Pu, S.; Pan, Y.; Zhang, Q.; You, T.; Yue, T.; Zhang, Y.; Wang, M. Endoplasmic Reticulum Stress and Mitochondrial Stress in Drug-Induced Liver Injury. *Molecules* **2023**, *28*, 3160. [CrossRef]
26. Gotama, K.T.; Soetikno, V.; Louisa, M.; Arozal, W. Hepatoprotective Effects of L-Citrulline against Doxorubicin-Induced Liver Damage in Rats: An Analysis of Serum Biomarkers. *Int. J. Appl. Pharm.* **2019**, *11*, 230–233. [CrossRef]
27. Wu, Y.-Z.; Wang, K.-X.; Ma, X.; Wang, C.-C.; Chen, N.-N.; Xiong, C.; Li, J.-X.; Su, S.-W. Therapeutic Effects of Atorvastatin on Doxorubicin-Induced Hepatotoxicity in Rats via Antioxidative Damage, Anti-Inflammatory, and Anti-Lipotoxicity. *J. Biochem. Mol. Toxicol.* **2023**, *37*, e23329. [CrossRef]
28. Villanueva-Paz, M.; Morán, L.; López-Alcántara, N.; Freixo, C.; Andrade, R.J.; Lucena, M.I.; Cubero, F.J. Oxidative Stress in Drug-Induced Liver Injury (DILI): From Mechanisms to Biomarkers for Use in Clinical Practice. *Antioxidants* **2021**, *10*, 390. [CrossRef]
29. Saleh, D.O.; Mahmoud, S.S.; Hassan, A.; Sanad, E.F. Doxorubicin-Induced Hepatic Toxicity in Rats: Mechanistic Protective Role of Omega-3 Fatty Acids through Nrf2/HO-1 Activation and PI3K/Akt/GSK-3β Axis Modulation. *Saudi J. Biol. Sci.* **2022**, *29*, 103308. [CrossRef]
30. Li, S.; Li, H.; Xu, X.; Saw, P.E.; Zhang, L. Nanocarrier-Mediated Antioxidant Delivery for Liver Diseases. *Theranostics* **2020**, *10*, 1262–1280. [CrossRef]
31. Meng, Z.; Wang, L.; Liao, Y.; Peng, Z.; Li, D.; Zhou, X.; Liu, S.; Li, Y.; Nüssler, A.K.; Liu, L.; et al. The Protective Effect of Heme Oxygenase-1 on Liver Injury Caused by DON-Induced Oxidative Stress and Cytotoxicity. *Toxins* **2021**, *13*, 732. [CrossRef] [PubMed]
32. Yarmohammadi, F.; Rezaee, R.; Haye, A.W.; Karimi, G. Endoplasmic Reticulum Stress in Doxorubicin-Induced Cardiotoxicity May Be Therapeutically Targeted by Natural and Chemical Compounds: A Review. *Pharmacol. Res.* **2021**, *164*, 105383. [CrossRef] [PubMed]
33. Kopsida, M.; Clavero, A.L.; Khaled, J.; Balgoma, D.; Luna-Marco, C.; Chowdhury, A.; Nyman, S.S.; Rorsman, F.; Ebeling Barbier, C.; Bergsten, P.; et al. Inhibiting the Endoplasmic Reticulum Stress Response Enhances the Effect of Doxorubicin by Altering the Lipid Metabolism of Liver Cancer Cells. *Mol. Metab.* **2024**, *79*, 101846. [CrossRef]
34. Hughes, D.; Mallucci, G.R. The Unfolded Protein Response in Neurodegenerative Disorders—Therapeutic Modulation of the PERK Pathway. *FEBS J.* **2019**, *286*, 342–355. [CrossRef]
35. Ong, G.; Logue, S.E. Unfolding the Interactions between Endoplasmic Reticulum Stress and Oxidative Stress. *Antioxidants* **2023**, *12*, 981. [CrossRef]
36. Lin, R.-W.; Ho, C.-J.; Chen, H.-W.; Pao, Y.-H.; Chen, L.-E.; Yang, M.-C.; Huang, S.-B.; Wang, S.; Chen, C.-H.; Wang, C. P53 Enhances Apoptosis Induced by Doxorubicin Only under Conditions of Severe DNA Damage. *Cell Cycle* **2018**, *17*, 2175–2186. [CrossRef]
37. Khafaga, A.F.; El-Sayed, Y.S. All-Trans-Retinoic Acid Ameliorates Doxorubicin-Induced Cardiotoxicity: In Vivo Potential Involvement of Oxidative Stress, Inflammation, and Apoptosis via Caspase-3 and P53 down-Expression. *Naunyn Schmiedebergs Arch. Pharmacol.* **2018**, *391*, 59–70. [CrossRef]
38. Mustafa, M.; Ahmad, R.; Tantry, I.Q.; Ahmad, W.; Siddiqui, S.; Alam, M.; Abbas, K.; Moinuddin, N.; Hassan, M.I.; Habib, S.; et al. Apoptosis: A Comprehensive Overview of Signaling Pathways, Morphological Changes, and Physiological Significance and Therapeutic Implications. *Cells* **2024**, *13*, 1838. [CrossRef]
39. Gadelmawla, M.H.A.; Alazzouni, A.S.; Farag, A.H.; Gabri, M.S.; Hassan, B.N. Enhanced Effects of Ferulic Acid against the Harmful Side Effects of Chemotherapy in Colon Cancer: Docking and in Vivo Study. *JOBAZ* **2022**, *83*, 28. [CrossRef]
40. Dhara, A.; Kaur, R.; Chattopadhyay, R.; Das, S.; Pal, S.; Sen, N. Role of Apoptosis in Cancer: War of the Worlds, Therapeutic Targets and Strategies. In *Apoptosis and Human Health: Understanding Mechanistic and Therapeutic Potential*; Jana, K., Ed.; Springer Nature: Singapore, 2024; pp. 169–205, ISBN 978-981-97-7905-5.

41. Özturk, N.; Ceylan, H.; Demir, Y. The Hepatoprotective Potential of Tannic Acid against Doxorubicin-Induced Hepatotoxicity: Insights into Its Antioxidative, Anti-Inflammatory, and Antiapoptotic Mechanisms. *J. Biochem. Mol. Toxicol.* **2024**, *38*, e23798. [CrossRef] [PubMed]
42. AlAsmari, A.F.; Alharbi, M.; Alqahtani, F.; Alasmari, F.; AlSwayyed, M.; Alzarea, S.I.; Al-Alallah, I.A.; Alghamdi, A.; Hakami, H.M.; Alyousef, M.K.; et al. Diosmin Alleviates Doxorubicin-Induced Liver Injury via Modulation of Oxidative Stress-Mediated Hepatic Inflammation and Apoptosis via NfkB and MAPK Pathway: A Preclinical Study. *Antioxidants* **2021**, *10*, 1998. [CrossRef] [PubMed]
43. Wang, L.; Wei, C.; Jing, J.; Shao, M.; Wang, Z.; Wen, B.; Lu, M.; Jia, Z.; Zhang, Y. The Effects of Polyphenols on Doxorubicin-Induced Nephrotoxicity by Modulating Inflammatory Cytokines, Apoptosis, Oxidative Stress, and Oxidative DNA Damage. *Phytother. Res.* **2025**, *39*, 2147–2164. [CrossRef]
44. Poteet, E.; Winters, A.; Yan, L.-J.; Shufelt, K.; Green, K.N.; Simpkins, J.W.; Wen, Y.; Yang, S.-H. Neuroprotective Actions of Methylene Blue and Its Derivatives. *PLoS ONE* **2012**, *7*, e48279. [CrossRef] [PubMed]
45. Duicu, O.M.; Privistirescu, A.; Wolf, A.; Petruş, A.; Dănilă, M.D.; Raţiu, C.D.; Muntean, D.M.; Sturza, A. Methylene Blue Improves Mitochondrial Respiration and Decreases Oxidative Stress in a Substrate-Dependent Manner in Diabetic Rat Hearts. *Can. J. Physiol. Pharmacol.* **2017**, *95*, 1376–1382. [CrossRef] [PubMed]
46. Jiang, Z.; Watts, L.T.; Huang, S.; Shen, Q.; Rodriguez, P.; Chen, C.; Zhou, C.; Duong, T.Q. The Effects of Methylene Blue on Autophagy and Apoptosis in MRI-Defined Normal Tissue, Ischemic Penumbra and Ischemic Core. *PLoS ONE* **2015**, *10*, e0131929. [CrossRef]
47. Mantawy, E.M.; El-Bakly, W.M.; Esmat, A.; Badr, A.M.; El-Demerdash, E. Chrysin Alleviates Acute Doxorubicin Cardiotoxicity in Rats via Suppression of Oxidative Stress, Inflammation and Apoptosis. *Eur. J. Pharmacol.* **2014**, *728*, 107–118. [CrossRef]
48. Gholami Jourabi, F.; Yari, S.; Amiri, P.; Heidarianpour, A.; Hashemi, H. The Ameliorative Effects of Methylene Blue on Testicular Damage Induced by Cisplatin in Rats. *Andrologia* **2021**, *53*, e13850. [CrossRef]
49. Abdallah, O.A.; Risha, E.F.; Elshopakey, G.E. Hepatoprotective and Antioxidant Effects of Artichoke against Carbon Tetrachloride-Toxicity in Rats. *Life Sci. J.* **2013**, *10*, 1436–1444.
50. Abdel-Wahhab, K.G.; Ashry, M.; Hassan, L.K.; El-Azma, M.H.; Elqattan, G.M.; Gadelmawla, M.H.A.; Mannaa, F.A. Hepatic and Immune Modulatory Effectiveness of Lactoferrin Loaded Selenium Nanoparticles on Bleomycin Induced Hepatic Injury. *Sci. Rep.* **2024**, *14*, 21066. [CrossRef]
51. Arab, H.H.; Alsufyani, S.E.; Ashour, A.M.; Gad, A.M.; Elhemiely, A.A.; Gadelmawla, M.H.A.; Mahmoud, M.A.; Khames, A. Targeting JAK2/STAT3, NLRP3/Caspase-1, and PK2/PKR2 Pathways with Arbutin Ameliorates Lead Acetate-Induced Testicular Injury in Rats. *Pharmaceuticals* **2024**, *17*, 909. [CrossRef]
52. Abdel-Wahhab, K.G.; Ashry, M.; Hassan, L.K.; Gadelmawla, M.H.A.; Elqattan, G.M.; El-Fakharany, E.M.; Mannaaa, F.A. Nano-Chitosan/Bovine Lactoperoxidase and Lactoferrin Formulation Modulates the Hepatic Deterioration Induced by 7,12-Dimethylbenz[*a*]Anthracene. *Comp. Clin. Pathol.* **2023**, *32*, 981–991. [CrossRef]
53. Alrashdi, B.M.; Askar, H.; Germoush, M.O.; Fouda, M.; Massoud, D.; Alzwain, S.; Abdelsater, N.; Salim, L.M.S.; Gadelmawla, M.H.A.; Ashry, M. Cardioprotective, Anti-Inflammatory, and Antioxidative Outcome of Costus against Bleomycin-Induced Cardiotoxicity in Rat Model. *J. Genet. Eng. Biotechnol.* **2025**, *23*, 100466. [CrossRef] [PubMed]

MDPI AG
Grosspeteranlage 5
4052 Basel
Switzerland
Tel.: +41 61 683 77 34

Pharmaceuticals Editorial Office
E-mail: pharmaceuticals@mdpi.com
www.mdpi.com/journal/pharmaceuticals

www.ingramcontent.com/pod-product-compliance
Lightning Source LLC
Chambersburg PA
CBHW061519210926
51850CB00008B/1974
9783725860814